Carbon Superstructures

This book covers how the understanding, as well as controllability, of the quantum electronic properties of carbon structures can be improved through a combined study of structural geometry, electronic properties, and dynamics of resonating valence bonds. It elaborates on varied properties such as growth mechanism, exotic transport properties, namely unusual geometry of microstructures mixed with electron distribution and spin properties in carbon. Transport mechanisms and new applications including hybrid quantum technology based on the superconducting diamond and diamond nitrogen-vacancy (NV) centers are discussed.

Features:

- Includes the experimental aspects of carbon physics, various carbon nanostructures, and simulations.
- Covers growth of carbon superstructures and various applications of their tunable electronic properties.
- Discusses how nanocarbon systems can be used in emerging technologies, including spintronic and quantum computing.
- Focuses on spin-related features and spin transport including the Kondo effect, spin-charge separation, spin-phonon coupling, anomalous Hall effect, and Luttinger liquid features.
- Explores carbon superstructure growth and their tunable electronic properties.

This book is aimed at students, researchers in physics, chemistry, engineering, materials science, electronics, and quantum technology.

Carbon Superstructures
From Quantum Transport to
Quantum Computation

Somnath Bhattacharyya

CRC Press
Taylor & Francis Group
Boca Raton London New York

CRC Press is an imprint of the
Taylor & Francis Group, an **informa** business

Designed cover image: ©Shutterstock

First edition published 2024
by CRC Press
2385 NW Executive Center Drive, Suite 320, Boca Raton FL 33431

and by CRC Press
4 Park Square, Milton Park, Abingdon, Oxon, OX14 4RN

CRC Press is an imprint of Taylor & Francis Group, LLC

© 2024 Somnath Bhattacharyya

Library of Congress Cataloging-in-Publication Data
Names: Bhattacharyya, Somnath, author.
Title: Carbon superstructures : from quantum transport to quantum
computation / Somnath Bhattacharyya.
Description: First edition. | Boca Raton : CRC Press, 2024. |
Includes bibliographical references and index. |
Identifiers: LCCN 2023049435 (print) | LCCN 2023049436 (ebook) |
ISBN 9781032327259 (hbk) | ISBN 9781032327266 (pbk) | ISBN 9781003316411 (ebk)
Subjects: LCSH: Electronics—Materials. | Carbon—Structure. |
Electron transport. | Quantum computing—Materials. | Microphysics.
Classification: LCC TK7871.15.C35 B53 2024 (print) |
LCC TK7871.15.C35 (ebook) | DDC 661/.0681—dc23/eng/20240129
LC record available at https://lccn.loc.gov/2023049435
LC ebook record available at https://lccn.loc.gov/2023049436

ISBN: 9781032327259 (hbk)
ISBN: 9781032327266 (pbk)
ISBN: 9781003316411 (ebk)

DOI: 10.1201/9781003316411

Typeset in Times
by codeMantra

*Dedicated to my supervisors, students, researchers, and colleagues
who contributed to the development of carbon science*

Contents

About the Author xiii
Acknowledgments..xiv
Introduction: Carbon: Harmony of Beauty, Strength, and Intelligence.................xvi

Chapter 1 Emergent Quantum Processes in Nature: Search for a New Material 1

    1.1   Ancient Carbon: Carbon Concerns the Creation of the Universe and Describing the Order and Disorder 1

    1.2   Carbon in Space: Carbon in the Extreme Environment on the Earth and in Deep Space (Soot to Diamond) 1

        1.2.1   First Clear Detection of Carbon Dioxide 2

        1.2.2   Carbon and Fire.. 3

    1.3   Carbon Cycle 4

    1.4   Dualism in Carbon Microstructures: Growth Mechanism and Structure 5

        1.4.1   Growth of Carbon Atoms and Molecules 5

        1.4.2   Atomic and Electronic Structure of Carbon 6

    1.5   Resonant Valence Bonds in Conducting Polymers Hydrocarbon Chain Conducting Polymer 8

        1.5.1   Conducting Carbon: Resonant Valence Bond 9

    1.6   Beauty in Carbon: Unusual Nanostructures: Pi-Electrons and Hybridization of S-P Orbitals in Carbon...........................10

    1.7   Carbon and Its Allotropes: Fullerenes, Nanotubes, Graphene, Nanocrystalline Diamond, etc. 10

        1.7.1   Fullerene 11

        1.7.2   Carbon Nanotube 11

        1.7.3   Graphene 12

        1.7.4   Diamond..13

    1.8   Superconductivity in Carbon Nanostructures: Topological Features: Spin-Orbit Coupling (SOC) 16

        1.8.1   The Periodic Table of Nanotubes 16

        1.8.2   SOI and Spin Selectivity for Tunneling Electron Transfer in DNA..19

    1.9   Computation Based on Biomolecules; DNA Structure–Based Computation 20

        1.9.1   Future Directions: Quantum Computer in Complex Carbon Structure Research Leading to Bioinformatics 20

    1.10   Carbon-Based Technology for Storing Energy (Sustainability): Charcoal, Carbon Dots, and Q-Carbon 21

        1.10.1   Charcoal 21

        1.10.2   Energy Storage Applications of Carbon Nanomaterials 21

1.10.3   Carbon Dots: A New Type of Carbon-Based
         Nanomaterial with Wide Applications...................... 23
1.10.4   Quantum Dots: Innovative Materials for
         Various Applications... 24
1.10.5   Q-Carbon... 25
1.10.6   Amorphous (Diamond-Like) Carbon....................... 26
Bibliography.. 28

Chapter 2   Synthesis of Carbon Nanostructures: Growth Model and
               Microstructure...31

2.1   Gas-Phase Reaction: Synthesis of Graphene
      Mechanical Exfoliation, Chemical Vapor Deposition,
      and Laser Ablation 31
      2.1.1   Synthetic Techniques for Graphene 31
      2.1.2   Micromechanical Cleavage.....................................31
      2.1.3   Chemical Vapor Deposition (CVD) 32
      2.1.4   The Mechanism of Graphene Vapor-Solid
              Growth on Insulating Substrates............................... 34
      2.1.5   Graphene Production by Laser Ablation....................35
      2.1.6   Doping of Graphene...35
      2.1.7   Synthesis of Graphene Oxide...................................35
2.2   Synthesis of Nanotubes: Arc Discharge, Chemical Vapor
      Deposition, and Laser Ablation... 36
      2.2.1   CNT Synthesis: Arc Discharge 36
      2.2.2   Nanotube Growth by CVD.......................................37
      2.2.3   Pulsed Laser Ablation 37
2.3   Functionalization of Carbon Nanostructures by Metals 38
      2.3.1   Functionalization of CNT 38
      2.3.2   Filling with GdCl$_3$...39
      2.3.3   Different Methods of Functionalization....................39
      2.3.4   Chelate System Form with Functionalized
              MWCNTs 39
2.4   Doping with Light Elements (Nitrogen, Boron, and
      Other Elements)... 40
2.5   Growth Mechanism(s) of Fullerenes and Nanotubes 41
      2.5.1   The Pentagon Road 42
      2.5.2   Growth Mechanism: Vapor-Solid-Solid....................43
      2.5.3   Vapor-Solid-Solid (VSS) Growth Process 46
      2.5.4   Imaging of CNT Growth.. 46
2.6   Growth of a Quantum Dot (QD).. 49
      2.6.1   Chirality 53
2.7   Synthesis Techniques and the CVD (for Diamond Growth)..... 53
      2.7.1   Growth Model of Nanodiamond Films 59
Bibliography.. 61

Chapter 3   Confined Low-Dimensional Structures.. 64

     3.1   Quantum Dots in Carbon Nanotubes 64
           3.1.1   Strong Localization 64
     3.2   Carbon Nanotubes Overview: Quantum Wires in
          Carbon systems.. 66
           3.2.1   Electronic Properties of SWNTs............................. 66
           3.2.2   Electronic Transport Properties of Carbon
                    Nanotubes...70
           3.2.3   Carbon Nanotubes as HF Devices 76
     3.3   Electrons in Graphene 79
           3.3.1   Dirac Dispersion of Electrons 84
     3.4   Deformation through Nanomanipulation in
          Multilayered Graphene 85
           3.4.1   Nanomanipulation Device Fabrication Technique 85
           3.4.2   Quantum Linear Magnetoresistance and Hall Effect 87
           3.4.3   Shubnikov de Haas Oscillations............................... 88
     3.5   Applications of CNT Hybrid Circuits and Graphene
          QDs at High Frequencies... 90
           3.5.1   Graphene-Based Qubits.. 90
     Bibliography.. 93

Chapter 4   Quantum SPIN Tunneling in Carbon Nanostructures and Devices... 95

     4.1   Spintronics Overview 95
           4.1.1   Magnetic Tunnel Junction Effect 96
     4.2   Magnetism in Pure Carbon?.....................................101
           4.2.1   Fullerenes 101
           4.2.2   Graphite and Graphene 102
           4.2.3   Electronic Transport....................................107
           4.2.4   Nitrogen Doping of Graphene 120
           4.2.5   Magnetism Induced by In-Plane Manipulation in
                    Graphene 121
     Bibliography.. 123

Chapter 5   Carbon Superstructures: Diamond Meets Graphene 126

     5.1   Diamond Transistors...126
           5.1.1   Defects in Graphene...................................127
           5.1.2   Graphene Superstructures: Transport in
                    Defective Graphene 132
           5.1.3   Model: Transport through Carbon Clusters 133
           5.1.4   Quantum Well 136
           5.1.5   Motivation for Carbon Superlattice.........................136
           5.1.6   Multilayered Superlattice Structures of the
                    Disordered Carbon 137
           5.1.7   Methodology: Tight-Binding Calculation 139

5.2   Microscopic Model of Correlated Disorder in
       Nitrogenated Amorphous Carbon 143
       5.2.1   Disorder Model of Growth of Carbon and
                Corresponding Electronic Structures 145
       5.2.2   LDoS Description 147
5.3   Summary 152
Bibliography 153

Chapter 6   Mesoscopic Phenomena: Electronic Transport in
             Low-Dimensional Carbon Films 157

6.1   A-B Effect 158
6.2   Geometric Phase 159
6.3   WL Process 160
6.4   Weak Localization in Graphene Films 161
       6.4.1   Zero-Field Conductivity 162
6.5   Activated and Hopping Conduction in Nanodiamond Films 163
       6.5.1   WL in Nanodiamond Films 164
       6.5.2   Three-Dimensional SL Structures Applied to
                N-UNCD Films 168
       6.5.3   Proposed Transport Model: Diffusive Transport
                in Low-Dimensional Disordered Carbon Films 173
       6.5.4   Experiment: Mobility and [n] vs. Temperature,
                Hopping, and Delocalized Transport 173
6.6   Summary 180
Bibliography 182

Chapter 7   Superconductivity in Boron-Doped Diamond and Related Systems 187

7.1   Superconductivity 187
       7.1.1   BCS Superconductivity 188
7.2   Non-BCS Superconductivity: Possible Pairing Mechanisms 189
       7.2.1   Vortex 189
7.3   Superconducting Carbon: Diamond 190
       7.3.1   Transport Features of Boron-Doped Diamond 190
       7.3.2   Microstructure Analysis 192
7.4   Spin–Orbit Coupling 196
       7.4.1   Josephson Tunneling 201
7.5   Andreev Reflection (AR) and Andreev Bound States (ABS) 202
       7.5.1   Electrical Conductivity 203
       7.5.2   Evidence for RSOC 203
       7.5.3   MR Transition Below and Above the Critical Point 207
7.6   Summary 211
Bibliography 211

Chapter 8   Carbon Hybrid System Odd-Frequency Order Parameter and
Vortex Phase..216

8.1   Duality in Superconductivity (Mixed with Impurities)..........216
       8.1.1   Ginsburg–Landau Theory and Vortex Structure 218
       8.1.2   Quantum Spin Liquid..218
       8.1.3   Anderson's RVB Model...219
       8.1.4   Superconductivity in Organic Conductors 222
8.2   Superconductivity in Twisted Graphene (TBC)................... 224
8.3   Berezinskii–Kosterlitz–Thouless Transition 226
       8.3.1   Field-Induced Superconductor–Insulator Transition.....226
8.4   Superconducting Carbon: A New Spin-Triplet
       Superconductor?...233
       8.4.1   ZBCP...233
       8.4.2   Topological Superconductor................................... 236
       8.4.3   FFLO and the SSH Model 238
       8.4.4   Vortex and Oscillatory Δ: FFLO and Odd
                Frequency 239
       8.4.5   Topology and the SSH Model 239
Bibliography.. 241

Chapter 9   Hybrid Quantum States: Diamond NV Center and Qubits.............. 244

9.1   Quantum Bits (Qubits)... 244
9.2   Diamond NV Centers 245
       9.2.1   Nitrogen-Vacancy Center (NVC) in Diamond 245
       9.2.2   The Electronic Structure 248
9.3   Manipulation of the NV Center Spin 250
       9.3.1   Methods of Detecting the Spin State of NV
                Centers: Spin Manipulation of the NV Center in
                the Ground State... 250
       9.3.2   Relaxation Times...251
       9.3.3   Using Quantum Entanglement to Entangle
                NV Centers...252
       9.3.4   Simulating One NV Center (Coupled with
                Superconducting Qubit)..253
       9.3.5   Evaluation of Highly Entangled States in
                Asymmetrically Coupled 3 NV Centers by
                IBM QE 254
       9.3.6   Geometric Phase 256
9.4   Hybrid Superconducting and Photonic Qubits 262
       9.4.1   Coupling Diamond NV Centers to a
                Superconducting Flux Qubit 262
       9.4.2   Implementation of a Hybrid Qubit System.............. 263
       9.4.3   Hybrid Quantum States: Superconducting
                Diamond and NV Center... 264

        9.4.4   Nanodiamond Quantum Sensor for Rapid
                Diagnosis of Viral Diseases 265
    9.5 Summary 268
    Bibliography 268

Chapter 10 Quantum Simulation of Carbon Structures: Close and Open
               Quantum Systems 270

    10.1 Superconducting Qubits 270
          10.1.1 Josephson Junction and Cooper-Pair Tunneling 270
          10.1.2 Josephson Relation 270
    10.2 Superconducting Quantum Interference Devices (SQUID) ... 271
          10.2.1 Intuitive Charge Qubits and Flux Qubits 272
          10.2.2 Nanotube-Based Qubits 273
          10.2.3 Graphene-Based Qubits 274
    10.3 Molecular Carbon, Biomolecules, and DNA-Based Qubits 278
          10.3.1 Future Directions: Quantum Computer in
                  Complex Carbon Structure Research Leading to
                  Bioinformatics 278
          10.3.2 Superconducting Phase-Slip Qubits 278
          10.3.3 Spin Qubits Based on Spin-Triplet Properties 282
    10.4 Basics of Quantum Simulations Applied to
          Hybrid Materials 283
          10.4.1 Quantum Tunneling and Coherence
                  Backscattering by Superconducting Qubits 283
          10.4.2 WL to WAL Geometric (AB) Phase 285
          10.4.3 Simulation of Spin–Orbit Coupling and
                  Spin–Flip Effect 287
          10.4.4 Spin–Orbit Coupling and Four-Level System 288
          10.4.5 Simulating Vortex and Topological
                  Insulators by Superconducting Qubits 290
    10.5 Shockley Model Simulation 291
          10.5.1 ABS and Vortex 291
          10.5.2 Topological Superconductor 292
          10.5.3 Geometric Phase Acquired by the GB 293
    Bibliography 295

Chapter 11 Conclusion: New Directions 299

Index 303

About the Author

Somnath Bhattacharyya, PhD is a Professor in the School of Physics at the University of the Witwatersrand, Johannesburg, South Africa. His research is focused on the areas of condensed matter physics, nanotechnology, and quantum computation. In 1997, Somnath Bhattacharyya completed his doctoral degree from the Indian Institute of Science, Bangalore in Condensed Matter Physics. He worked as a researcher in the United States, Germany, France, and England. His major interest is in the transport properties of carbon, and his major achievements include multilayer carbon-based resonant tunnel devices, high-speed transport and spintronics in nanostructured carbon devices, n-type doping of nanocrystalline diamond films, spin-triplet superconductivity in boron-doped diamond films, theoretical models for transport in disordered carbon, and recently, quantum simulations of many body systems. Prof. Bhattacharyya published two books, four book chapters, and over one hundred papers in peer-reviewed journals. In the book entitled *Evolution: Classical Philosophy Meets Quantum Science*, he presented a model of 'creation' and connected it with the quantum science. Some of the ideas on quantum transport described in that book are extended in the present book. At present, Prof. Bhattacharyya is engaged in developing a new infrastructure for a wider range of nanotechnology that will include quantum bits, quantum matter, and quantum simulators based on hybrid carbon structures.

Acknowledgments

The story of carbon in this book was developed as a result of my intense curiosity about the birth of the universe, primordial energy, intelligence, and the universe. I have studied carbon for more than 25 years. This book is written based on the works of a large number of scientists which enriched the science of carbon. I sincerely thank to the people whose experimental results and calculations are displayed in this book. I had the good fortune to collaborate with (late) Prof. S. V. Subramanyam, Prof. G. Turban, Prof. F. Richter, Dr. A. Kraus, Dr. D. Gruen, and Prof. Ravi Silva. Former colleagues, S. Ahmed, A. Sumant, C. Cardinaud, C. Speath, and L. Gomez Rojas are acknowledged for assistance and discussions. Over time I have benefited from interactions with the pioneers in the field of carbon materials, condensed matter physics, nanotechnology, and quantum technology. Discussions with Kostya Novoselov, Andrea Ferrari, David Tomanek, Mildred Dresselhaus, Mikhail Katsnelson, Paul May, ACH Castro Neto, Gehan Amaratunga, Ping Sheng, and Sense Jan van der Molen have been very inspiring.

In quantum technology discussions with William Oliver (MIT), Lloyd Hollenberg (Melbourne), John Martinis (UCSB, Google), Nick Bronn (IBM), Rupak Biswas (NASA), Meya Meyappan (NASA), and Francesco Petruccione (UKZN) were very helpful. For electron microscopy, I received generous support from Angus Kirkland (Oxford). Milos Nesladek offered me superconducting diamond samples which were used to fabricate diamond-based resonators at Zeiss lab (USA) and measured by Ilya Besedin at MiSIS.

In condensed matter physics and superconductivity discussions with Ping Sheng (HKUST), Satoshi Okuma (Tokyo), Y Takano (NIMS), T.V. Ramakrishnan, N. Kumar, A. Taraphdar, Jan Aarts (Leiden), and Brian Hickey (Leeds) were very useful.

I conducted experimental as well as theoretical research on everything from quantum computation to quantum electronics, which greatly aided me in comparing the findings of many studies with the principles of basic physics. During my time at MISIS in Moscow, Russia in 2019, I had the chance to give seminars explaining the beauty of intricate carbon structures that served as the inspiration for this piece.

I would like to express my profound gratitude to the University of the Witwatersrand in Johannesburg, South Africa, for supporting my research where I worked on this book. Some part of the book was written in 2022 (Laboratoire de Chimie des Mat´eriaux Inorganiques, CNRS, Universit´e Paul-Sabatier, Toulouse, France), Gangtok, Sikkim, India, and 2023 (University of Surrey, Guildford, United Kingdom) while I was on sabbatical. My research on carbon-based nanotechnology was supported by the National Research Foundation, South Africa, DST-NRF CoESM, and CSIR-NLC.

I got substantial inspiration from my former students and research fellows. George Chimowa (Nanotubes, N-doped diamond), Dmitry Churochkin (Superconductivity, weak localization), Mikhail Katkov (superlattice calculations), Ross McIntosh (calculations), Christopher Coleman (graphene, superconductivity), Siphephile Ncube & Mosse Sav (nanotubes), Davie Mtsuko (Superconductivity), Farai Mazhandu & Declan Mahony (NV center), and Shaman Bhattacharyya (quantum computation) produced the results which were used in this book. C. Coleman and V.R. Sodisetti provided me with some diagrams and text that are used in this book. Alvaro de Sousa and Andre Strydom (UJ) are acknowledged for synthesis of metal-functionalized CNT samples and transport property measurements. I have used some of the calculations and data analysis of Mr. Shaman Bhattacharyya in this book, who critically revised the work. I am very thankful to the young scholar Mr. Samya Mukherjee (Kalyani, India) for supplying me with plenty of information, which acts as an introduction to my writings. Prabin Pykurel (Gangtok, India) is acknowledged for improving some parts of the text. Angela Simon (Vacutec, Johannesburg) offered me assistance to improve the text of some parts of the manuscript.

My colleague Neil Coville (Wits University, Johannesburg) very carefully read the manuscript and made suggestions about the content of this book. At Wits, I enjoyed discussions with my colleagues Rudolph, Daniel, Alex, Robert, Joao, Andrew, Yorick, Arthur, and Darell. I thank Emmanuel Flahaut and Wolfgang Bacsa for stimulating discussions on nanotubes during my visit to Toulouse in 2022. Close interactions with Victor Koledov and Svetlana von Gratowski (Moscow, Russia) inspired me to write the book.

Special credit should go to the publisher Dr. Gagandeep Singh, who invited me to write this book in 2021. Thanks to all reviewers for carefully reviewing the synopsis and proposal of this book and providing some of the most valuable suggestions. I am thankful to Karthik Orukaimani from Codemantra for proofreading.

Finally, I must thank my parents H.P. Bhattacharyya & Anjali Bhattacharyya, my wife, Monali and sons Shaman & Aryaman, for supporting my work.

Introduction
Carbon: Harmony of Beauty, Strength, and Intelligence

Carbon is well-regarded as an energy reservoir as well as for storing a wealth of information within nanostructures exhibiting outstanding electronic transport properties which have been mimicked in emergent quantum technology for the artificial neural network. The complexity of carbon bonds combines the most fascinating and rigid chain-like and ring structures yet fast interconnectivity appeared to be the basis of any biological unit which is believed due to its *'dual'* character in bonds named 'resonating valence bonds'. Condensed matter physics is developed based on the *resonating valence band model* originally found in benzene ring structures having a *'topological'* feature. This resonance model explaining interacting spin systems like in a magnet can be applied to solve the fascinating electronic properties of conducting carbon systems, which can pave the path connecting *carbon physics* and quantum biology. We elaborate on how the understanding, as well as controllability, of the quantum electronic properties of carbon structures can be largely improved through a combined study of structural geometry, spintronic properties, and dynamics of resonating valence bonds. Pure carbon forms several beautiful nanostructures, and a good model explaining their growth mechanism is addressed in the book.

Due to the *'dual'* characteristics of carbon bonds, hybrid carbon superstructures combining diamond and graphene have been constructed to study resonant tunneling in a carbon system which also explains the high-speed transmission of information in a carbon structure. A new addition to this family of beautiful yet strong nanostructured carbon is a novel topological insulating and superconducting phase that can be made in diamond and graphene structures. We focus on this superstructure which can be useful not only for building intelligent devices but also mimicking carbon structures available in nature or biological systems that store information through intelligent processing.

Diamond crystals absorb nitrogen and boron atoms to create beautiful color centers and exhibit very interesting optical and vibrational properties which need deep insight. The doping of diamond and other carbon nanostructures by incorporating nitrogen and boron is discussed since controlled doping in carbon has not achieved yet. Heavily boron-doped diamonds showing spin-triplet-like superconducting transition can offer a new class of quantum bits or spin qubits which is expected to play a key role in the development of artificial intelligence. By accumulating nitrogen atoms, carbon can construct the elements of life to the smallest magnetic center, the well-celebrated nitrogen-vacancy (NV) color center of the diamond which is the heart of quantum technology. We focus on the fascinating high-speed and spin transport in carbon nanostructures and particularly on 'superconductivity' in boron-doped diamonds.

Perhaps we cannot treat carbon as a semiconductor since the bandgap cannot be defined in the very defective structure of carbon. From an insulating state it can show very high conductivity in association with dopants followed by a superconducting transition. Overall, carbon is not an electron-rich system. However, a hybrid structure of carbon may work well, which can be found in nature. A natural object takes up as many different forms as possible. It can absorb (attract) other species from nature, and after some time, it desorb (repel) materials. The cycle continues based on attraction and repulsion. This can be reproduced in a quantum system (simulator). As a classical device, a transistor is operated by the application of a voltage that produces an ON/OFF state. A spintronic device works based on the duality of spin-polarized states; however, it does not work well in carbon, which is a highly defective material. Due to the duality or uncertainty of states in carbon, classical or linear transport in carbon seems to be impossible. Quantum transport associated with a dual characteristic, either in structure or in the electronic/spintronic state, is common in carbon. This can create cycles or a quantum system, a topologically protected system. A cycle can absorb and reject by deforming itself, a carbon cycle.

Things start from a dot or an atom. A shell structure is developed based on the strength of the nucleus or the size of the positively charged core. For a heavy nuclear system, the core is compensated by negatively charged electrons arranged in different shells. For a heavy metal, free electrons can be arranged from different shells. Light materials do not have lots of electrons which can be used for an electronic device such as a transistor. How do you make a shell structure which is curved? Well, the bonds will try to arrange themselves so that a ring structure of a closed loop can be formed. So, the linear space becomes curvilinear. An sp^3 bond is split into an sp^2 or a double-bonded structure which varies periodically. Hence, a resonant valence structure is created in a carbon system which can provide extra electrons so that transport in carbon can be comparable to metals. An insulator state of carbon can be transformed into a metallic state and even a superconducting state in curvilinear space by introducing a small defect, that is, duality.

In this book, we explain (i) dualism in carbon explaining mass-energy transfer through alternative resonant valence bond structures, (ii) resonant tunneling to Josephson tunneling, (iii) static beauty (geometry) mixed with dynamics (electron distribution), (iv) stability in complex structures offering extreme speed, and (v) application in quantum technology based on topologically protected beautiful structures with very low noise.

In Chapter 1, we introduce this unique element through the synthesis, fundamentals of the electronic properties, and applications in electronic devices extending to emergent quantum technology. In Chapter 2, we explain a special growth model for carbon which includes both ordered and disordered phases through the optical and vibrational properties of carbon. We introduce the concept of a strong localization or close quantum system of carbon in Chapter 3. The system starts to interact with others through the tunneling process and creates an open quantum system as described in Chapter 4. The examples of an open/closed system in a real superstructure are described in Chapter 5. Transport in the superstructure will be described by mesoscopic physics in Chapter 6. In Chapter 7, addition of spin-orbit coupling presents topological features of the carbon system. Unconventional superconductivity in the carbon system is discussed in Chapter 8. A hybrid system of boron-doped superconducting diamond and diamond NV center is suggested in Chapter 9. Finally, the application of qubit in constructing the hexagonal structure of carbon to simulate quantum transport features is discussed in Chapter 10.

1 Emergent Quantum Processes in Nature
Search for a New Material

1.1 ANCIENT CARBON: CARBON CONCERNS THE CREATION OF THE UNIVERSE AND DESCRIBING THE ORDER AND DISORDER

After hydrogen, helium, and oxygen, carbon is the fourth most abundant chemical element in the observable universe by mass. Carbon is abundant in stars, the Sun, comets, and most planets' atmospheres. Carbon, in the form of carbide, can be found at the Earth's core alongside iron. Carbon on Earth may have originated extraterrestrially from the solar wind, comets, or meteorites that collided with the newly formed Earth. However, extra-terrestrial carbon structures such as diamonds and fullerenes are transported to Earth.

We'd like to know when carbon was created. The answer is that carbon predates fire. According to some calculations, the Big Bang produced 1064 carbon atoms (compared to 1080 atoms of hydrogen). It is also proposed that carbon and oxygen were formed much later in stars rather than during the Big Bang. All carbon and oxygen in all living things are produced in nuclear fusion reactors known as stars. The early stars are massive and fleeting. They produce heavier elements by consuming hydrogen, helium, and lithium. Carbon forms the first crystal with four bonds in the form of a diamond around 4.5 billion years ago because of the dissociation of methane at extremely high temperatures and pressure.

A trigonal or double-bonded graphic structure is formed at a slightly lower temperature. The nuclear fusion process by which hydrogen burns and produces helium in a star or the Sun; carbon is formed from fusion of Helium. Fred Hoyle proposed the concept of "nucleosyntheses," as well as a mechanism known as "the Triple-alpha" process. Carbon, like fire, transforms or recycles energy, but it also stores information in the microstructures.

1.2 CARBON IN SPACE: CARBON IN THE EXTREME ENVIRONMENT ON THE EARTH AND IN DEEP SPACE (SOOT TO DIAMOND)

The complexity of carbonaceous molecules in space, their abundance, and the time scale they form are critical questions in cosmochemistry. Despite the great diversity of galactic and interstellar regions, organic chemistry appears to follow common pathways throughout the universe. The majority of the carbon in space is thought to exist in the form of large molecules known as Polycyclic Aromatic Hydrocarbons (PAHs) – these

DOI: 10.1201/9781003316411-1

molecules are abundant in space but have not been directly observed (according to circumstantial evidence of 1980). PAHs were thought to form efficiently only at high temperatures; on Earth, they occur as by-products of fossil fuel combustion and can also be found in char marks on grilled food. However, the interstellar cloud where the researchers observed them has not yet begun to form stars, and the temperature is approximately 10° above absolute zero. These findings imply that these molecules can form at much lower temperatures than previously thought, prompting scientists to reconsider their assumptions about the roles of PAH Chemistry in the formation of stars and planets. Astronomers used telescopes to detect infrared signals indicating the presence of aromatic molecules, which are molecules with one or more carbon rings. PAHs, which contain at least two carbon rings, are thought to contain 10%–25% of the carbon in space, but the infrared signals were not distinct enough to identify specific molecules. Their existence in interstellar and circumstellar meteorite regions is either unknown or debatable. In any case, such delicate species are easily destroyed by UV radiation, shocks, and thermal processing, and they are unlikely to survive incorporation into the solar system without being harmed. The more refractory material, in particular macromolecular carbon, may more faithfully retain an interstellar heritage.

1.2.1 FIRST CLEAR DETECTION OF CARBON DIOXIDE

NASA's James Webb Space Telescope captured the first clear evidence of carbon dioxide in the atmosphere of a planet outside the solar system (Figure 1.1). A small hill between 4.1 and 4.6 μm in the resulting spectrum of the exoplanet's atmosphere provides the first clear, detailed evidence of carbon dioxide ever detected in a planet outside the solar system. This observation of a gas giant planet orbiting a Sun-like star 700 light-years away provides important insights into the composition and formation of the planet. Access to this part of the spectrum is critical for measuring the abundance of gases such as water,

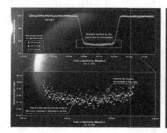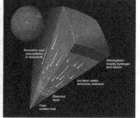

FIGURE 1.1 Figure 1.1 (left) Clear evidence for carbon dioxide in the atmosphere of a planet outside the solar system as observed by NASA's James Webb Space Telescope. A series of light curves from the Near-Infrared Spectrograph shows the change in brightness of three different wavelengths of light from the WASP-39 star system over time as the planet transited the star on July 10, 2022. [First Clear Detection of Carbon Dioxide NASA's James Webb telescope; https://webbtelescope.org/contents/media/images/2022/042/01GB31JJZT8NSEZQCJ8BE5EY56?news=true]. A similar claim can be found in the paper [JWST Transiting Exoplanet Community Early Release Science Team. Identification of carbon dioxide in an exoplanet atmosphere. Nature 614, 649–652 (2023). (middle) The NASA/ESA Hubble Space Telescope made a striking observation of the carbon star CW Leonis, which resembles a baleful orange eye glaring from behind a shroud of smoke. [https://esahubble.org/news/heic2112/] (right) Neptune's cutaway reveals the planet's several layers, one of which might be the source of diamonds. According to models, the diamonds seem to gather in a layer somewhat above the planet's core. [25].

methane, and carbon dioxide, which are thought to exist in a variety of exoplanets. From the NASA/ESA Hubble Space Telescope, the carbon star CW Leonis is observed, which resembles a menacing orange eye peering out from behind a shroud of smoke. In this Hubble Space Telescope image, CW Leonis glows from deep within a thick shroud of dust. CW Leonis, located approximately 400 light-years from Earth in the constellation Leo, is a carbon star – a luminous type of red giant star with a carbon-rich atmosphere. As the outer layers of CW Leonis itself were thrown out into the void, dense clouds of sooty gas and dust engulfed this dying star.

1.2.2 CARBON AND FIRE

Carbon is derived from the Latin carbo, which means coal or charcoal. Carbon was discovered in prehistory and was known to the earliest human civilizations in the forms of soot and charcoal. Fire has been used for over 400,000 years. To generate heat, ancient people used fire to burn organic materials (wood or fossilized rocks). Controlling fire was critical for the development of the devices, so supplies of organic sources or carbon were sought. Carbon was the first element to be used in the construction of civilization, and it was particularly revered as a God in ancient Indian culture. Carbon has been identified as God's abode in the form of fire. Agni, or fire, sleeps in a carbon-rich material for a very long time, but it can be awakened as visible flame and light when the carbon-rich material is excited. By fracturing the materials or carbon, fire and light can be liberated from the carbon system. The Sanskrit word "Angara," which means "charcoal or source of fire," is related to the name of a sage named "Angiras," who was one of the main composers of the most ancient book "Rig Vedas," written around 1500 BCE. This book describes the carbon cycle and the role of fire in producing our carbon-based life cycle. Carbon in the form of charcoal was created around the time of the Romans using the same chemistry as it is today, by heating wood in a pyramid covered with clay. Carbon was undoubtedly studied in ancient Africa for heating and melting minerals. Diamonds were previously known to be found in India and China. From the time of their discovery in the 9th century BCE to the mid-18th century AD, India was the leading country in the world in diamond production, but the commercial potential of these sources had been exhausted by the late 18th century. In 1725, diamonds were discovered in Brazil and later in many other countries.

As a result, fire and carbon are inextricably linked. Carbon absorbs heat (fire) and emits light (fire). It is an excellent material for dealing with fire, such as holding the fire for an extended period and raising the temperature of its body to extremely high levels, which is required for metal smelting. Carbon is flammable. The atomic number of elements should be 2, 10, 18, or 36 for inertness. Carbon has six electrons that are halfway between 2 and 10, and can move to the lower or higher side by releasing or accepting four electrons. This implies that carbon has a dual nature and can take any configuration by using single to double and triple bonds. Let's add hydrogen to carbon until we reach the magical number of 10, at which point the hydrocarbon becomes flammable. Methane has a total of ten electrons. Ethane is made up of two carbon atoms and six hydrogen atoms, for a total of 18 electrons. Propane, butane, and octane are also flammable, with three, four, and five carbon atoms, respectively. Carbon acts as the Earth (solid), air (gas), fire (plasma), and water (photosynthesis) by absorbing heat and light from the Sun. Carbon dioxide and water combine to form glucose and oxygen.

"There is not a law under which any part of the universe is governed which does not come into play and is touched upon in the chemistry of a candle," writes Michael Faraday. By radiation and conduction, the flame's heat melts the wax; capillary action drags the wax up the wick; the wax then vaporizes; chemical reactions create the flame; heated solid carbon particles glow; and convection currents remove the combustion byproducts.

1.3 CARBON CYCLE

Recycling Energy in Carbon = Recycling Information; Control Disorder Level = Cleaning up the System

The carbon cycle is a biogeochemical cycle that exchanges carbon between the Earth's biosphere, pedosphere, geosphere, hydrosphere, and atmosphere. Carbon is the most abundant element in biological compounds, as well as in many minerals such as limestone. The carbon cycle, such as the nitrogen and water cycles, consists of a series of events that are critical to the formation and maintenance of life. It describes life, carbon, and the movement of carbon throughout the biosphere, as well as long-term processes of carbon sequestration to and releases from carbon sinks. Carbon sinks on land and in the ocean currently absorb roughly one-quarter of anthropogenic carbon emissions each year. Because of dissolved carbon dioxide, carbonic acid, and other compounds, the acidity of the ocean surface has increased by about 30%, fundamentally altering marine chemistry. Most of the fossil carbon has been extracted over the last half-century, and rates of extraction continue to rise rapidly as a result of climatic change.

Humphry Davy popularized the carbon cycle, which was first described by Antoine Lavoisier and Joseph Priestley. The global carbon cycle is now commonly divided into several major reservoirs of carbon with interconnected pathways of exchange: the atmosphere, terrestrial biosphere, ocean, sediments, and the Earth's interior. These carbon stores interact with the other components through geological processes. Carbon exchanges between reservoirs take place because of a variety of chemical, physical, geological, and biological processes.

The carbon cycle is a natural method of recycling carbon atoms, which travel from the atmosphere into organisms on Earth and then back into the atmosphere. Most of the carbon is stored in rocks and sediments, with the remainder in the ocean, atmosphere, and living organisms. Carbon enters the atmosphere as CO_2, which is absorbed by autotrophs like green plants. Animals consume plants, absorbing carbon into their systems. Animals and plants die, and their bodies decompose, releasing carbon back into the atmosphere. Carbon is the chemical foundation of all lives on Earth. Carbon compounds regulate the Earth's temperatures, make up the food that sustains us, and provide energy that fuels the global economy. Depending on the pressure and temperature at different levels below the surface of the Earth, carbon is transformed into diamond and graphite (Figure 1.2). Thus, the carbon cycle maintains a balance between different life forms and helps to keep the Earth's temperature stable. Plants absorb carbon dioxide and sunlight during photosynthesis to produce fuel – glucose and other sugars – for building plant structures. The fast biological carbon cycle is built on this process. Carbon dioxide aids in the retention of solar energy, preventing it from escaping back into space; without carbon dioxide, the oceans would be frozen

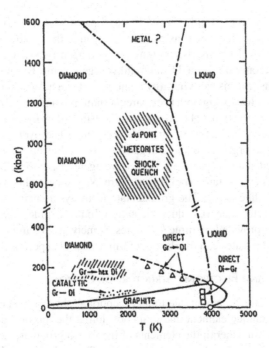

FIGURE 1.2 One recommended variation of the carbon phase diagram. In addition to diamond and graphite, other phases exist, including hexagonal diamond, a high temperature-high pressure phase, meteorites, and shock-quench, which have not received much attention but might be related to carbynes (du Pont in the diagram). Liquid carbon that has not been thoroughly researched [26].

solid. In response to the effects of climate change, there are three main mitigation strategies: reducing carbon emissions through low-carbon technology by prioritizing renewable energy resources, recycling, minimizing energy use, and implementing energy-conservation measures.

1.4   DUALISM IN CARBON MICROSTRUCTURES: GROWTH MECHANISM AND STRUCTURE

1.4.1   GROWTH OF CARBON ATOMS AND MOLECULES

To find real applications for an element like carbon, the structure must be grown in a highly controlled environment. A two-level (or higher) system is required to store quantum information in the form of a superposition of states, such as a "dual." In the following chapters, we will explain how carbon exhibits this "dual" character in every way to distinguish itself. The application of carbon as a source of energy since ancient times to modern quantum technology is discussed. In Chapter 2, a plausible mechanism for the creation and transformation (rearrangement) of carbon will be proposed based on electric and magnetic dipole interactions and their dualistic nature.

A vortex or a blackhole model can be used to explain the formation of carbon. In philosophy, hydrogen atoms emerge as a fireball-like vortex. Positive and negative particles, ±energies, and even space with positive and negative curvatures will be created from a neutral (and infinitesimally small particle or a void). They should be separated by a gap so that electrons can revolve around the proton (mass). Due to its dualistic nature, two electrons, two protons, and two neutrals are produced. Following that, two helium atoms are formed from four hydrogen atoms. Because of duality, two helium and two hydrogens can **virtually** combine to form a carbon atom. Carbon has dual properties, such as being both transparent (diamond) and absorbent (graphite), as well as a mixture of the two. (The growth mechanism will be discussed in Chapter 2.)

The atomic structure has a strong dualistic nature. The hydrogen atom is made up of a positively charged core and a negatively charged electron, which represent a steady and excited state, respectively. A reactive agent or an excited state is required in addition to the neutral state to build a structure. Near the absolute zero temperature, a pure or ground state exists. The ground state is always associated with or covered by an excited (pi) state due to its dualistic nature. In three dimensions, the full coverage of the state is provided by three "p" states, namely, p_x, p_y, and p_z. It's fascinating to learn that s and p states can overlap and form a state of superposition (Figure 1.3).

1.4.2 ATOMIC AND ELECTRONIC STRUCTURE OF CARBON

Carbon is a fascinating element due to its exceptional physical and chemical properties. Carbon is a unique element due to its position at the top of the fourth group in the periodic table. In general, the elements of the fourth group are neither metals nor insulators; however, carbon is chemically active. The carbon atom has six electrons distributed as $1s^2$, $2s^2$, and $2p^2$ in its electronic structure. These atomic orbitals can hybridize in a variety of configurations, including sp^1, sp^2, and sp^3. As a result, carbon can be found in a variety of stable forms. Carbon, for example, is the most abundant element in all organic compounds. Furthermore, carbon can form covalent bonds,

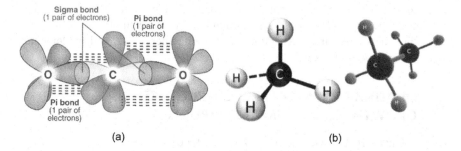

(a)   (b)

FIGURE 1.3 (a) The Lewis structure of carbon dioxide. Carbon is making 2s and 2p bonds to the oxygen atoms. The 2s bonds indicate that there are two equivalent molecular orbitals formed. To form two hybrid molecular orbitals, we need to mix two atomic orbitals, an s orbital and a p orbital. The resulting hybrid orbitals are called sp hybrids. The two unhybridized p-orbitals on carbon form p bonds to the oxygen atoms. The carbon atom in the molecule of methane is the one in the middle (black), while hydrogen atoms are the white ones. The orbitals between the carbon atom and the hydrogen are sp^3-hybridized, with the geometry of tetrahedral shapes. (b) The molecule of ethane, the carbon atoms are sp^2-hybridized. The bond between the two carbon atoms is σ-type one, while the p_z orbital is perpendicular to the plane of the molecule, forming a π-bond. (http://butane.chem.uiuc.edu/cyerkes/Chem102AEFa07/Lecture_Notes_102/Lecture%2015%27-102.htm.)

which can result in the formation of one of nature's strongest bonds. Carbon's richness stems from its electronic structure. The valence shells, $2s^2$ and $2p^2$, have four electrons available to form bonds, whereas the core electrons, $1s^2$, have no chemical activity. When the low energy configuration of the valence electrons, $2s^2$, $2p^1$, and $2p^1$, is examined, it is discovered that the $2p^z$ orbital is empty. This enables carbon to form as many bonds as possible by forming excited states, resulting in the formation of $2s^1$, $2p^1$, $2p^1$, and $2p^1$ orbitals. As a result, carbon could form four bonds. Consider the compound shown in Figure 1.3 to better understand carbon atom hybridization. In this case, the carbon atom is surrounded by four hydrogen atoms, which means that all orbitals must be mixed to form one unpaired electron in each of the four orbitals, and each hydrogen group contributes one electron to the electron pair. In this case, however, the s orbital and the three p-orbitals have all contributed to the formation of the sp^3 hybrid orbital. In terms of geometry, the orbitals arrange themselves in three dimensions to minimize the repulsive force between them. The s orbital has three components, one along each axis, whereas the p orbital has four (x, y, and z). It has two lobes, one in the positive direction of the axis and the other in the negative, forming a tetrahedral geometry with an angle of $109.5°$ between each orbital in the case of the sp^3 hybrid.

Ethane is another example of sp^2 carbon hybridization. As in the previous example, only the s, p_x, and p_y orbitals are mixed this time. The p_z orbital contains an electron that does not mix with the other orbitals, resulting in the formation of a sp^2 hybrid (one s and two p-orbitals are contributing). The orbitals are arranged geometrically to minimize electric repulsion, forming a trigonal planner geometry with an angle of $120°$ between the orbitals that lie in the plane, while the p_z orbital remains perpendicular to the plane. When sp^2-hybridized carbon atoms are bonded together, the sp^2 orbitals overlap, forming the so-called saturated bond shown in Figure 1.3. The p_z orbitals overlap side by side, forming the bond.

The sp^2 hybridization of carbon atoms results in lattice geometry, structural flexibility, and optical properties. In terms of electronic structure, the saturated bond formed by the sp^2 hybrid's overlap forms a deep valence band that does not affect the electronic properties. When the half-filled p_z orbitals bind covalently, the bond is formed. These electrons describe most of the electronic properties of graphene, which will be discussed in the following section.

The formation of bonds can be explained from the interactions of forces. It is possible when things are repelled a potential or attractive force is required to control the repulsive force. This duality is represented by a hand and a self-matter. A transparent-colored object is another good example. This dual property in carbon is the result of a single (hand) and a double (soft resonating) bond. Carbon uses its two "s" and two "p" electrons to redistribute four hands (or heads) and creates either four equal hands directed in four directions or three directions in a 2D plus one normal to the plane, which can be liberated. The free one can move over a long distance in an extremely stable structure built by three other hands. It's like a triangle with a unique center that controls the triangle's property without breaking the symmetry (Figure 1.4).

Take two strings, separate them by two fixed obstacles, and then vibrate them. If they are not touching, resonance is created. As a result, energy is transferred from one side to the other via resonance. Consider a unique structure, such as a hexagon, in which three double bonds are separated from insulating single bonds. This structure

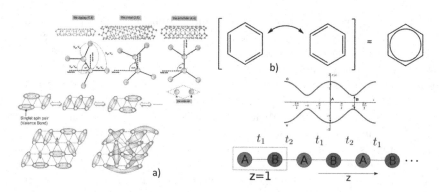

FIGURE 1.4 (a) Two equivalent resonance structures of hexagons which are equal to a circle. A chiral structure has a mirror symmetry. It consists of left-handed and right-handed rotations similar to a DNA structure [27]. (b) Resonance Valence Bond model used in a strongly correlated system in condensed matter explains quantum phase transition and superconductivity. Exotic structure can be suggested. Sketch of the 1D chain of atoms in consideration. The integer unit cell coordinate is denoted by z, and the intracell and intercell hopping amplitudes are denoted by t_1 and t_2, respectively [28].

can generate a ring current that can flow both clockwise and counterclockwise. Several hexagons can be fused to form a large superstructure. This resonance can spread throughout the structure, and thus life begins. A lifeless material, carbon comes to life.

The speed of a process or the requirement for a one-of-a-kind structure – which can be extremely stable or dense while accommodating speed, such as a vortex surrounded by spin liquid. It is a diamond with a wide and well-defined band gap surrounded by a graphitic structure or ring. It has resonating bonds that are single (diamond-like) and double (graphite-like), representing strong repulsion and attraction, respectively. As a result, a band gap appears and disappears. Carbon has 2s and 2p electrons that form isotropic and anisotropic orbitals that correspond to stable (or massive) and unstable (or mass-less) fermions, respectively. The p-orbitals protrude from the flat plane of "s" orbits as if they have no mass. Two "p" electrons rotate in a plane that is perpendicular to the plane of rotation of "s" electrons. This is like a spin-orbit interaction (SOI) configuration, which can arrange and form sp³, sp², and sp¹ bonds to provide various bonding configurations.

1.5 RESONANT VALENCE BONDS IN CONDUCTING POLYMERS HYDROCARBON CHAIN CONDUCTING POLYMER

Conducting polymers, also known as "Intrinsically Conducting Polymers" (ICP), are organic polymers that conduct electricity. Such compounds may be metallic conductors or semiconductors. The most significant advantage of conductive polymers is their processability, primarily through dispersion. Organic synthesis and advanced dispersion techniques can be used to fine-tune the electrical properties. Contiguous sp² hybridized carbon centers form the backbones of conducting polymers. Each center has one valence electron in a p_z orbital, which is orthogonal to the other three sigma bonds. All p_z orbitals combine to form a molecule with a widely delocalized set of orbitals. When the material is doped by oxidation, which removes some delocalized electrons, the electrons in these delocalized orbitals have high mobility. Thus, conjugated p-orbitals form a one-dimensional electronic band, and when this band is partially emptied, the electrons within it become mobile. A tight-binding model can be used to easily calculate the band structures of conducting polymers. In theory, these same materials can be doped by reduction, which introduces electrons into an otherwise empty band. In theory, these same materials can be doped by reduction, which introduces electrons into an otherwise empty band. Most organic conductors are oxidatively doped to produce p-type materials. Undoped conjugating polymers, on the other hand, are either semiconductors or insulators. The energy gap in such compounds can be less than 2 eV, which is too large for thermally activated conduction. Undoped conjugated polymers, such as polythiophenes and polyacetylenes, have low electrical conductivities ranging from 10^{-10} to 10^{-8} S/cm. Subsequent doping of the conducting polymers will result in conductivity saturation at values ranging from 0.1 to 10 kS/cm for different polymers.

1.5.1 CONDUCTING CARBON: RESONANT VALENCE BOND

Linus Pauling proposed the benzene hexagonal ring resonant valence bond structure, in which a single and double bond alternate and produce a ring current. The Su–Schrieffer–Heeger (SSH) model is a one-dimensional lattice model with topological properties in condensed matter physics. It was developed in 1979 by W-P Su, J R Schrieffer, and A J Heeger to describe the increase in electrical conductivity of a polyacetylene polymer chain due to the presence of solitonic defects. It is a quantum mechanical tight-binding approach that describes spinless electron hopping in a chain with two alternating types of bonds. Electrons in one location can only hope to travel to nearby locations. The system can be in a metallic (conductive) or insulating phase depending on the ratio of the hopping energies of the two possible bonds. Depending on the boundary conditions at the chain's edges, the finite SSH chain can behave as a topological insulator. There is an insulating phase for the finite chain that is topologically non-trivial and allows for the existence of edge states that are localized at the boundaries. The model depicts a half-filled one-dimensional lattice with two sites per unit cell, A and B, each representing a single electron. Each electron in this configuration can either hop within the unit cell or hop to an adjacent cell via the nearest neighbor sites. The dispersion relation will have two bands, as with any 1D model with two sites per cell (usually called optical and acoustic bands). A band gap exists when the bands do not touch. If the gap is at the Fermi level, the system is classified as an insulator. Phil Anderson built a broken symmetry model using this structure. This is the modern condensed matter theory of superconductivity's foundation. Although there is no single theory of high-temperature superconductivity, the Anderson–Higgs model is very effective at explaining many aspects of phase transition in magnetic and superconducting systems. We will explain the dualistic theory of structure and conduction in the carbon system, which developed the condensed matter theory of systems, in the later part (Chapters 7 and 8) of this book. The change of electrical properties in carbon structures based on the variation of the band structures (band gap) will be discussed in Chapters 3 and 4.

1.6 BEAUTY IN CARBON: UNUSUAL NANOSTRUCTURES: PI-ELECTRONS AND HYBRIDIZATION OF S-P ORBITALS IN CARBON

Carbon's complexities are still unknown. It can create some stunning structures by arranging hexagons, pentagons, heptagons, and other large superstructures. By alternating single and double bonds, it can form chains and even beautiful tubes – carbon nanotubes. A cubic diamond of many different tubes can be formed from a planar graphene structure. In comparison to metals and semiconductors, these structures remain extremely complex (Silicon). Carbon's beauty and intelligence are determined by unresolved chirality and topological features rather than its crystal structure. One of the reasons for this is the presence of pi-electrons in carbon microstructures. Carbon is not a metal or a traditional semiconductor similar to the other seven elements that brought about a revolution in human civilization's history. It is not a metal like gold, copper, silver, and titanium; however, its electrical conductance can outperform copper via a different mode of transport. It can be transformed into a superconductor by doping it with other elements such as boron and potassium. However, none of these elements are superconductors, including carbon. It may be capable of high-temperature superconductivity. Carbon can have a variety of structures, including planar graphitic and three-dimensional diamond-like carbon. While silicon has a diamond structure, it is an excellent semiconductor with a relatively small band gap or window that can be easily opened and closed. Carbon opens a wide range of windows that cannot be easily closed in the structure of a diamond. It makes plenty of space within the gap. The graphitic structure, on the other hand, does not open the window at all because it lacks an energy gap.

We recognize carbon as black coal, carbon dioxide, all organic compounds, and gleaming diamonds. The beauty of carbon nanostructures lies not only in their gleaming diamonds but also in their low dimensionality, from the smallest sphere, the Nobel Prize–winning fullerene, to the narrowest carbon nanotube (CNTs) and, more recently, the thinnest layer of carbon, the Nobel Prize–winning graphene. All these nanostructures show a superconducting transition. Carbon combined with hydrogen resulted in the Nobel Prize–winning conducting polymer structures, which are much more electrically conducting than copper. Finally, superconductivity can be seen by incorporating boron into diamond, which is initially more insulating than glass or quartz (see Chapter 7). Carbon nanostructures are made up of resonating valence bonds that exhibit duality in mechanical (soft to hard), electrical (super-insulator to superconductor), and optical properties (fully absorbent to reflecting and transparent). Incorporation of graphitic layers or dopants (e.g., nitrogen and boron) the transport properties of diamond as well as disordered carbon films can be controlled over a very wide range (see Chapters 5 and 6). By accumulating nitrogen carbon, the elements of life (amino acids) can be constructed to the smallest magnetic center, the well-known nitrogen-vacancy (NV) color center of the diamond, which is the heart of the emergent quantum technology (see Chapter 9).

1.7 CARBON AND ITS ALLOTROPES: FULLERENES, NANOTUBES, GRAPHENE, NANOCRYSTALLINE DIAMOND, ETC.

The fourth family of the periodic chart includes carbon, the sixth element on the list. It has two s-orbitals and two p-orbitals for each of its four valence electrons. The amount of exterior shell electrons makes it chemically similar to silicon and germanium. Atomic orbitals within the valence band have a very tiny energy difference, enabling open-hand development to hybridization. Thus, four distinct kinds of carbon allotropes can be categorized as follows: diamond, fullerenes, graphite (graphene), and single- and multi-walled CNTs. The amorphous material known as coal, charcoal, and candle black is excluded from this. Amorphous carbon can be distinguished from other allotropes of carbon based on its physical attributes, mechanical, and electronic characteristics.

Buckyballs and nanotubes are two examples of the nanostructured forms of fullerene, a novel carbon allotrope identified in 1985. Their discoverers, R. Curl, H. Kroto, and R. Smalley, received the 1996 Chemistry Nobel Prize. More exotic allotropes, like glassy carbon, were found as a consequence of the rekindled interest in novel forms, and it was also realized that "amorphous carbon" is not strictly amorphous. The discovery of single-walled CNTs in Japan in 1991 had a big effect on science and technology. The Nobel Prize for CNT was expected by many academics. A single isolated layer of graphene with exceptionally high electronic properties was first identified in 2004 by K. Novoselov and A. Geim. In 2010, they received the Physics Nobel Prize.

1.7.1 FULLERENE

Fullerene is a carbon allotrope whose molecule is made up of carbon atoms linked together by single and double bonds to form a closed or partially closed mesh with fused rings of five to seven atoms. Closed mesh fullerenes are denoted informally by the empirical formula C_n, where n is the number of carbon atoms. However, there may be more than one isomer for some values of n – the family is named after Buckminsterfullerene (C_{60}). Fullerrite is the bulk solid form of pure or mixed fullerenes. Fullerenes had been predicted for some time, but they were discovered in nature and space only after their accidental synthesis in 1985. The discovery of fullerenes significantly increased the number of known carbon allotropes, which were previously limited to graphite, diamond, and amorphous carbons such as soot and charcoal. They have sparked intense research, both for their chemistry and for their technological applications, particularly in materials science, electronics, and nanotechnology. Because each carbon atom is connected to only three neighbors rather than the usual four, the bonds are commonly described as a mix of single and double covalent bonds.

1.7.2 CARBON NANOTUBE

CNTs and buckytubes are other names for cylindrical fullerenes. CNTs are distinguished by their small diameter, a few centimeters in length, and extremely thin materials. CNTs are cylindrical rolled-up graphene sheets with diameters of a few nanometers and lengths ranging from a few microns to centimeters. CNTs are further classified as Single-Walled Carbon Nanotubes (SWCNTs), which are made of a single rolled-up graphene sheet, and Multi-walled Carbon Nanotubes (MWCNTs), which are made of multiple stacked rolled-up graphene sheets. The tube layers of MWCNTs are held together weakly by van der Waals' forces. This force is estimated to be 0.8 eV/nm of contact between two nanotubes in a crystal, which is enormous when we consider that a nanotube typically measures a few micrometers. As a result, it is critical to separate these bundles and individualize CNTs.

SWCNTs are an allotrope of carbon with diameters in the nanometer range that are intermediate between fullerene cages and flat graphene. SWCNTs can be imagined as cut-outs of a two-dimensional hexagonal lattice of carbon atoms rolled up along one of the hexagonal lattice's Bravais lattice vectors from a hollow cylinder. MWCNTs are made up of nested single-wall CNTs that are weakly bound together in a tree ring-like structure by van der Waals' interactions; these nanotubes are also sometimes used to refer to double- and triple-walled CNTs. CNT surfaces can be chemically modified by coating spinel nanoparticles with hydrothermal synthesis and used for water oxidation. CNTs are represented in the same way as traditional composites, with a reinforcement phase surrounded by a matrix phase. Popular ideal models include cylindrical, hexagonal, and square shapes. While the ideal model implementation is computationally efficient, it does not replicate the micro-structural features observed in scanning electron microscopy of actual nanocomposites. These MWCNTs are frequently used to improve the mechanical and electrical properties of composite materials, as well as for field emission devices, whereas SWCNTs are more commonly used for the fabrication of transistors due to their smaller size dimension and appreciable electrical properties (Chapter 3).

1.7.3 GRAPHENE

Graphene is a single graphite layer composed of a hexagonally arranged sp^2-bonded, stable two-dimensional allotrope of carbon with a variety of distinct properties. It has a single atomic layer of highly oxidized carbon atoms formed by the chemical oxidation of graphite, a cheap and abundant material. The oxidized forms of graphene are laced with oxygen-containing groups. A single-layered graphene oxide is produced using a modified number method and has a functional group content of 40%–60%, resulting in high solubility in water and other solvents that behave similar to water. Graphene, a single layer of carbon atoms, has been touted as the strongest material known to exist, 200 times stronger than steel, lighter than paper, and with exceptional mechanical and electrical properties. When bi-layer graphene was twisted at a specific angle, it demonstrated superconductivity. However, it has been claimed that a two-dimensional carbon topological insulator superior to graphene exists.

Carbon in two dimensions has great applications in field-effect electronics. The density of 2d electron gas in carbon can be modulated by gate bias which is compared with other strongly correlated materials. Hence, the superconductivity transition in the carbon system can be explained (Figure 1.7).

Since its successful synthesis in 2004, graphene has become the focus of research in the material science fields due to its unique properties, attracting a large amount of research for its application. Graphene's extraordinary properties suggest some promising applications, including optical electronics, photovoltaic systems, and composites. However, it has been discovered that the production of single-layered graphene for commercial applications is still being researched to improve its quality, size, and quantity. It has been reported that single-layered graphene can be synthesized using a variety of methods, but Chemical Vapor Deposition (CVD) is the most promising method for producing high-quality, single-layered graphene.

1.7.4 DIAMOND

Since their use as religious icons in ancient India, diamonds have been prized as gemstones. Their use in engraving tools dates to the dawn of time. Diamonds' popularity has grown since the 19th century because of increased supply, improved cutting and polishing techniques, global economic growth, and innovative and successful advertising campaigns. In 1772, French scientist Antoine Lavoisier used a lens to focus the Sun's rays on a diamond in an oxygen atmosphere and demonstrated that the only product of the combustion was carbon dioxide, proving that a diamond is composed of carbon. Diamond is a solid form of the element carbon, with its atoms arranged in a diamond cubic crystal structure. Another solid form of carbon known as graphite is the chemically stable form of carbon at room temperature and pressure, but diamond converts to it extremely slowly. Diamond has the highest hardness and thermal conductivity of any natural material, and these properties are used in major industrial applications such as cutting and polishing tools. It has the greatest thermal conductivity and sound velocity. It has low adhesion and friction, as well as a very low coefficient of thermal expansion. It has high optical dispersion and optical transparency extending from the far infrared to the deep ultraviolet. It is also the most electrically resistant. It is chemically inert, meaning it does not react with the most corrosive substances, and it has high biological compatibility. Natural minerals and oxides are responsible for the presence of impurities in natural diamonds. Diamond is an ideal material for biological applications due to its excellent electrical, mechanical, and biological properties. Diamond is an excellent electrode material for sensing applications due to its ability to tune its electrical properties from insulating to conducting by tuning its band gap energy through doping. Furthermore, because of its chemical stability, it can be used as a coating material for implants and prostheses. Significant advances in CVD technology over the last two decades have made the production of low-cost, high-quality diamond coatings a scientific and commercial reality (Chapter 2).

Most diamonds are formed in the Earth's mantle. Furthermore, as the crust thickened, some blocks of the crust or terranes were buried deeply enough to undergo ultra-high-pressure metamorphism. These have been evenly distributed as micro-diamonds with no evidence of magma transport. Furthermore, when meteorites collide with the ground, the shock wave can generate high temperatures and pressures, allowing micro- and nanodiamonds to form. Micro-diamonds of impact origin can be used to identify ancient impact craters. The Popigai crater in Russia, formed by asteroid images, may have the world's largest diamond deposits, estimated in trillions of carats. Diamonds are dated by analyzing inclusions using radioactive isotope decay. Most gem-quality diamonds are found at depths of 150–250 km in the lithosphere. Such depths are found in mantle keels, the thickest part of the lithosphere, beneath cratons. These regions have enough pressure and temperature to allow diamonds to form, and they are not converted, so diamonds can be stored for billions of years until they are sampled by a volcanic kimberlite eruption.

Diamonds form in the mantle because of a metasomatic process in which a C-O-H-N-S fluid or melt dissolves minerals in rocks and replaces them with new minerals. Diamonds can be formed from this fluid by either reducing oxidized carbon (e.g., CO_2 or CO_3) or oxidizing a reduced phase like methane. A series of growth zones in diamonds can be identified using probes such as polarized light, photoluminescence, and cathodoluminescence. The characteristic pattern in lithosphere diamonds is a nearly concentric series of zones with very thin luminescence oscillations and alternating episodes where the carbon is resorbed by the fluid and then grown again. Diamonds below the lithosphere have a more irregular, almost polycrystalline texture, owing to higher temperatures and pressures, as well as convection transport. In 2021, another was discovered in the Ellendale Diamond Field in Western Australia. Although diamonds are rare on Earth, they are abundant in space. About 3% of the carbon in meteorites is in the form of nanodiamonds with diameters of a few nanometers. Small diamonds can form in sufficient quantities in space because their lower surface energy makes them more stable than graphite (Chapters 5 and 6). Some nanodiamonds have isotropic signatures, indicating that they formed outside of the solar system in stars. Experiments at high pressure predict that large amounts of diamonds will condense from methane into a "diamond rain" on the ice-giants Uranus and Neptune. Some extrasolar planets may be made almost entirely of diamonds. Diamonds could be found in carbon-rich stars, especially white dwarfs. Carbonado, the hardest form of diamond, is thought to have originated in a white dwarf or supernova. The first minerals may have been diamonds formed in stars. A wide range of carbon nanostructures show a superconducting transition by doping (Figure 1.5) by twisting layers of graphene (Figure 1.6), by increasing the n_{2D} (Figure 1.7) or by applying pressure in boron-doped diamonds (Figure 1.8). There are claims of unconventional (spin-mediated) superconductivity in carbon; however, in most of the cases, the superconducting pairing mechanism was found to be unclear. The overlapping of the atomic orbitals in carbon nanostructures due to curvature can induce SOI. SWCNTs show chiral structures which can be corrected with the spin-orbit coupling (SOC).

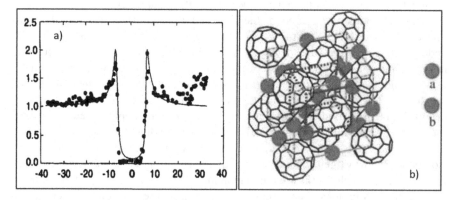

FIGURE 1.5   A superconducting Rb_3C_{60} sample's dI/dV vs. bias voltage V (meV) plot at 4.2 K in temperature. (a) I-V measurements were used to determine numerically the experimental conductance values. (b) Model showing how alkali metal ions are arranged in the fcc M_3C_{60} lattice's tetrahedral and octahedral holes. Alkali metal ions are represented by smaller, light gray spheres, while the cluster is represented by larger, gray-shaded spheres [29].

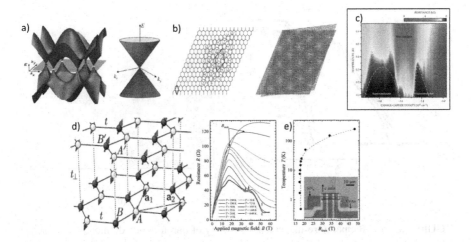

FIGURE 1.6 (a) Electronic band structure of graphene and Dirac cone. (b) Twisted graphene bi-layer shows superconductivity with (c) a variation of temperature vs. charge carrier density [30]. (d) Antiferromagnetism in the Bernal-stacked graphene [31]. (e) Superconducting transition at high temperatures in graphite was claimed [32].

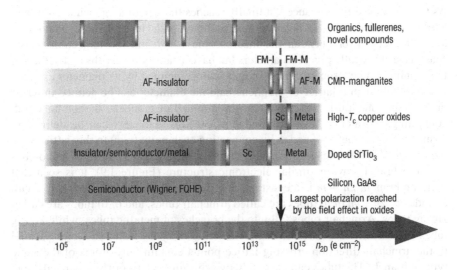

FIGURE 1.7 Carbon needs optimum/low $[n]_{2D}$ for field modulation and active device application. The 2D carrier density of carbon materials are compared with semiconductors and superconductors. New field-effect device: special electrostatic modulation of carriers is needed for carbon-based devices, like high gate bias in strongly correlated systems [33].

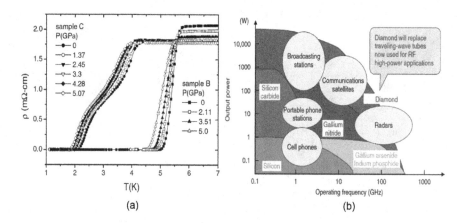

(a)   **(b)**

FIGURE 1.8 (a) Temperature-dependent resistivity of boron-doped diamond shows super-conducting transition at different pressure. Superconductivity in boron-doped diamond at high pressure [34]. (b) Diamond used as high-power electronics at high frequencies since it outperforms other semiconductors in speed [35].

1.8 SUPERCONDUCTIVITY IN CARBON NANOSTRUCTURES: TOPOLOGICAL FEATURES: SPIN-ORBIT COUPLING (SOC)

1.8.1 THE PERIODIC TABLE OF NANOTUBES

As the size of a solid is reduced, it finally reaches the size of a molecule. However, while molecules have no defects, macroscopic solids have defects. Nanotubes are at the crossover between the macroscopic world and the molecular world. They have atomically precise diameters but have a variety of lengths and defects at their endings. The structure of a nanotube can be defined by its diameter as given in the rhombohedral lattice of graphene, and the diameter can be defined in terms of the rhombohedral lattice. Note that the unit cell of graphene is rhombohedral, containing two atoms. Within a wedge of 30°, all diameters can be defined in terms of the parameters n and m by connecting each point (n,m) in this lattice with the origin. Vertical lines in this table define tube branches or families. Nanotubes within a branch have a similar electronic structure (Figure 1.9). It is seen that within a branch, $2n + m$ is a constant, and thus, each branch can be labeled with an index of $2n + m$. (n,n) tubes are called armchair tubes, and $(n,0)$ tubes are called zigzag tubes. Armchair tubes fall into the branches of metallic tubes, while zigzag tubes fall into the branches of semiconducting of types I and II. Lines parallel to the armchair direction crossing lattice points contain only atoms of the same type of branch. The tubes can have only two chiralities: left- or right-handed. Note that in literature, helicity is often confused with chirality. Typical diameters of single-wall CNTs fall between the two lines, between branches 17 and 27. Thus, based on symmetry, nanoparticles can be described by two or more parameters, from which periodic tables can be constructed and families of nanoparticles of similar electronic structure can be identified.

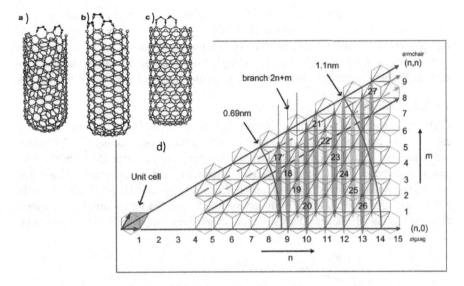

FIGURE 1.9   (a)–(c) Chiral structures of carbon nanotubes (top left). (d) (bottom) Periodic table of carbon nanotubes: rhombohedric lattice of graphene as a function of the helicity parameters *n* (horizontal axis) and *m* (vertical axis): connecting one point from the rhombohedral lattice with the origin defines a roll up vector or the diameter of the tube. Vertical lines show the tube branches or family of nanotubes with similar electronic structure. (*n,n*) tubes are called armchair tubes, and (*n,*0) tubes are called zigzag tubes. Armchair tubes fall into the branches of metallic tubes, while zigzag tubes fall into the branches of non-metallic tubes. (Adapted from Bacsa [36].)

1.8.1.1 SOC in Curved Graphene, Fullerenes, Nanotubes, and Nanotube Caps

The interaction between the spin of an electron and its orbital motion around the nucleus is known as SOC (or SOI). The SOC in quantum physics is a relativistic interaction of a particle's spin with its motion inside a potential – this SOC is one of the relativistic effects that occurs whenever a particle with non-zero spin moves around a region with a finite electric field. SOC can be thought of as a type of effective magnetic field seen by electron spin in the rest frame. Based on the concept of the effective magnetic field, it is simple to imagine SOC as a natural, non-magnetic method of generating spin-polarized electron current. A "Vortex," which can also be found in magnetic and superconducting materials, can generalize this model (Figure 1.10). In a vortex structure, the interaction of charged particles (electric fields) and spin ensembles (magnetic fields) can be broadly understood (see in Chapter 7). This concept will be compared to vortexes (in Chapter 10).

Chirality, also known as handedness, is a digital relationship between three vectors that differentiates an object from its mirror image, such as the spread fingers of the right and left hands. A strong relativistic SOI can fix the chirality of ground-state magnetic textures defined by magnetization vectors, gradients, and an electric field from broken inversion symmetry. We focus on the chirality observed in excited states of the magnetic order, dielectrics, and conductors with evanescent transverse spins. Even in the absence of a relativistic effect, the evanescent waves' transverse spin is locked to the momentum and the surface normal to their propagation plane. This chirality acts as a generalized SOI, leading to the discovery of various chiral interactions in spintronics that mediate the excitation of quasiparticles into a single direction, resulting in phenomena, for example, chiral spin and phonon pumping, chiral spin Seebeck, spin skin, magnonic trap, magnon Doppler, and spin diode effects.

1.8.1.2 Chirality

Carbon has been established as a strongly correlated system based on detailed electrical transport, particularly from the superconducting transition, and magnetic properties (either intrinsic or in a doped superstructure). A strong SOC can be produced by the resonant valence bond. The curvature of carbon structures results in a high SOC. Strong SOC is also produced by chiral or twisted carbon structures (Figure 1.11). SOC is linked to the NV center. As a result, the SOC can serve as the foundation for quantum computation. One can simulate the helical DNA structure, which guides the chiral structure of CNTs.

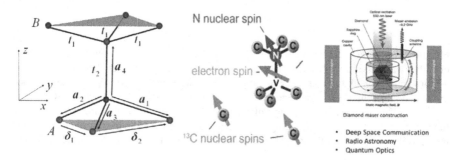

FIGURE 1.10 Sketch of the diamond lattice structure, with the hopping amplitudes (left). The A (bottom) and B (top) are sublattice planes [28]. Diamond NV center consisting of a nitrogen atom and a vacancy center in a diamond lattice (middle). Interactions between the nuclear and the electron spin determine the properties of a spin qubit which also has a spin-triplet state. Applications of diamond sensors for deep space communication, radio astronomy, and quantum optics (right) [37].

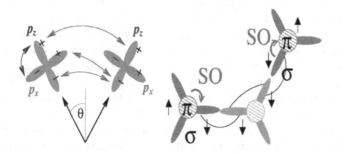

FIGURE 1.11 Sketch of the relevant orbitals, p_x and p_z needed for the analysis of spin-orbit effects in a curved nanotube (left). The combined effects of the intra-atomic spin-orbit coupling, curvature, and applied electric field has been analyzed by using perturbation theory (right). For nanotube caps, spin-orbit coupling causes spin-splitting of the localized states at the cap, which could allow spin-dependent field-effect emission [38].

1.8.2   SOI AND SPIN SELECTIVITY FOR TUNNELING ELECTRON TRANSFER IN DNA

Electron transfer (ET) in biological molecules such as peptides and proteins involves electrons tunneling between well-defined localized states (donors to acceptors). An analytical model for ET based on DNA tunneling in the presence of SOI was proposed, resulting in a strong spin asymmetry with intrinsic atomic SO strength in the MeV range (Figure 1.12). The ET behavior as a function of injection state momentum, SOC, and barrier length and strength from a Hamiltonian is consistent with charge transport through π orbitals on DNA bases. Two concomitant mechanisms for spin selection emerge in a highly consistent scenario: spin interference and differential spin amplitude decay. If the SO coupling is realistic, high spin filtering can occur at the expense of reduced amplitude transmission. The spin filtering scenario is completed by addressing the spin-dependent torque under the barrier using a conserved definition for the spin current.

The strength of the SOI pertinent to transport in a low-dimensional structure is significantly influenced by the relative geometrical arrangement of current-carrying orbitals. Controlling the spin selectivity seen in experiments, which is a defining characteristic of the Chiral-Induced Selectivity Effect, depends critically on understanding the origin of the enhanced SOI in chiral systems (CISS). Recent tight-binding orbital models for spin transport in DNA-like molecules postulate that the angular relationships between adjacent bases' orbitals on the helical chain are what induce the band SOC. By straining the molecule in a conductive probe AFM/Break junction

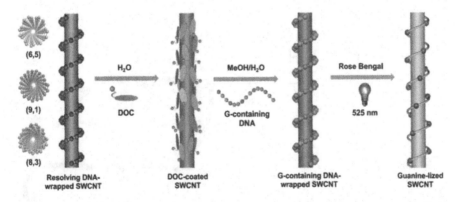

FIGURE 1.12 Covalent modification of carbon nanotubes is a promising strategy for engineering their electronic structures. However, keeping modification sites in registration with a nanotube lattice is challenging. A solution using DNA-directed, guanine (G)-specific cross-linking chemistry is reported. Through DNA screening we identify a sequence, $C_3GC_7GC_3$, whose reaction with an (8,3) enantiomer yields minimum disorder-induced Raman mode intensities and photoluminescence Stokes shift, suggesting ordered defect array formation. Single-particle cryo–electron microscopy shows that the $C_3GC_7GC_3$ functionalized (8,3) has an ordered helical structure with a 6.5 Å periodicity. Reaction mechanism analysis suggests that the helical periodicity arises from an array of G-modified carbon-carbon bonds separated by a fixed distance along an armchair helical line. These findings may be used to remodel nanotube lattices for novel electronic properties [39].

type configuration, such arrangements could be investigated. When a double-strand DNA model is compressed or stretched, the strain-dependent kinetic and SO coupling in two experimentally feasible setups with unusual deformation properties can be seen. Mobility and SO coupling can be greatly affected by strain, and the analytical model supports the qualitative trends of the experiments.

1.9 COMPUTATION BASED ON BIOMOLECULES; DNA STRUCTURE–BASED COMPUTATION

1.9.1 FUTURE DIRECTIONS: QUANTUM COMPUTER IN COMPLEX CARBON STRUCTURE RESEARCH LEADING TO BIOINFORMATICS

The fundamental unit of a biologically complicated molecule must be comprehended through a slightly skewed structure and connectivity. We must comprehend how the atoms in the structures can entangle and send an electrical signal coherently (data). Long coherence lengths (time) and carrier mobility have been proven in carbon nanostructures such as nanotubes and graphene. Some organic molecule compounds exhibit spin-triplet superconductivity (with a protein-like structure) (Figure 1.13). In the grain boundary (GB) areas of boron-doped superconducting diamond films, long coherence times and spin-triplet superconductivity are also noted.

Detailed microscopic investigations reveal a structure resembling a neural network at the grain boundaries. As a consequence, a quantum simulator can be used to examine carbon nanostructures that resemble biological complexes. Simulations of small molecules

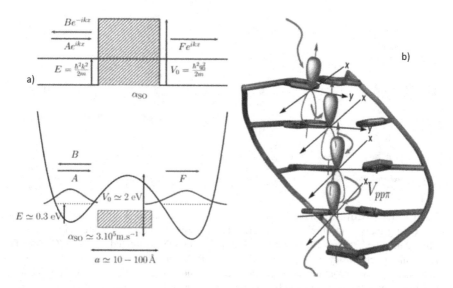

FIGURE 1.13   (a) Scattering potential barrier model with SO interaction (hatch). The label for the incident (A) and scattered (B and F) wave functions amplitudes are indicated. The well parameters are estimated in the text based on polaron transport. (b) Orbital model for transport in DNA. The figure depicts the electron carrying orbitals (p_z orbital perpendicular to the base planes) coupled by V_{pp} Slater-Koster matrix elements. It is well known that any transport mechanism occurs by ET through these orbitals [40].

or chemical reactions are expected to be some of the first practical issues that quantum computers will address. The hunt for novel drugs could then be sped up using computers, as could the creation of catalysts that use less energy during chemical reactions.

We can continue our quest to create specific cancer treatments by exploring the mysteries of DNA's proteins. We could map proteins in their totality using quantum computing, just like we can map genes. A fundamental mechanism in biology, the binding of gene regulatory proteins to the genome, could be predicted using a quantum processor. On a D-Wave Two X device, the research was conducted. However, ten qubits can be used to build the DNA structures, which can then be simulated on an International Business Mechanics (IBM) Quantum Experience (QE) if a suitable mathematical model is provided. Future carbon studies will use quantum computers to investigate some of the most crucial issues: Examples of nanostructure growth processes at the atomic level include comprehending chirality, functionalizing carbon nanostructures, storing data in nanostructures, and transporting (or transferring) charges between nanostructures. We talk about carbon-based quantum technology in Chapter 10.

1.10   CARBON-BASED TECHNOLOGY FOR STORING ENERGY (SUSTAINABILITY): CHARCOAL, CARBON DOTS, AND Q-CARBON

Carbon is one of the oldest and most prevalent materials in the universe, and it can be found in a wide variety of unusual structures, including soot, which includes carbon nanospheres, nanorods, and ultra-thin graphene. In addition to the basic types of carbon, we also observe strong structures for energy storage, such as charcoals.

1.10.1   CHARCOAL

Charcoal is a lightweight black carbon residue made by intensely burning wood (or other animal and plant materials) in a low-oxygen environment to completely evaporate all water and volatile components (Figure 1.14). In the traditional form of this pyrolysis process, known as charcoal burning, which is frequently accomplished by creating a "charcoal kiln," the heat is provided by burning a portion of the initial substance itself while using a constrained amount of oxygen. In a close retort, the substance can also be heated. Non-volatile compounds should not be present in the source substance to produce coal with high purity. By destructively distilling wood, wood tar, and pyroligneous acid are produced as liquid by-products, wood gas is produced as a gaseous by-product, and wood charcoal is produced as waste.

1.10.2   ENERGY STORAGE APPLICATIONS OF CARBON NANOMATERIALS

Duality also applies to carbon structures, which show pliable, absorbent, conducting graphite in addition to a superdense, shiny, and insulating diamond. Graphite's hexagonal ring structures can produce interlayer areas where gas molecules can be absorbed to store energy. By incorporating nitrogen (or boron) into a cubic diamond, a tiny vacancy can also be produced. This vacancy can store a significant amount of information that is helpful for quantum technology that uses light-matter interaction.

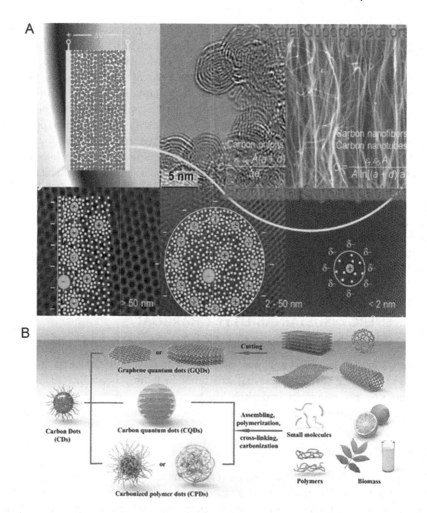

FIGURE 1.14 A schematic model for the microstructure of activated carbon fibers: (A): (top) for a high surface area fiber where the basic structural units are randomly arranged, (bottom) for a fiber after some heat treatment, showing partial alignment of the basic structural units. Tangled structure within a grain proposed for many polymer-derived graphitic carbons including glassy carbon. The presence of voids accounts for the low mass density of glassy carbons, and the voids are likely sources of mechanical weaknesses. Schematic diagram of the carbon aerogel microstructure. Each shaded circle represents an amorphous carbon particle. (left) Idealized structure of a partially graphitized carbon black particle. (right) Another schematic view of a carbon black particle showing short graphitic segments [42].
(B) Carbon dots can be found in graphene and layered polymer dots which can be synthesized by dissociating or higher dimensional carbon nanostructures or assembling small molecules or polymers or biomass.

We will expand on how quantum information can be stored in carbon-based materials after describing the fundamentals of quantum dots (Chapter 3) and quantum wells (Chapter 5) in semiconductors. Resonant states of quantum wells and quantum dots can be reached by electromagnetic fields. Metallic graphene and graphite carbon can create a deep potential state from a zero potential state that can be applied to supercapacitors.

Carbon can hold energy as quantum dots, quantum wells, or quantum boxes, such as the crossing points of filaments in Q-carbon (carbon dots). The energy is kept as a deep potential that ought to be simple to reach. Similarly, knowing layered structure and connectivity is necessary for data storage. CNTs are utilized in cutting-edge energy storage systems, particularly electrochemical supercapacitors, and lithium-ion batteries. Researchers have looked into the novel uses of CNTs in bendable and stretchable energy storage devices, micro-scaled current collectors, and binder-free electrodes. The capabilities of CNTs in these energy storage technologies have greatly increased in recent years.

Carbons can be made with many beautiful shell-like structures such as carbon onions (Figure 1.15a) as shown in transmission electron microscope (TEM) images. Figure 1.15b shows hollow carbon spheres. The spheres were made by firstly making a removable template and covering the template with carbon. Removal of the template left "balloon"-like empty structures with a shell made of carbon. In this set of images, small carbon fibers can be seen. These were grown inside the spheres by a "ship in a bottle" approach. Figure 1.15c shows carbon helices. The dark regions show copper catalyst particles. In the presence of a carbon precursor, the carbon deposits on the copper and two symmetrical carbon fibers grow away from the particle with a helical shape.

1.10.3 CARBON DOTS: A NEW TYPE OF CARBON-BASED NANOMATERIAL WITH WIDE APPLICATIONS

Researchers were inspired to create ultrafine materials like carbon dots, which are smaller than 10 nm and were found due to their fluorescent properties, because of the widely recognized broad application of carbon nanomaterials. Numerous uses for carbon dots are still being discovered in a variety of industries, including biology, medicine, electronics (sensing and imaging tools, added materials for optoelectronic devices and solar cells), and energy (light-emitting diode, supercapacitors, and voltaic cell). The beginning materials and synthesis techniques (such as top-down and bottom-up routes) have an impact on the structure and surface functional groups of carbon dots. Due to time, material, and equipment limitations, ultrasonic synthesis is advised for mass production. Other top-down routes include arc discharge, laser ablation, electrochemical oxidation, and ultrasonic synthesis. Microwave synthesis, thermal decomposition, hydrothermal treatment, and plasma treatment are examples of bottom-up methods that use quick-processing, low-cost tools. It typically appeared in liquid or suspension form with quasi-spherical particles made of carbon atoms that had undergone sp^2/sp^3

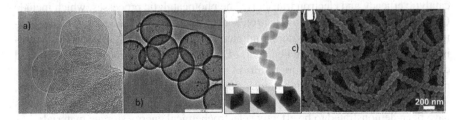

FIGURE 1.15 (a) High-resolution transmission electron microscopy shows image of a carbon onion [42]. (b) Hollow carbon spheres [43]. (c) Helical and Twisted carbon filaments are formed which can store energy [44].

hybridization and had a lot of nitrogen and oxygen in their structures. The properties and structures of carbon dots, which are influenced by the synthesis routes, decide their uses.

There are mainly three identified properties namely optical properties due to the π-conjugated electrons in sp^2 atomic structures (which include UV-absorption, photoluminescence, and fluorescence behaviors), electronic properties (with higher ET abilities, which are enhanced by its surface functional group and heteroatom dopant), and catalytic properties (by improving the activities of catalysts).

These characteristics make carbon dots a possible candidate for a range of applications, including sensing, imaging, solar cells, and catalysis. They are particularly well suited for use in biological environments due to their chemical stability, non-toxicity, and biocompatibility. Due to their surface area and active surface sites, carbon dots have been found as a decorative hybrid electrode agent that could enhance the conductivity and functionality of the energy storage device. Three different kinds of carbon dots – graphene quantum dots, carbon nanodots, and carbonized polymeric dots can be found under the general term "carbon dots," which refers to ultrafine dots-sized carbon allotropes.

Graphene quantum dots are made of mono/multi-layers of nano graphite sheet with π-conjugated bonds. Their sizes are larger than the carbon nanodots and range below 100 nm with their lateral dimensions larger than their height. Carbonized polymeric dots hold π-conjugated bonds in the polymeric structure and do not show quantum confinement effects due to their particle size.

1.10.4 QUANTUM DOTS: INNOVATIVE MATERIALS FOR VARIOUS APPLICATIONS

The semiconductors known as Qdots have been found to have light-emitting properties. They can replace optical fiber or light-emitting diodes and can improve full-color output in a flat television. This material has a wide variety of light emission wavelengths, is low in production costs, and operates sustainably over an extended period without degrading. In terms of electrical compatibility with optical devices, low loss of optical feedback resonance, and low emission threshold frequency, it can be compared to other emitters. Changes in Qdot size, form, composition, and structure may have an impact on the luminescence process. Due to Qdots' high rate of photogenerated electron-hole pairs, stable light reaction, programmable band gap, and long quantum lifetime, they were particularly effective in photocatalytic applications. The structural alterations of Qdots, such as heteroatomic doping and morphology control, may increase photocatalytic effectiveness. By manipulating the Qdot size or creating different kinds of heterojunctions, the band gap can be changed, enhancing both photocatalytic efficacy and electron transport. The photocatalytic processes, which include Qdots' capacities to function as electron acceptors or donors (Lewis acid/base), photosensitizers, electron accelerators, and electric field generators, may provide a satisfactory explanation for these behaviors. The resonance behavior of the sp^2-bonded carbon Qdots within its structure results in greater electrical conductivity and may be preferred over some metals. In batteries, supercapacitors, and micro-supercapacitors, where they are used as electrode materials or combined with support agents, Qdots have been demonstrated to be effective electrochemical energy storage materials.

With high ET and resistance to degradation during use, these cutting-edge materials answer the call for creating an effective, reliable energy storage system. They are the ideal materials for electrical devices because of their surface area, active sites, electrical conductivity, and solubility in different liquids. When the electrodes were combined with Qdots, it was found that the specific capacity and cyclic stability of lithium batteries significantly increased. An attempt to market environmentally friendly devices is necessary because heavy and toxic metals make Qdots devices unstable and hinder their mass manufacturing. If the intended use necessitates a secure environment rather than hazardous materials, synthetic strategies and metal selection become crucial steps.

1.10.5 Q-CARBON

New carbon allotrope with minimal sp-orbital hybridization and its lithium-ion battery-related nucleation and lithiation processes: Q-carbon, a new metallic carbon allotrope, was found through first-principles computations (Figure 1.16). The so-called Q-carbon had a cage structure made of three compounds and three carbon atoms in three dimensions. Q-carbon has a minimal level of s-p orbital hybridization, according to the energy distribution of electrons in various orbitals. Li^+ aggregation inside Q-carbon during lithiation was indicated by the calculated Li^+ binding energies. Therefore, a typical two-phase transition was implied by the formation and gradual expansion of the Li_8C_{32} phase in Q-carbon. As a result, the theoretical voltage on Q-carbon could remain constant at 0.40 V, successfully preventing the growth of Li dendrites. Stress-strain analysis verified a stable Li_8C_{32}/C_{32} two-phase interface, and a calculated Li^+ diffusion barrier of 0.50 eV guaranteed efficient Li^+ diffusion along a 3D pathway. This research offered new insight into the design of new carbon materials for energy storage applications and was of great importance for our understanding of the two-phase transition of Li^+ storage materials. Chapter 7 discusses Q-carbon's superconductivity.

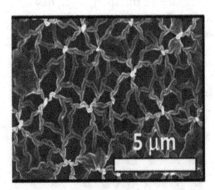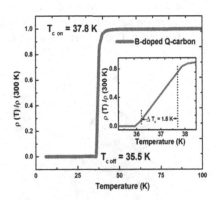

FIGURE 1.16 High-resolution FESEM images showing filamentary structures of B-doped Q-carbon showing dense and (some) disconnected filamentary structures of B-doped Q-carbon (indicated by white arrows) formed due to interfacial instability, respectively (left). Temperature-dependent (normalized) resistivity measurements of B-doped Q-carbon thin films showing the onset temperature of superconducting transition temperature at 37.8 K (right). The inset depicts the enlarged view of the superconducting transition showing the transition width to be 1.5 K [45].

1.10.6 AMORPHOUS (DIAMOND-LIKE) CARBON

So far, we have discussed crystalline and nanostructured forms of carbon. In carbon world, a large area is occupied by (soft) amorphous carbon and a special type of hard carbon called "diamond-like carbon (DLC)" or "tetrahedral amorphous carbon (t-aC)" was deposited by applying high-energy ion beam or high bias in a CVD. The properties of ta-C are compared with a-C films. Figure 1.17 shows the classification of amorphous carbon films by making a ternary diagram from 74 types of amorphous carbon analyzed. In general, ta-C films are presumably free of hydrogen or have minimal amounts because, by definition, the tetrahedral structure predominantly contains σ bonds. Therefore, ta-C was considered the region where the sp^3 ratio exceeded 50% and the hydrogen content was <~5 at%. Further, a-C was classified as the region where the sp^3 ratio and hydrogen content were ≤50% and ≤5 at%, respectively. Figure 1.17 also depicts a clear separation between ta-C and a-C in terms of the nanoindentation hardness. The large region with an sp^3 ratio and hydrogen content of <50% and >5 at%, respectively, was labeled as a-C:H if a large amount of hydrogen was introduced into a-C. Four regions are all DLC may be classified as polymer-like carbon films in the cases where the hydrogen content is high (>~40 at%) and a linear chain structure is dominant. These films also exhibit small hardness values. However, there are also films with hardness values >9 GPa, even when the hydrogen contents exceed 40 at%. We summarize this chapter by classifying carbon in different dimensions (Figure 1.18).

- Main problem discussed in this chapter.

  A cosmic picture of carbon starting from deep space to the carbon cycle on this planet, which fosters life in presented. A wide range of carbon in terms of structure and (electrical) properties is discussed. We try to find the elementary structure of carbon which gives elegant transport properties in a carbon sample which include quantum tunnel as well as superconducting transport. Two atoms can form four-level systems. The electronic structure

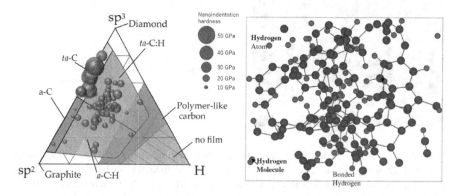

FIGURE 1.17 (Left) Distribution of 74 types of amorphous carbon films on the ternary diagram. The diameter of the circle corresponds to the nanoindentation hardness of each amorphous carbon film [46]. (Right) Molecular dynamic simulation of atomic structure of a hydrogenated DLC film consists of sp^2 and sp^3 bonded carbon atoms and hydrogen atoms [47].

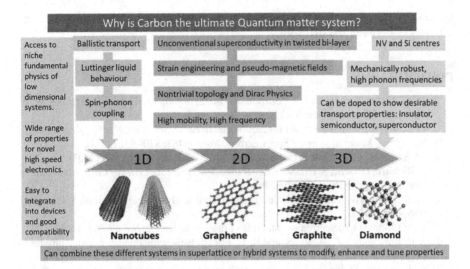

FIGURE 1.18 Carbon is the ultimate quantum matter. Classification of carbon systems. These properties can be combined in a hybrid carbon system. It can show a wide range of outstanding transport properties in 1D, superconductivity in 2D, and optical properties in 3D, which can be used in high-speed electronics as well as quantum computation.

has a symmetric (ground) s-state and an antisymmetric part or an excited p-state which can be considered a two-level (a qubit) system.

- What has been achieved?

Two s electrons and two p electrons in a carbon atom consisting of can hybridize which is similar to the superposition of quantum states in the qubits. The superposition can be obtained in several ways. Four electrons can be distributed equally in space, connect the neighboring carbon atoms, and form a sp^3-bonded (diamond) structure. The electric fields originating from the strain of the bonds are distributed all over the space equally. The superposition of 2s and 2p electrons can also yield a planar structure of graphite where one carbon atom is linked with three neighboring carbon atoms. For a linear chain, one single bond (p-state, orbit-like, spin 1) alternates with a double bond (s-like, spin zero). A unit of the single and double bond can be considered a two-level system or a qubit. Therefore, a resonating structure can be described as a multi-qubit system. By introducing a dopant, the symmetry of a structure can be broken and a spin-qubit or a spin-triplet state evolves. A resonating valence band structure was described as the SSH model (particularly, in conducting polymers) and the Fulde-Ferrel and Larkin-Ovchinnikov (FFLO)-type superconductivity.

- What has not been achieved?

Chapter 1. Why carbon has so many allotropes in all dimensions from 0d to 3d? We know carbon cycles and the transformation of energy associated with carbon structures. We don't know why and how to control the cycle and the transformation of energy. For example, the mechanism for photosynthesis (classical or quantum) is not known. A quantum processor will be useful to simulate the process.

Q. Construct structures of various carbon allotropes and find the similarities and differences between the structures.

BIBLIOGRAPHY

1. R. M. Hazen, *Symphony in C: Carbon and the Evolution of (Almost) Everything*, W. W. Norton & Company, New York city, 2019.
2. G. A. J. Amaratunga, A dawn for carbon electronics?, *Science* 297, 1657–1658 (2002).
3. N. Savage, Material science: Super carbon, *Nature* 483, S30–S31 (2012).
4. M. Stoneham, Electrons in carbon country, *Nature Materials.* 3, 3–5 (2003).
5. S. Richiiro, D. Gene, S. Dresselhaus Mildred, *Physical Properties of Carbon Nanotubes*, Imperial College Press, London, 1998.
6. K. S. Novoselov, et al., Electric field effect in atomically thin carbon films,. *Science* 306, 666–669 (2004).
7. L. F. Lindoy, Optoelectronics: Marvels of molecular device, *Nature* 364, 17–18 (1993).
8. J. H. Burroughes, C. A. Jones, and R. H. Friend, New semiconductor device physics in polymer diodes and transistors, *Nature* 335, 137 (1988).
9. S. J. van der Molen, and P. Liljeroth, Charge transport through molecular switches, *Journal of Physics: Condensed Matter* 22, 133001–133030 (2010).
10. Z. Ezziane, DNA computing: Applications and challenges, *Nanotechnology* 17, R27–R39 (2006).
11. J. Narayan, Dislocations, *Journal of Materials Research.* 5, 2411 (1990).
12. S. Bhattacharyya, and S. V. Subramanyam, Conducting carbon films: Structure, properties and applications. In: D. L. Wise (Ed.), *Electrical and Optical Polymer Systems: Fundamentals, Methods and Applications*, Marcel Dekker, New York, 1998, pp. 201–296.
13. A. Shaikjee, and N. J. Coville, The synthesis, properties and uses of carbon materials with helical morphology, *Journal of Advanced Research.* 3, 195–223 (2012).
14. C. Cheng, Q. Liang, M. Yan, Z. Liu, Q. He, T, Wu, S. Luo, Y. Pan, C. Zhao, and Y. Liu, Advances in preparation, mechanism and applications of graphene quantum dots/ semiconductor composite photocatalysts: A review, *Journal of Hazardous Materials* 424, 127721 (2022).
15. N. Munyebvu, E. Lane, E. Grisan, and P. D. Howes, Accelerating colloidal quantum dot innovation with algorithms and automation, *Materials Advances.* 3, 6950–6967 (2022).
16. A. Nurmikko, What future for quantum dot-based light emitters? *Nature Nanotechnology.* 10, 1001–1004 (2015).
17. A. P. V. K. Saroja, M. S. Garapati, R. Shyiamala Devi, M. Kamaraj, and S. Ramaprabhu, Facile synthesis of heteroatom doped and undoped graphene quantum dots as active materials for reversible lithium and sodium ions storage, *Applied Surface Science* 504, 144430 (2020).
18. J. S. Steckel, J. Ho, and S. Coe-Sullivan, *Photon Spectra* https://go.nature.com/mMfCaf (September 2014).
19. W. Yang, X. Li, L. Fei, W. Liu, X. Liu, H., Xu, and Y. Liu, A review on sustainable synthetic approaches toward photoluminescent quantum dots, *Green Chemistry* 24, 675–700 (2022).
20. M. Wu, H. Chen, L. P. Lv, and Y. Wang, Graphene quantum dots modification of yolk-shell Co3O4@ CuO microspheres for boosted lithium storage performance, *Chemical Engineering Journal* 373, 985–994 (2019).
21. N. Zahir, P. Magri, W. Luo, J. J. Gaumet, and P. Pierrat, Recent advances on graphene quantum dots for electrochemical energy storage devices, *Energy & Environmental Materials* 5, 201–214 (2022).

22. Z. Lin, L. C. Beltran, Z. A. De los Santos, Y. Li, T. Adel, J. A. Fagan, A. R. H. Walker, E. H. Egelman, and M. Zheng, DNA-guided lattice remodelling of carbon nanotubes, *Science* 377, 535–539 (2022); S. Iqbal, H. Khatoon, A. H. Pandit, S. Ahmad, Recent development of carbon based materials for energy storage devices, *Materials Science for Energy Technologies* 2, 417–428 (2019).

23. L. Sun, X. Wang, Y. Wang, and Q. Zhang, Roles of carbon nanotubes in novel energy storage devices, *Carbon* 122, 462 (2017).

24. S. V. Salazar, V. Mujica, and E. Medina, Spin-orbit coupling modulation in DNA by mechanical deformations, *Chimia* 72, 411–417 (2018).

25. D. Kraus, On Neptune, it's raining diamonds, *American Scientist* 106, 285–287 (2018). https://news.mit.edu/2021/space-complex-carbon-molecules-0318.

26. R. E. Hanneman, H. M. Strong, and F. P. Bundy, Hexagonal diamonds in meteorites: Implications, *Science* 155, 995–997 (2017).

27. J. Charoenpakdee, O. Suntijitrungruang, and S. Boonchui, Chirality effects on an electron transport in single-walled carbon nanotube, *Scientific Reports* 10, 18949 (2020).

28. S. S. Perchgova and V. M. Yakovenko, Schockley model description of surface states in topological insulators, *Physical Review B* 86, 075304 (2012).

29. Y. Iwasa, Superconductivity: Discoveries of the fullerenes, *Nature* 466, 191–192 (2010).

30. J. L. Miller, Unconventional superconductivity discovered in graphene bilayers, *Physics Today* 71, 15 (2018).

31. T. C. Lang, Z. Y. Meng, M. M. Scherer, S. Uebelacker, F. F. Assaad, A. Muramatsu, C. Honerkamp, and S. Wessel, Antiferromagnetism in the Hubbard model on the Bernal-stacked honeycomb bilayer, *Physical Review Letters* 109, 126402 (2012).

32. P. Esquinazi, T. T. Heikkilä, Y. V. Lysogorskiy, et al., On the superconductivity of graphite interfaces, *JETP Letters* 100, 336 (2014).

33. M. Stoneham, Electrons in Carbon Country, *Nature Materials* 3, 3 (2004); C. H. Ahn, et al., Electric field effect in correlated oxide systems, *Nature* 424, 1015 (2003).

34. E. A. Ekimov, Superconductivity in diamond induced by boron doping at high pressure, *Physica Status Solidi B* 246, 667 (2009).

35. M. Kasu, *Frontiers of Materials Research* (Special Report). 8, 1 (2010).

36. W. Bacsa, One and two dimensional crystalline materials: Industrial applications, *NSTI Nanotech Conference Proceedings, Washington DC, USA* (2017).

37. J. D. Breeze, et al., Continuous-wave room-temperature diamond maser, *Nature* 555, 493–496 (2018).

38. D. Huertas-Hernando, F. Guinea, and A. Brataas, Spin-orbit coupling in curved graphene, fullerenes, nanotubes, and nanotube caps, *Physical Review B* 74, 155426 (2006).

39. Z. Lin, et al., DNA-guided lattice remodeling of carbon nanotubes, *Science* 377, 535–539 (2022).

40. S. Varela, I. Zambrano, B. Berche, V. Mujica, and E. Medina, Spin-orbit interaction and spin selectivity for tunneling electron transfer in DNA, *Physical Review B* 101, 241410 (2020).

41. S. Iqbal, H. Khatoon, A. H. Pandit, and S. Ahmad, Recent development of carbon based materials for energy storage devices, *Materials Science for Energy Technologies* 2, 417–428 (2019); L. Sun, X. Wang, Y. Wang, and Q. Zhang, Roles of carbon nanotubes in novel energy storage devices, *Carbon* 122, 462 (2017).

42. D. Ugarte, Curling and closure of graphitic networks under electron-beam irradiation, *Nature* 359, 707 (1992)

43. S. Li, A. Pasc, V. Fierro, and A. Celzard, Hollow carbon spheres, synthesis and applications – a review, *Journal of Materials Chemistry A* 4, 12686 (2016).

44. A. Shaikjee, P. J. Franklyn, and N. J. Coville, The effect of substituted alkynes on nickel catalyst morphology and carbon fiber growth, *Carbon* 49, 2950 (2011).

45. A. Bhaumik, R. Sachan, and J. Narayan, High-temperature superconductivity in boron-doped Q-carbon, *ACS Nano* 11, 5351 (2017).
46. N. Ohtake, M. Hiratsuka, K. Kanda, H. Akasaka, M. Tsujioka, K. Hirakuri, A. Hirata, T. Ohana, H. Inaba, M. Kano, et al., Properties and classification of diamond-like carbon films, *Materials* 14, 315 (2021).
47. A. Erdemir and C. Donnet, Tribology of diamond-like carbon films: Recent progress and future prospects, *Journal of Physics D: Applied Physics* 39, R311–R327 (2006).

2 Synthesis of Carbon Nanostructures
Growth Model and Microstructure

2.1 GAS-PHASE REACTION: SYNTHESIS OF GRAPHENE MECHANICAL EXFOLIATION, CHEMICAL VAPOR DEPOSITION, AND LASER ABLATION

2.1.1 SYNTHETIC TECHNIQUES FOR GRAPHENE

To date, various synthetic techniques, exist for the synthesis of graphene. Namely, micromechanical cleavage, chemical vapor deposition (CVD), laser ablation, arc discharge, epitaxial growth on SiC, chemical reduction of graphite, liquid phase exfoliation, and electrochemical exfoliation of graphite. In this chapter, we expand and elaborate on some of the synthesis techniques for the synthesis of graphene.

2.1.2 MICROMECHANICAL CLEAVAGE

Graphene was first isolated in its free-form using the "scotch-tape" method developed by Geim and Novoselov in 2004. This method is only suitable for research purposes only and not suitable for large-scale production of single-layered graphene. The method includes peeling off layers of graphene from highly oriented pyrolytic graphite (HOPG) using adhesive tape, and the tape is repeatedly pressed together and cleaved to yield single-layer graphene. The simplicity of this method has motivated researchers to develop a new method to produce graphene devices that are contamination-free and do not require further chemical treatments or lithographic techniques. The novel method includes mechanically peeling off a few layers of graphene sheets from bulk graphite, the layers are placed onto prefabricated electrodes (see Figure 2.1).

The scotch-tape exfoliated graphite flakes were placed on SiO_2 substrates and investigated via SEM with Kleindiek nanoprobes of tip radius ~100 nm housed inside the instrument. During this process, thin strips of graphene were peeled off from larger flakes and transferred to a different substrate containing prefabricated electrodes. The multilayered strips were secured using platinum tabs fabricated from a combination of electron beam and gas injection (OmniGIS). The method also ensures that the graphene layers are suspended, thus ensuring electron transport without interacting with the substrate material. The magnetic field-dependent resistance measurements of the device fabricated through this method were investigated since

DOI: 10.1201/9781003316411-2

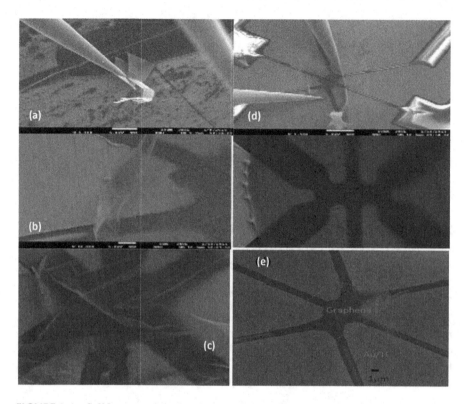

FIGURE 2.1   Solid-state exfoliation of graphene: SEM images illustrating the main steps followed in the development of suspended multilayered graphene devices. (a) Thin strips of layered graphene are peeled off of larger flakes. (b) The strips are transported to the desired location using a nanomanipulator probe tip. (c) Smooth, thick, and non-wrinkled multilayer graphene device with Hall bar electrodes. The tungsten tabs keep the suspended graphene in place: (d) wrinkled graphene and (e) smooth, thin, and transparent graphene. The dotted line shows the edge of the graphene layer [9].

the defect level is expected to be different from graphene synthesized through other techniques (see Chapter 3). The comparison of the two results showed the change in the electron transport characteristics from single-layer graphene to graphite, showing the influence of the number of layers.

2.1.3   CHEMICAL VAPOR DEPOSITION (CVD)

- C-containing gases on catalytic metal surfaces.
- C-atoms diffuse on the surface and form a solid solution, C-precipitates and forms graphene when cooling.
- Process difficult to control on polycrystalline metal foils, single crystals give better results.

CVD is a synthetic method generally used to produce thin-film samples. It involves introducing a vaporous precursor chemical into the CVD chamber containing a substrate material. At high temperatures, the vaporous chemical decomposes and leads to the growth of a layered material on the substrate (see Figure 2.2). This method is a well-established technique used to grow graphene on metal substrates such as Copper or Nickel. It can be scaled up easily and has been reported before for forming large samples of graphene. This method can also be used for doping graphene samples by changing the input gases and allows for comparative analysis between samples of different compositions.

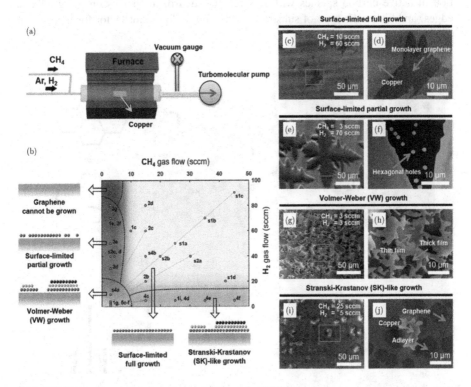

FIGURE 2.2 Graphene development through CVD phase diagram with various CH_4 and H_2 gas fluxes. Diagram of the growth chamber in (a) schematic form. (b) Graphene development phase diagram for various CH_4 and H_2 gas fluxes. The experimental conditions employed in this work are represented by light gray dots in the plot. (c, d) SEM picture of graphene formed using the surface-limited full growth mode and matching enlarged image of the box in (c) Dashed lines are the recommendations that link the light gray dots that have the same gas flow ratio. SEM images of graphene produced using the surface-limited partial growth mode are shown in (e, f), along with a magnified version of the image of the box in (e). Graphene developed in the VW growth mode is shown in the SEM images (g, h), along with an enlarged version of the box in (g). (i, j) A magnified picture of the box in (i) and an SEM image of graphene developed in an SK-like growth mode [10].

2.1.4 THE MECHANISM OF GRAPHENE VAPOR-SOLID GROWTH ON INSULATING SUBSTRATES

High-performance graphene-based devices are usually built on wafer-scale single-crystal graphene films on insulating substrates via the CVD method. However, the lack of understanding of the growth mechanisms of insulating substrates greatly hinders the progress of this method. This also includes amorphous 2D carbon materials containing topological defects (see Figure 2.3). Absorption studies of various carbon species CH_x ($x = 0, 1, 2, 3, 4$) on three insulating substrates, that is, h-BN, sapphire, and quartz were carried out using first-principles calculations. The study revealed that graphene growth on an insulating substrate was mainly due to the reaction of active carbon species with less reactive hydrogen species on the graphene edges, and hence the type of substrate used is not a dominant factor for the growth

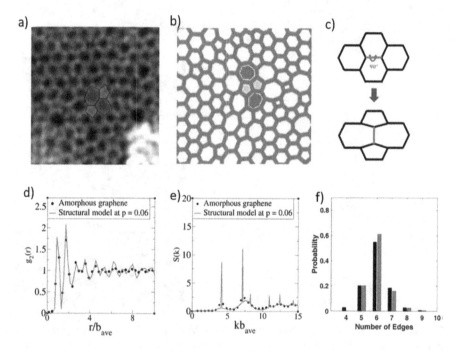

FIGURE 2.3 Amorphous 2D materials with SW topological flaws are shown. 2D amorphous graphene imaged by TEM in (a). (b) Structural model of a perfect honeycomb network with SW flaws added to create a disordered hyperuniform 2D material. Illustration of a SW flaw that alters the topology of the local network and results in a cluster of two pentagons and two heptagons (c). (d) Amorphous graphene sample from experiments and our structural model at $p = 0:06$, where the distance is scaled by the mean bond length of the relevant system. (e) Structure factor $S(k)$ of an experiment-derived amorphous graphene sample and our structural model at time $p = 0:06$. (f) Polygonal shape distributions of our structural model at $p = 0:06$ (black on the left side) and the experimental graphene sample (gray on the right side). Stone–Wales defects preserve hyperuniformity in amorphous two-dimensional networks [11].

mechanism. The gas precursor, CH_3 plays two key roles in graphene CVD growth on an insulating substrate and is explained as follows:

  i. To feed the graphene growth.
  ii. Removal of excessive hydrogen atoms at the edge sites of the graphene.

The threshold reaction barriers for the growth of graphene armchair (AC) and zigzag (ZZ) edges were calculated as 3.00 and 1.94 eV, respectively. Thus, from this observation, the ZZ edge grows faster than the AC one. We have successfully explained why the circumference of a graphene island grown on insulating substrates is generally dominated by AC edges, which is a long-standing puzzle of graphene growth. Furthermore, the sluggish graphene growth rate on an insulating substrate was calculated which agreed well with existing experimental observations. The understanding of the mechanisms of graphene growth on insulating substrates at the atomic scale provides insights into the experimental design of high-quality graphene growth on insulating substrates.

2.1.5 GRAPHENE PRODUCTION BY LASER ABLATION

This method offers the advantage of being energy efficient since it uses minimal energy by only heating the targeted area. The drawback is the reduction in production time. This method is already intensively adopted in the production of CNTs and fullerenes and is important for carbon studies.

2.1.6 DOPING OF GRAPHENE

There are various claims of doping of graphene such as doing holes, doing electrons, and substitutional doping. Charge transfer through hole doping can be achieved with the incorporation of H_2O, NO_2, Br_2, I_2, and O_2 molecules. Organic molecules such as F_4-TCNQ, TCNE, BPO, diazonium salts, and C_{60} can be added to graphene to induce charge. There were attempts to add metals with work function higher than graphene i.e., bismuth, antimony, and gold, etc. However, metals with weak interaction such as Al, Cu, Ag, Au, and Pt have been added to graphene. By arc discharge and in presence of boron, hole doping of graphene was claimed.

Electron doping of graphene can be achieved by incorporating ammonia, ethanol, SiC, SiO_2, potassium (reactive, reduces mobility), polyethyleneimine, etc. Metals having work function lower than graphene has also be used. Different bonding configurations lead to defects (ripples) in graphene structure. It can lower the mobility of carriers and opens a band gap.

2.1.7 SYNTHESIS OF GRAPHENE OXIDE

Bulk graphite consists of multiple packed layers of 2D graphene sheets; hence, chemically exfoliating the sheets from graphite can extend their applications. In the synthesis of graphene oxide (GO), the oxidant reactant $NaNO_3$ was used with K_2FeO_4 replacing $KMnO_4$. This approach yielded promising results for the production and application of both graphite oxide and its derivatives. The conventional method for synthesizing GO was modified to achieve this outcome. Furthermore, $NaNO_3$ was not utilized in this process and GO has been used for wastewater post-treatment purification. Another approach for producing graphene involved the reduction of GO using L-ascorbic acid, which is a form of vitamin C. This resulted in stable suspensions of rGO, which could be prepared in water, as well as in solvents such as N, N-dimethylformamide (DMF) and N-methyl1-2-pyrrolidone (NMP).

2.2 SYNTHESIS OF NANOTUBES: ARC DISCHARGE, CHEMICAL VAPOR DEPOSITION, AND LASER ABLATION

2.2.1 CNT SYNTHESIS: ARC DISCHARGE

In earlier days, most of the work reported in the literature utilized the carbon arc method to produce CNTs and fullerenes. In a typical synthesis, a current between 50 and 100 A and a potential difference of 20–25 V is applied in an inert atmosphere of Helium (see Figure 2.4). The magnitude of the z current required must scale with the electrode diameter. The composition of fused nanotubes and nanoparticles forms a hard gray outer shell, while a soft fibrous core only contains two-thirds nanotubes and one-third of nanoparticles. Solvents such as ethanol, methanol, and isopropanol

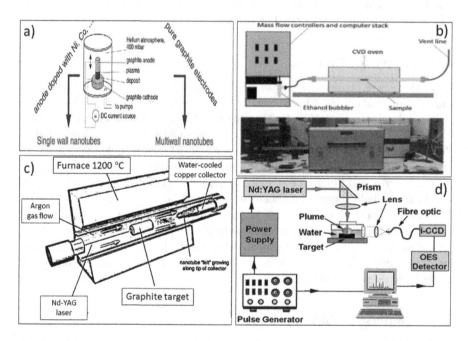

FIGURE 2.4 (a) Schematic diagram of the CVD oven in the Hone lab as well as two digital pictures of the setup. (b) The results of SWNTs come from patterned catalyst islands that were grown by the CVD method [21,34]. (c) A graphite target placed in a furnace can be ablated to produce CNTs. (d) Schematic diagram of the pulse laser deposition system [35].

are used during ultrasonication to separate the nanotubes. The transmission electron microscopy (TEM) image of the carbon arc deposit core shows both nanoparticles and nanotubes of a diameter of $1/zm$ with a current of 20–200 A.

2.2.2 Nanotube Growth by CVD

The production and quality of CNTs synthesized using CVD depends on various factors e.g. catalyst size, catalytic concentration, temperature, and gas flow. Like graphene growth via CVD, the formation of CNTs takes place on the surface of catalytic materials. The size can be controlled by carefully varying the size of the catalysts and this can be performed at low temperatures. The synthetic procedure occurs in two steps: (i) catalysts are first prepared by sputtering, PVD, dip coating, and so on, and (ii) the catalytic substrate is heated in a carbon-rich gaseous environment at 500°C–1000°C.

The CVD method is a low-cost and simple technique for producing CNTs at low temperatures and ambient pressure. Furthermore, it can use a range of hydrocarbons in any state (gas, solid, or liquid), and allows for the use of various substrates and growth can occur in a variety of forms. The growth forms can be powders, thin or thick films, aligned or entangled, straight or coiled, or even a desirably twisted form. Multi-walled CNTs can be grown without a metal catalyst in arc discharge and laser ablation methods. For CVD growth a metal catalyst is required, generally; however, it can be grown from NaCl crystals as well. In contrast, metal catalysts are required for the growth of single-walled CNTs in CVD, arc discharge, and laser ablation methods.

The working principle of CVD includes the decomposition of hydrocarbon feedstock at certain temperatures and allows the growth of carbon materials on transition metal substrates. It utilizes precursor gas at lower temperatures compared to laser ablation temperatures of 500–1100 K to produce CNTs. Since this method can be modified in many ways, it is a preferable method to produce CNTs. The chamber consists of a furnace with operating temperatures of 400–1000 K, a temperature controller, mass flow controllers, and a quartz reaction tube (see Figure 2.4a and b). Carbonaceous gaseous compounds such as methane, ethylene, and acetylene are broken down in the oven to feed into the CVD chamber. Pre-patterned substrates are used for the growth of aligned CNTs in which the size is determined by the diameter of the patterned catalysts. MWCNT, DWNT, and SWCNT are mostly produced via CVD and the differences are determined by the type of catalyst and reaction conditions (Figure 2.5).

2.2.3 Pulsed Laser Ablation

Pulsed laser ablation was first discovered by a scientific group at Rice University and this method is currently mostly used to produce SWCTNs. As displayed in Figure 2.4c this technique utilizes a laser beam to ablate a solid graphite target with a small amount of metal catalyst at elevated temperatures (800–1200 K). An inert carrier gas is used to mobilize the atomic plume from the point of ablation to the cold region where it condenses and then self-assembles around the metal center to form SWCNTs. Highly crystalline SWCNTs with low defect density (90% purity) can be produced through this technique. The quality and quantity of the SWCNT are controlled by the catalyst ratio, pressure, gas flow rate, temperature, and laser parameters.

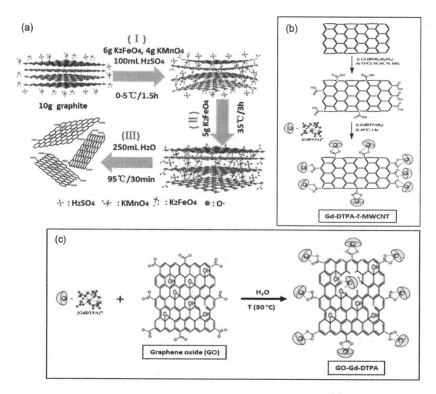

FIGURE 2.5 (a) Illustration of preparation of GO from graphite material. (b) Schematic diagram of functionalization process of a multi-walled carbon nanotube at different temperatures using a mixture of nitric and sulfuric acid [36]. (c) Schematic diagram of the loading of magnetic molecular Gd-DTPA onto the surface of GO [12].

2.3 FUNCTIONALIZATION OF CARBON NANOSTRUCTURES BY METALS

2.3.1 FUNCTIONALIZATION OF CNT

Chemical functionalization is commonly utilized to modify CNTs and enhance their already remarkable characteristics, which in turn extends their range of potential applications. Due to their remarkable features such as high mechanical strength, stiffness, and 1D structure, CNTs are typically insoluble in most solvents, making them suitable for rigorous chemical reactions required for functionalization. In non-covalent functionalization, the structure of the CNT is maintained while covalent functionalization interferes with the CNTs introducing defects. The route chosen usually depends on the intended application of the CNT. Since this study investigates electronic transport, covalent functionalization has been selected as the method for firmly attaching Gadolinium-diethylenetriamine penta-acetic acid (Gd-DTPA) to the outer walls of CNTs to enhance the interaction between the CNTs and magnetic nanoparticles. Covalent functionalization enables a dependable modification of the chemical and electronic properties of CNTs, which can lead to the development of new functions.

Ideally, CNTs should be chemically inert and not have any functional groups, but this is not always the case due to limitations in synthesis and purification techniques.

2.3.2 FILLING WITH GDCL₃

Gadolinium chloride (GdCl₃) was used in this study for encapsulation within the core of the CNTs. GdCl₃ is white crystals in a solid state with a hexagonal structure, like other lanthanide chlorides at ambient conditions. Its structure features a 9-coordinate system with a tricapped trigonal prismatic shape. The preparation of multi-walled carbon nanotubes (MWNTs) filled with GdCl₃ was achieved using an established methodology.

2.3.3 DIFFERENT METHODS OF FUNCTIONALIZATION

The chemical functionalization of MWCNTs can change their electrical, structural, and magnetic properties by creating or breaking bonds during the chemical reaction process. This modification can lead to improved or defective properties of the CNTs. However, it is possible to modify CNTs both internally and externally. In endohedral functionalization, which is the functionalization on the inside, the CNTs are suspended in a mixture containing the desired nanoparticles for functionalization. The nanoparticles enter the inner tube of the CNT through capillary action. After some time, the suspension is evaporated leaving the filled CNT with nanoparticles. The size of the nanoparticles to be enclosed inside the CNT should be smaller than the diameter of the tube. Exohedral functionalization which is the functionalization on the outer sides of the CNT, achieves the introduction of defects onto the outer surface of the tube by treating it with an acid mixture. There exist various techniques for the functionalization of CNTs with organic groups; however, it is challenging to modify graphene through those methods besides graphene being a carbon allotrope. The structure of graphene can be modified through chemical reactions that involve chemical reactants. Different organic functional groups such as the carboxylic, carbonyl, epoxy, and hydroxyl groups can be formed on the structure of GO (Figure 2.6).

2.3.4 CHELATE SYSTEM FORM WITH FUNCTIONALIZED MWCNTS

Doping: Change of the position of the Fermi level can be achieved by substitution of atoms such as nitrogen and boron. It changes sp² hybridization. Doping can also be achieved by the application of electric fields or charge transfer. In the charge transfer doping the Raman G structure remains unaffected.

Different organic groups such as hydroxyl, carbonyl, or sulfonyl can be attached to the MWCNTs depending on the application of the material. The attachment is required for enhancements of chemical or physical properties such as improving water solubility, and dispersion, reducing toxicity, and aiding the transport of macromolecules into cells. The carbonyl group can form a strong chemical bond with MWCNT through the C–C bond, but pristine MWCNTs are usually unreactive chemically and not soluble in water. Macromolecules can be attached to MWCNTs

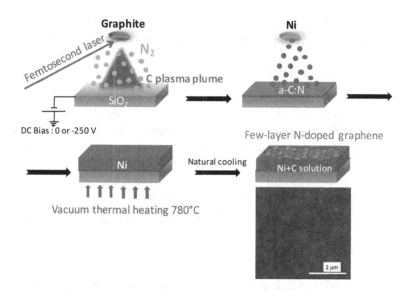

FIGURE 2.6 Doping: Few-layer N-doped graphene films produced by femtosecond pulsed laser deposition are synthesized in this way. To create a carbon-based plasma that expands in a gaseous nitrogen environment, a femtosecond laser beam is focused on a graphite target in a vacuum. The deposited a-C:N layer is then followed by a nickel film (150 nm) by vacuum thermal evaporation. The thickness of the nickel film depends on the N_2 pressure and substrate bias. To create a few-layer N-doped graphene on the top surface, the a-C:N/Ni sandwich is next heated to 780°C in vacuum for 30 minutes with a heating ramp of 4°C/min. This is followed by natural vacuum cooling for 5 hours. The NG2 sample (2.0%N) is relevant to the SEM image [13].

either by aggregation or affinity. Modifying MWCNTs allows for the chelation of the CNT and a chelating molecule. This modification enables MWCNTs to be coupled with proteins, oligonucleotides, peptides, or other macromolecules.

The oxidation of MWCNT with –COOH and –OH groups can be achieved through the use of acids, ozone, or plasma. The carbonyl group is a useful site for surface modification of MWCNT. Covalent linking of the metal with the CNT through chemical reactions using several chemical solvents is commonly used. Furthermore, unzipped MWCNT unlike graphene can potentially possess different properties. Graphene can be functionalized first and then have the macromolecules attached to it after.

2.4 DOPING WITH LIGHT ELEMENTS (NITROGEN, BORON, AND OTHER ELEMENTS)

Doping is used to alter the properties of materials. In graphene, doping affects the sp^2 hybridization which alters the electronic and chemical properties of the carbon atoms. A special type of doping called N-doping is used to introduce n-type charge carriers in carbon systems such as graphene. While some attention has been given to the synthesis and applications of N-doped graphene, the atomic configurations of such dopants in the sublattices of graphene are not well known. The nature of the dopants is equally important. While both monolayer and entangled multilayered NG sheets were prepared and a model for the substitution of carbon atoms by dopants had been proposed further research regarding the configurations was. One experiment which involved the synthesis of carbon on a copper sheet showed that if N_2 is used as the dopant then two nitrogen atoms are found to substitute two carbon atoms in the graphene sheet. These atoms are only separated by a single carbon atom and both atoms lie within the same sublattice. Moreover, most of the nitrogen atoms are incorporated in the same sublattice namely the A sublattice. Moreover, it is favorable for a secondary defect to be introduced in the A sublattice rather than introducing a new defect in the B sublattice in terms of energy. Using Raman spectroscopy, it is observed that the incorporation of nitrogen atoms competes with the annealing process if ammonia is present. Moreover, the incorporation of nitrogen only occurs at high temperatures. At low temperatures, incorporation is reduced. The local density of states (LDOS) is also affected by doping. From dI/dV measurements and measurements involving the LDOS, nitrogen doping results in the n-type doping of graphene. Additional localized states are formed due to the electronic coupling between the nitrogen atom and the nearest neighbor carbon atoms. Charge localization also occurs around the nitrogen atom.

2.5   GROWTH MECHANISM(S) OF FULLERENES AND NANOTUBES

In the previous section, we discussed the basics of carbon growth. Here we present a detailed study. We start with the growth of C_{60}. Fullerenes are occasionally formed by the arrangement of pentagons and hexagons in the graphene forcing it to close into a hollow cage. Isotopic studies and mixed samples of ^{12}C- and ^{13}C- have demonstrated that during the formation of fullerenes, the reacting material undergoes a breakdown process to make atomic carbon or small fragments such as C_2 or C_3 before recondensing into fullerenes. It has been widely debated the way cylindrical fullerene tubules form. One school of thought assumes that the tubules are always capped and that the growth mechanism involves C_2 absorption assisted by pentagonal defects on the caps. Another school of thought assumes that the tubules are open during the growth process and that carbon atoms are added at the open ends of the tubules. One of the concepts focuses on tubule growth at low temperatures (~1100°C) and assumes growth is nucleated at active sites of a vapor-grown carbon fiber of about 1000 A diameter. The absorption of a C_2 dimer near a pentagon in the tubule cap is thought to be linked to the growth of the CNT (Figure 2.7).

The procedure entails a sequence of C_2 dimer additions that add a row of hexagons to a carbon tubule, as well as projection mapping of the cap atoms and the atoms on the cylinder of a carbon tubule, where each pentagon defect is indicated by a light gray shaded hexagon bearing a number. The tubules grow at their open ends when CNTs are grown by the arc discharge method. If the tubule has chirality, the absorption of a single C_2 dimer at the active dangling bond edge site will add one hexagon to the open end. Pentagons are introduced, causing positive curvature, while heptagons are used to cap, changing the size and direction of the tubules. The toroidal surface forms when six heptagons are added to the outside of an open tube to maintain the tubule's consistent diameter.

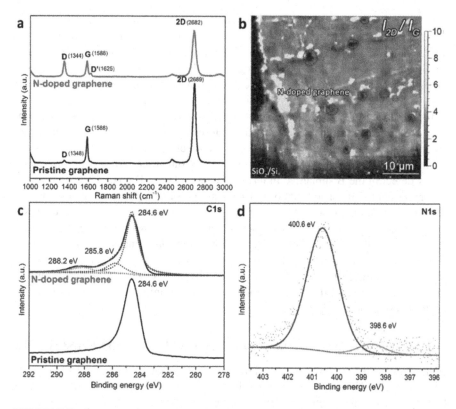

FIGURE 2.7 Raman spectroscopy and XPS of as-synthesized N_G and pristine graphene samples (a) Typical Raman spectra of N-doped and pristine graphene on SiO_2/Si substrate. The wavelength of Raman laser line is 514 nm. (b) 2D-band to G-band intensity ratio (I_{2D}/I_G) mapping of NG_{on} SiO_2/Si substrate. The NG sample is composed of single-layer sheets and some bi-layer or tri-layer islands. (c) XPS C1s line scan of N-doped and pristine graphene. The main peak at 284.6 eV corresponds to the graphite-like sp^2 C, indicating most of the C atoms are arranged in honeycomb lattice. Peaks at 285.8 and 288.2 eV can be attributed to C-N bonding and oxygenated group structures. (d) This confirms the presence of substitutional (400.6 eV) and pyridine-like (398.6 eV) N dopants [14].

2.5.1 The Pentagon Road

Smalley and colleagues modified their original party line idea in response to the finding of a bulk method for producing fullerenes with an appealing mechanism they called the "Pentagon Road." According to Smalley's hypothesis about carbon nucleation, any tiny graphitic fragment's lowest-energy form exhibits the following characteristics:

1. Only hexagons and pentagons are present in the fragments.
2. The maximum number of pentagons are present.
3. The pentagons are not next to each other (Figure 2.3).

These open graphitic cups are carbon clusters with more than 20–30 atoms that are too tiny to form closed fullerenes with isolated pentagon caps. These structures have a low energy as compared to other structures.

Bulk fullerene production has a slow-cooling environment which enables the annealing of carbon clusters to form structures of minimum energy. Thus, these graphitic cups form above other potential structures due to thermochemical stability, and kinetic reactivity requires that they continue to grow toward fullerenes. According to the original Pentagon Road idea, the primary mechanism for this growth is the incorporation of C_2 and other small carbon particles into the expanding graphitic network's reactive edges.

Once produced, the closed fullerenes have no open edges and typically do not continue to expand. Any nucleation path ends in the formation of C_{60}. Therefore, if this process is followed the cluster formation will not continue. The formation of the C_{60} is a closed structure and is not able to continue growing. It is hypothesized that an abundance of small carbon particles required for this technique must exist in the later phases of fullerene production, although there is no experimental evidence to support this idea. The Pentagon Road has nevertheless been modified to account for the reaction between developing graphitic cups and rings of intermediate size.

A continuous supply of C_2 and C_3 would be unnecessary in such reactions. These reactions should have low energy barriers because neither reactant would need to undergo a major reorganization. Uncertain outcomes have come from theoretical analysis of the proposed Pentagon Road intermediates and other hypothetical medium-sized carbon cluster isomers. Based on calculations using the density functional theory, pentagonal-shaped graphitic pieces will be more energetic than those with planar structures. Recent coupled-cluster calculations on C_{20} revealed that the bowl-shaped graphitic cup with a central pentagon has a lower energy than the monocyclic ring and is almost degenerate to the fullerene isomer.

Even though the corners of the polygon have significant curvatures, faceting reduces the free energy in the case of highly ordered carbon fibers by introducing interlayer long-range order. It appears that polygonal cross-sections are more common in large-diameter tubules, even though the majority of the carbon atoms in a faceted nanotube would be on flat edges. Neighboring nanotubes with polygonal cross-sections are potential hosts for intercalated guest species over planar regions.

The tiny size of the carbon atoms in a C_{60}-based tubule, resulting from their special atomic arrangement, prevents substitutional contaminants. In addition, the C_{60} tubule's monolayer structure could reduce the likelihood of the screw axis dislocation, which is the most common flaw in bulk graphite. The special geometrical configuration of C_{60} and related CNTs gives them greater strength and stiffness along the tubule axis as well as resistance to compression under hydrostatic pressure.

2.5.2   GROWTH MECHANISM: VAPOR-SOLID-SOLID

2.5.2.1   Why Carbon Nanotubes Grow

Scientists are still unable to provide a simple explanation for why CNTs grow. Since the encapsulation of a catalyst with graphitic carbon is energetically preferable to CNT growth in every way, CNTs shouldn't grow to solve this problem, a theoretical model has been developed using comprehensive first-principles calculations and molecular dynamics theory [15]. They demonstrated a hitherto unknown but essential characteristic of the CNT-catalyst interface: the interfacial energy of the CNT-catalyst edge is contact angle-dependent. By taking off the graphitic cap, the contact angle is increased, leading to an interfacial formation energy reduction of up to 6–9 eV/nm and overcoming van der Waals cap catalyst adhesion. Due to a map of this exceptional and straightforward relationship, this work enabled scientists to comprehend what causes CNT growth.

The role of temperature and gas in the formation of CNTs is not completely considered in the model. The closed tube technique would be favored since vapor phase growth only occurs at 1100°C, rendering any dangling bonds that might be involved in the open tubule growth mechanism unstable. Since the tubule structure is only quasi-stable, noble gas and other gases are typically utilized at 100 Torr to cool the carbon system in most studies. It is unclear how the quasi-stable tubule or fullerene phases grow as the carbon system is cooled by He gas (Figure 2.8).

In cases when the inner tubule does not have access to a carbon source, several organizations have documented the containment of a small-diameter carbon tubule inside a tubule of a larger diameter. Although nucleation of the cap from the terminated cap may be required, it appears that such a feature requires growth via an open tube mechanism. The addition of carbon atoms to the open ends, capping of the longest open end, and the start of new tubules are demonstrated in this example for the open tube growth of nanotubes from a carbon supply (Figure 2.9).

2.5.2.2 Cobalt-Encapsulated Nanoparticle Synthesis and Characterization

Two techniques for plasma-based synthesis include arc discharge and laser ablation. In thermal synthesis, only thermal energy is employed, and the hot zone of the process, including plasma accelerated CVD, never rises above 1200°C. Carbon feedstock almost always yields CNTs when active catalytic species like Fe, Ni, and Co are present. The best possibility to develop a controllable method for the deliberate creation of nanotubes with certain features appears to be CVD. Catalysts made of Ni, Fe, or Co are put on a substrate and crystallized through chemical etching or thermal annealing with ammonia. Carbon is introduced into the reaction chamber in the gaseous phase.

Using an energy source such as a plasma or a heated coil, carbon molecules are broken down to their atomic level. The substrate, which has been coated with a catalyst, attracts this carbon through dispersion. On top of this metal catalyst, nanotubes develop. The carbon source used is methane, carbon monoxide, or acetylene. MWNT

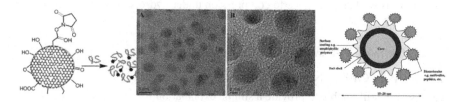

FIGURE 2.8 TEM images of as-synthesized C-QDs show crystalline core and the shell structures (at high resolution) [38]. The C-QDs can be functionalized and connected to each other to form a lattice-like structure (TEM at low resolution).

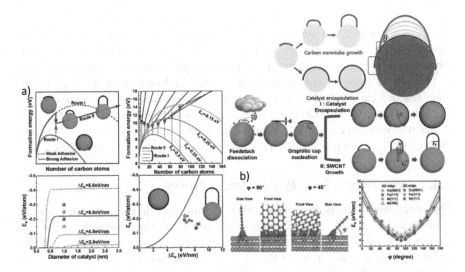

FIGURE 2.9 Left (a) The formation energy profiles for a developing graphitic structure on a catalyst nanoparticle for weak (black) and high (light gray) adhesion situations. Between the two paths of growth – route I being catalyst encapsulation and route II being graphitic cap lift-off and SWCNT growth – bold and dotted lines show energetically favorable and unfavorable growth, respectively. Due to a corresponding drop in formation energy and an increase in formation energy, the aligned arrows indicate favorable and unfavorable graphitic cap lift-off.

[Middle] (b) Energy profiles for the formation of graphitic structures on catalyst nanoparticles ($r_p = 0.5$ nm) with varying numbers of carbon atoms that either enclose the nanoparticle (way I) or lift off to generate SWCNTs with varying diameters (route II). Unfavorable and advantageous graphitic cap lift-off are indicated by aligned arrows. A graphitic SWCNT cap with numerous carbon atoms has a better chance of lifting off the catalyst surface if a point of the route II curve is below a point of the route I curve. Curves showing the preference for SWCNT growth or encapsulation for various catalyst sizes. The identical-hued route I curves are represented by a number of (four) spheres. Key requirements for SWCNT growth over encapsulation include having a high enough to outweigh Co, Fe, and Ni are frequent SWCNT growth catalysts, and their computed DFT energies are presented. The evolution of a graphitic cap structure on the catalyst nanoparticle surface is feasible via two different pathways: (I) catalyst encapsulation; and (II) graphitic cap lifting and SWCNT growth. The angles as well as the radii r_C, r_T, and r_P and the separation between graphitic carbon and the nanoparticle d used in the theoretical model. The charge density varies for graphene flake adsorbed on Ni(111) at contact angles of 90° and 45° which was studied for adsorbed armchair (AC) and zigzag (ZZ) graphitic flake edges [15].

field emitter towers and some SWNTs can be created using direct plasma-enhanced chemical vapor deposition (PECVD) methods. A glow discharge is generated in a chamber or reaction furnace by applying a (high frequency) voltage to both electrodes, a substrate is placed on the ground electrode, and the reaction gas is fed into the chamber from the opposite plate.

Using thermal CVD or sputtering, catalytic metals such as Fe, Ni, and Co are employed on a Si, SiO_2, or glass substrate. A crossover between plasma-based growth and CVD synthesis is PECVD and microwave plasma-assisted CVD (MWCVD). The carbon for PECVD synthesis originates from feedstock gases like CH_4 and CO, unlike arc discharge, laser ablation, and solar furnace; therefore, there is no requirement for a solid graphite supply. To enable development at low temperatures and pressures, argon-assisted plasma converts the feedstock gases into C_2, CH, and other reactive carbon species (C_xH_y). Large-scale, low-cost SWNT synthesis using alcohol catalytic CVD (ACCVD) which is done by methanol and ethanol vapors is passed over iron and cobalt catalytic metal particles supported with zeolite. With this method, CNT can be produced at a relatively low temperature of about 550°C.

2.5.3 VAPOR-SOLID-SOLID (VSS) GROWTH PROCESS

This mechanism entails four phases. First, a thin metallic coating is annealed or laser-ablated to create metallic nanoparticles on a substrate. Second, carbon and hydrogen are released from the hydrocarbon gas and dissolve in the metal catalyst particle. Third, carbon permeates the particle. The fourth step is to coat the substrate and stop the catalyst and nanotube growth.

"Tip growth" is a mechanism where particles detach from the body of the CNT and move to the head of the nanotube. The ongoing production of carbon atoms on the support side depicts the growth of CNTs with the metal lifted at the tip of CNTs. CNT is twice as heat resistant as diamond and is just as hard. Compared to copper, CNT can carry 1000 times more current and is thermally stable up to 4000 K. CNT can be metallic or semiconducting depending on diameter and chirality. The properties that have been predicted for an atomically ideal CNT are far from the CNTs that have been produced. This is because we still do not understand all the mechanisms of CNT growth.

2.5.4 IMAGING OF CNT GROWTH

This has been done by notable researchers using both HRTEM techniques and TEM. In 2005, Raty et al. showed their research by including in situ HRTEM images of a developing carbon nanofiber for one cycle in the elongation process [17]. They included drawings to help locate mono-atomic Ni step-edges at the graphene–Ni interface. The illustration of SWCNT growth on a Fe catalyst in ab initio simulations is also of interest in developing an understanding of the growth mechanism of CNTs under experimental conditions.

Hofmann et al. [18,19] showed two models of CNT formation showing in situ TEM images of MWCNT nucleation and growth as well as TEM images of SWCNT nucleation and growth [18,19]. Terrones showed how an electron beam can knock carbon atoms out of the side walls of MWCNTs and into the metal cluster it is encasing. A convex dome is formed out of the metal cluster's flat cross-section, and a carbon cap surrounds it. Atomic processes take place at the metal dome's base, and new MWCNTs appear coaxial to the original MWCNT. Wang et al. [20] reported on the CNT growth steps showing a sample with a thin-film catalyst. The steps include catalyst thin-film dewetting, catalyst dots formation, and CNT nucleation on catalyst dots. High density of catalyst dots leads to the growth of tall VACNT carpets, and low catalyst dots density leads to entangled CNT carpets. They also reported on ab initio simulations of SWCNT growth on a 1 nm Fe catalyst.

Reversible hydrocarbon adsorption, dehydrogenation processes on the surface, and the formation of a catalytic fibrous kind of carbon were demonstrated [17]. However, alternative routes for the formation of carbon on metal catalysts showing the complexity of hydrocarbon decomposition are also proposed.

Two other avenues for the creation of a graphitic cap structure on the catalyst nanoparticle surface are graphitic cap lifting and SWCNT growth. The aforementioned theoretical model accounts for the angles and, radii r_C, r_T, and r_P, in addition to the separation between graphitic carbon and the nanoparticle. The electron density of graphene flakes adhering to Ni (111) at contact angles of 90° and 45° differs from the electron density. For graphitic flake edges adsorbed in an armchair (AC) and zigzag (ZZ) configurations on various metal surfaces, the interface formation energy, e, as a function of angle. A diagram shows the energy profiles for the development of graphitic structures with low and high adhesion. In this work, the energy advantages and disadvantages of the two growth strategies – catalyst encapsulation and graphitic cap lift-off and SWCNT growth – were presented. It was also possible to discern between favorable and unfavorable graphitic cap lift-off due to formation energy. On a catalyst nanoparticle ($r_P = 0.5$ nm), graphitic structures either encapsulate (route I) or lift off to produce SWCNTs of various diameters (route II) and the corresponding formation energy profiles are shown. A graphitic SWCNT cap with many carbon atoms can lift off the catalytic surface if route II parallels route I.

2.5.4.1 Iijima's Research

The interference patterns for the parallel planes are used to calculate the chiral angle 0, which is then discovered from the orientation of the tubule axis concerning the closest zigzag axis, according to Iijima's report on his TEM observations of N coaxial carbon tubules with various inner diameters [21]. He also discovered that icosahedral C_{140} fullerene hemispheres cap single-wall tubules, and he demonstrated that CNTs have three common cap terminations: symmetric polyhedral, asymmetric polyhedral, and symmetrical fiat. Iijima's TEM image depicts a nanotube with a bill-like end, pentagons at "B" and "C," pentagon and heptagon responsible for "B" and "C" curvatures, and pentagon and heptagon causing tube form transitions.

The majority of CNTs have continuous carbon shell caps that fit on the ends of the long cylinder. Iijima's research on caps discusses their features. They displayed the potential number of caps that might be permanently linked to each (n, m) CNT, and it was found by projection mapping that only single pentagon caps – not permutations – are counted. The shortest capable diameter is found in the armchair tubule (5, 5), which joins neatly to a C_{60} hemisphere. The C_{60} hemisphere is perfectly connected to the (9, 0) zigzag tubule. The smallest chiral tubule (6, 5) satisfies the isolated pentagon rule because of its distinctive cap, which is made up solely of pentagons and hexagons. Thirteen different caps can be joined to the chiral tubule (7, 5). In addition, they displayed the projection mapping of an icosahedral cap on a (10, 5) tubule as well as the chirality of the (10, 5) tubule. A capped chiral graphene tubule is created by combining rows of hexagons. The same chiral tubule may be specified by a large number of caps that link smoothly to the same vector $C_h = n_{a1} + m_{a2}$. There aren't any pentagon- and hexagon-only caps that can fit continuously on nanotubes smaller than C_{60} (n, m). Therefore, it is unlikely to find CNTs with a diameter of less than 7 A.

The (4, 2) chiral vector is not thought to be a CNT because it lacks a cap. There are more methods to construct caps for larger tubules (> 10 Å), and therefore they might be more prevalent. This probability should rise quickly with tubule diameter, just as the number of isomers adhering to the isolated pentagon rule rises with fullerene carbon atom count. Most of the investigated tubules have closed caps. Furthermore, open-ended tubules have been seen in high-resolution TEM and STM images. Tubule open ends are highlighted as edges by STM because of hanging bonds there. Repeating cap formations are visible in closed-end tubules. Six pentagons, each having a disclination of $-7r/3$, form the cap of a carbon tubule according to Euler's theorem in w. They displayed CNT caps that are symmetric, asymmetric, and fiat, with larger tubules possessing flat tops and smaller ones having spheres. The "hemisphere" of a big fullerene (about C_{6000}) corresponds to the cap of a tubule with a diameter of $dt''-42$. Some caps, as explained in w, have a cone-like form. The placement of a pentagon at "B" of $2-r$, which introduces a $+r/3$ disclination, explains the bill-like cap. $2-r$ is the solid angle of a hemisphere. Using a method like Iijima's, researchers discovered that most of their nanotubes with a chiral structure.

Nonchiral tubules $(0 = 0$ or $r/6)$ are preferred in multi-walled nanotubes, based on TEM experiments carried out. On the surface of CNTs, the chiral angle of the cylindrical shell has been measured using atomic resolution STM measurements. There have been both chiral and nonchiral tubules found. It was demonstrated that the tubular cylinder and the adjacent inner cylinder were both helical with chiral angles of $0 = 5$ and $0 = -4$, respectively, generating a relative chiral angle, using a Moiré pattern technique that produces the interference pattern between the stacking of adjacent layers. To show networks that multilayer nanotubes can be composed of cylinders with different chiral orientations, atomic resolution STM research is needed.

The numerous additional cap configurations form the foundation of several theoretical justifications in favor of chiral nanotubes. Although rare, semi-toroidal tubule terminations are quite important. In this study, TEM images of a tube with six graphene shells and six pentagon-heptagon pairs in a hexagonal network, as well as images of inner cylinders with normal polyhedral terminations and six with semi-toroidal terminations, which produce a tubule cap lip, were presented. According to Iijima, most of the carbon tubules he observed through electron diffraction in a TEM had screw axes, and chirality (0 is neither 0 nor $7r/6$).

The chirality of adjacent cylindrical planes does not correlate in multiwall tubules (19.1,9). Some investigations have shown circular cross-sections normal to the nanotube axis for cross-sections with fivefold symmetry, even though the majority of efforts have focused on CNTs with circular cross-sections parallel to the tubule axis. It has been observed that vapor-grown carbon fibers with a heat treatment temperature of around 3000°C exhibit polygonal cross-sections.

2.5.4.2   Summaries of Works of Various Researchers and Propose a Growth Model of Carbon Nanostructures

From gaseous components, the carbon atoms or ions come together to form solid structures and experience both repulsive and attractive forces. The repulsive force is due to the Coulomb force, while the attractive force is due to magnetic dipole interactions. In the gas or plasma phase, carbon atoms group and form a compressible ball-like structure. The carbon atoms are confined by a strong magnetic field in a dense plasma ball, where they experience high pressure that forces the bonding of the atoms and forms a highly energetic and metastable structure. Furthermore, the structure formed has a very high compressive stress. Within the diamond structure, there exists a strong repulsive force between the atoms; however, the presence of an oscillatory or periodic electric field creates a closed path of electric field lines. The carbon atoms follow the closed path and form ring structures. The dual nature of the electric field in the diamond structure leads to the formation of two types of rings rotating in opposite directions. As a result, the diamond structure is made up of hexagonal rings in the cubic structure and forms a face-centered cubic unit crystal system.

The carbon atoms are forced to be recycled by the electromagnetic field, which increases the pressure within the plasma ball due to the confinements of the carbon particles by the plasma. However, the atoms that are not constrained by plasma do not recycle and do not contribute to the formation of diamond structures because they do not form a closed loop or do not return. Therefore, the growth and formation of the diamond crystals are explained by the Coulomb repulsion force between the carbon atoms and the atoms that are recycled within the plasma. The electric field directs the charged particles in a circular motion, and the associated magnetic field directs the charges perpendicular to the electric field lines. The hexagonal rings are formed at an angle due to the vertical force, resulting in a structure that looks like an inclined tube. These rings being parallel or as a stack form a helical structure. We consider the attractive forces between carbon atoms; a stable ring structure is formed as the carbon atoms try to occupy space. The rings rotate in opposite directions due to their dual nature, resulting in an extended graphitic structure. These rings then form layers that are stacked on top of each other. The net electric field created by the rings between adjacent layers is zero as it cancels out. Hence, a stable layered structure of graphite is formed. Due to the pressure, a single bond in the graphitic ring can be converted into a double bond and form an excited state. However, this conversion is balanced by the presence of a single bond or the ground state. The alternating s- and p-type bonds make the structure stable, and the minimum energy level is achieved in the ground state. Amorphous carbon is formed because of a combination of attractive and repulsive forces of carbon atoms, which prevent the formation of complete cycles.

2.6   GROWTH OF A QUANTUM DOT (QD)

In Chapter 1, we presented carbon quantum dot (Qdot) structure. Here we propose the growth of carbon from a generalized QD or a zero-dimensional structure. A localized charge is converted into a quantum wire, and this triggers the growth of one-dimensional structures like CNTs. Like atoms, these structures have a dual state. A QD is a spin triplet system that is strongly associated with SOIs. A set of three qubits can illustrate the impact of geometry on the system. It starts with a QD that possesses a two-level (or a multi-level) system. This can be seen as a zero-state connection with the spin–orbit coupling (SOC). It confines the charge resulting in a strong localized (zero) state and a gap which results in a topological state.

The chirality of the system is determined by the size of the QDs and the energy levels associated with them, which is influenced by the SOC. The addition of absorbers can then transform the QD into a quantum wire or a tube for a carbon structure, or a quantum well structure [21] (Figure 2.10).

We start with a QD-like atom that forms a 1D chain or 2D circuit by connecting with other atoms. A chain made of sp^3-bonded carbon is not electrically conductive, but by introducing alternating double sp^2 bonds, it becomes conductive, forming a resonating structure. Pentagonal and hexagonal structures are formed to minimize the energy of the structure. Each ring is associated with a ring current and a magnetic dipole moment (μ) which can also interact with each other and form a resultant (large) dipole moment. The arrangement of rings in the structure is due to the torque generated by the distributed orientation of dipole moments. The dipole moment of one ring influences the orientation of another ring, which is important in the formation of a new ring. The new ring is formed in a way that aligns its dipole moment in a specific direction to increase the net dipole moment of the overall structure. To attain stability the ring forms a curved structure that produces a pentagon. The pentagon is surrounded by six hexagonal rings and this introduces a curvature to the network structure. A maximum SOC implies a maximum curvature or maximum uplift. Chirality in this case refers to the mirror symmetry that is proportional to the energy of the SOC. The energy is dependent on the size of the QD or the level splitting. This type of structure exhibits alternating single and double bonds that are capable of resonating, known as resonating valence bonds (Figure 2.11).

In a fullerene structure, two pentagons are located on opposite sides of a small sphere, and the magnetic moment of the rings combines to form either a unit or a

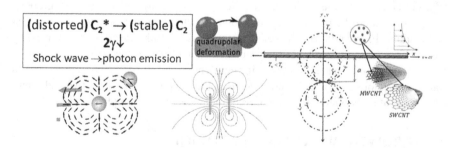

FIGURE 2.10   (top) Production of C_2 and associated valence bonds. Deformation of molecular orbital structure plays a key role to the formation of carbon structures. (below) C_2 dipoles become quadrupoles which is described as splitting of C_2 [24]. Quantum dot model of carbon based on dipolar interactions (right) of nanotubes.

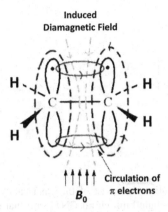

FIGURE 2.11 The interactions between molecular orbitals of carbon atoms (associated with H atoms) produce circulations of pi electrons. Interference between two layers of carbon creates an induced diamagnetic field [38]

large moment. A planar graphene structure lacks a large moment as the ring current in the adjacent rings cancels each other out. However, introducing a defect center such as a pentagon can break the time-reversal symmetry and result in a net magnetic moment. In principle, a structure resembling a vortex could be advantageous for generating a large moment. The core of this vortex structure resembles either a QD or a fullerene molecule. The curvature of this structure results in a large SOC and interactions between the magnetic moments of adjacent rings. CNTs, being the smallest hollow tubes, can be seen as resembling a small vortex that begins from a defect center or a catalyst. These nanotubes can grow by extending or elongating the fullerene structures (Figure 2.12).

During the growth of CNTs, high temperatures are required which causes the metal catalysts to melt and mix with carbon atoms. Small molten metal blobs can absorb C_2 radicals in the CVD or plasma deposition chamber, which are then precipitated on the surface of the metal drop, forming a ring around it. The size of the ring depends on the size of the catalyst, which serves as the building block of the CNTs. The carbon atoms then settle on the rings and the nanotube grows. Metal particles introduce a charge to the carbon atom, resulting in the formation of a high-charge state. This creates a QD localized state or a void-like object, which acts as a seed for the creation process. It acts like a vacuum state that repels other incoming charges due to a significant pressure difference or potential difference. The zero space is created by removing objects with high pressure (by bending the space). The CNTs are formed by starting with a void or a zero, which is a neutral state located at the outer layer of the vortex. This state extends to infinity and is in a steady or continuous state. At the center of this state, there is an extremely attractive force that represents a dynamic zero. During the gas-phase synthesis of CNTs, the rings are arranged in a way that their dipoles are aligned in the form of a vortex. The orientation of these rings forms a chiral structure of the CNTs. The chiral structure of CNTs is formed due to the interactions between the dipoles of the rings that make up the tube. These interactions cause the rings to align in a specific manner, which

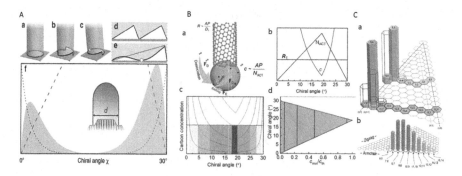

FIGURE 2.12 [Left] (A) Continuum model of the CNT-catalyst system: (a) achiral, (b) multiple- kink chiral, and (c) single-kink chiral CNTs on a flat substrate. Unrolled CNT–substrate interfaces for (d) two-kink and (e) single-kink nanotubes show the nanotube tilt off the vertical, reducing the edge-substrate gap; the white dot in c and e marks the contact point. A nascent CNT of diameter d is shown on a solid catalyst [19].
(B) SWCNT growth mode without enough etching agents. (a) Growth model for SWCNT growth in the absence of sufficient etching agents. (b) In the growth regime, the SWCNT growth rate (RT), active sites for carbon incorporation, and catalyst surface carbon concentration (c) plots against the tube chiral angle. (c) Carbon concentration as a function of the SWCNT chiral angle at different fluxes of carbon deposition. The threshold carbon concentration (c_{th}) for catalyst encapsulation is indicated by the gray line. (d) The chirality-selective growth of SWCNTs as a function of c_{min}/c_{th}, where the dark zone indicates the SWCNTs can survive under different carbon fluxes [25].
(C) Predicted CNT-type distributions. (a) Distributions calculated directly based on atomistic computations for two CNT sets (0.8 nm and 1.2 nm). The empty bars show chirality distributions for liquid-catalyst case. (b) Full (n,m) distribution based on an analytical fit to MD interface energies [26].

leads to the development of a chiral structure. Moreover, the magnetic moment of the rings is aligned perpendicular to the magnetic moment of the tube. This model of magnetic moment interactions can explain both the chiral structure and the SOC. The movement of charge or spin on the surface of the tube creates a circular flow around the highly reactive metal particle that acts as a catalyst. The charged atoms, such as C_2, move along the vortex core in an upward direction (for tip growth) or a direction toward the orbits. A QD can be described as a localized state, which is also referred to as a 'zero' or a dipole. It has the potential to grow into larger structures such as a benzene ring, a sphere (fullerene), and even a hollow tube (CNT), all of which possess a void structure and a big zero. To initiate the growth process, disordered carbon, which may be in a liquid state, is utilized as the source material along with precursors like C_2H_4, C_2H_6, and C_4H_8. However, since this material has a cage structure, the atoms are bonded as pentagons and hexagons, creating an egg-like structure that serves as a seed for growth. Due to its curved surface, a carbon structure can have a significant overlap of its orbitals, resulting in a high degree of SOC, which will be further discussed in the following text.

The addition of C_2 to a fullerene structure initially causes the egg-like structure to be stable and deform. The target is to achieve the maximum excited state or SOC.

Chiral tubes (m unequal n) with angles between 0° and 30° have the highest SOC and result in more tube growth. Slow growth is observed in (0, 0), (m, 0), or (0, n) due to the equal forces on the metal particle. A metal particle can rotate and produce chiral CNTs if the forces at its ends are different. Chirality can also be produced from the random breakup or fragmentation of a fullerene, like the shell of an egg. The broken shell of the egg acts as a seed for the growth of nanotubes with chiral structures, resulting in a mixture of chirality.

2.6.1   CHIRALITY

In the studies of Li et al., it has been proposed that low-temperature SWCNT growth can occur based on non-equilibrium quantum chemical molecular dynamics (QM/MD) simulations and density functional theory (DFT) calculations. A self-assembly process for these SWCNTs was developed from a [6] cycloparaphenylene ([6] CPP) precursor via ethynyl (C_2H). This process is more advantageous than a previously proposed Diels-Alder-based growth mechanism because it preserves the (6, 6) armchair chirality of an SWCNT fragment throughout the growth phase. In addition to extracting hydrogen from the SWCNT fragment, C_2H radicals also operate as the growth's primary carbon source, as demonstrated by QM/MD simulations and DFT calculations. Simulations show that macrocyclic hydrocarbon seed molecules with pre-selected edge structures can develop chirality-controlled SWCNTs when the reaction conditions are properly chosen to allow radical species to efficiently remove hydrogen throughout the growth process (Figure 2.13).

The structure of double-walled carbon nanotubes (DWNTs) has been identified using the latest generation of transmission electron microscope. This has made it possible to do a statistical analysis based on numerous nanoobjects. It has been seen that the DWNTs' inner and outer tubes are not orientated arbitrarily, which raises the possibility of mechanical interaction between the two concentric walls. With the help of atomic-scale modeling, the existence of incommensurate domains, whose inner tubes attempt to attain a local stacking orientation to lessen strain effects, and whose structures rely on the diameters and helicities of both tubes can be deduced.

2.7   SYNTHESIS TECHNIQUES AND THE CVD (FOR DIAMOND GROWTH)

Ever since 1954 when the first confirmed and reproducible synthetic diamond was achieved, many methods of diamond synthesis have come into application. The various methods used depend on the use and type of diamond quality required as given below.

a. **High pressure–high temperature (HPHT):** This is a method that tries to replicate the natural process of diamond formation. A pure form of carbon, for example, coke or graphite is placed under very high pressure using an anvil and heated to very high temperatures. By so doing the carbon transforms into

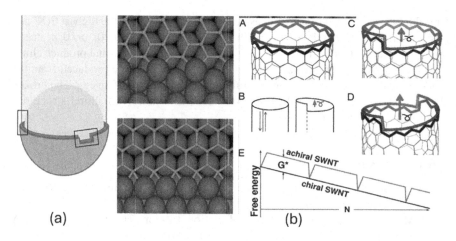

(a)   (b)

FIGURE 2.13 Model of dislocation growth. Screw dislocation along the axis is seen in chiral CNT. Experimental observations and growth rate proportional to the Burgers vector of screw dislocations are in agreement. A dislocation of an axial screw in the CNT. As a perfect crystal, an achiral zigzag $(n, 0)$ tube (A) can be sliced, moved by a Burgers vector (B–D), and then sealed again. This process produces a chiral tube (B). The axial screw dislocations with a single and double value of b, respectively, are present in the chiral $(n, 1)$ in (C) and $(n, 2)$ in (D) tubes; the associated kinks at the open tube end are shown in red. (E) Free energy profile of chiral or achiral nanotube development. Dislocation theory of chirality-controlled nanotube growth [5].

its more stable allotrope diamond. This method is used to make the main micro and single crystalline diamond which has found its place in abrasives and jewelry. Nanocrystalline diamonds (powder instead of films) can be produced by a detonation technique. Single-crystal diamonds can be produced by a high-temperature, high-pressure technique; however, doping diamonds with light elements, particularly with boron, is not easy (Figure 2.14).

Detonation of explosives: Nanocrystalline diamonds (powder instead of films) can be produced by a detonation technique. This method uses the same principle as the HPHT method. Some carbon-containing compound is detonated in a metal chamber and the high pressure and temperature created during the explosion converts the carbon into diamond. Rapid cooling prohibits the conversion of the diamond back to the more stable graphite. It is commonly used for industrial synthetic diamonds.

Laser ablation: In this chapter, we have discussed the synthesis of graphitic carbon (CNTs, graphene) by Nd-YAG laser deposition at high temperatures. However, deposition of diamond-like carbon (DLC) requires a very high-power laser (a 248 nm pulsed UV excimer laser) which can ablate a graphite target at low pressure ($\sim 10^7$ mbar). The fluency of the laser was varied to synthesize two types of ultra-thin a-C layers (~ 4 J/cm^2 for sp^2-rich and ~ 20 J/cm^2 for sp^3-rich DLC layers) of different sp^3 content with significantly different band gaps [85% sp^3, 2.8 eV (barrier-layer of DLC) and 50% sp^3, 1.5 eV (well-layer), Figure 2.15e]. DLC films can also

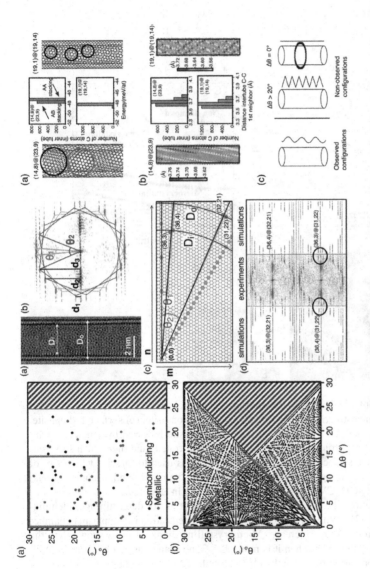

FIGURE 2.14 (Left) (a) HRTEM images of a DWNT and Fourier transform measurement find the layer line spacings d_2 and d_3, and the chiral angles (~0.5°). Distribution of possible chiral indices after the analysis of the layer lines which suggests four configurations (36,3)@(32,21), (36,4)@(32,21), (36,4)@(31,22), and (36,3)@(32,22). (b) Random distribution of $\Delta\theta$ for all (ni, mi) and (no, mo) in the ranges $1.5 < D_m < 4.0$ nm and $0.30 < \Delta r < 0.40$ nm. In all cases, favored (delimited area) and nonobserved (dashed gray area) configurations are marked. (Middle) Analysis of local energies of a (14,8)@(23,9) and (19,1)@(19,14) DWNT ($\Delta\theta = 22.46°$) in the form of histogram plots (middle). Different stackings (AA or AB) are highlighted by circles. (Right) Analysis of the C–C first-neighbors intertube distances in the form of histogram plots (middle) and spatial distribution along the tube (left and right). (c) Sketches illustrating the different types of roughness between walls. Structural Properties of Double-Walled Carbon Nanotubes Driven by Mechanical Interlayer Coupling.

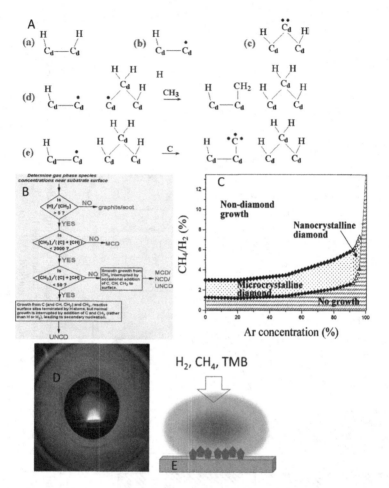

FIGURE 2.15 (Top) (A) Diagram illustrating the several surface carbon radicals that result from hydrogen abstraction. (a) Hydrogen ended diamond site, (b) surface radical site, (c) surface biradical site, (d) a different surface biradical site followed by a reaction with methyl to produce a CH_2 group, and (e) A C atom (not illustrated) and the radical site combine to form a reactive surface. (middle, left) A flowchart is used to forecast which sort of diamond will form depending on the active radicals. On the other hand, a compositional mapping for an Ar-H_2-CH_4 CVD system was proposed by Lin et al. [29].

(B) The flow chart shows the model put forward by May et al. for determining which diamond type will form [31]. With the increase of the Ar concentration in the plasma chamber the methane-to-hydrogen ratios increase. Microcrystalline diamond films grow when the ratio is greater than 1, however, when it is more than 3, no diamonds can be formed. With the increase of Ar concentration, more than 85% growth of NCD can be found. Films do not grow when the ratio is between 1 and above 3 and Ar% is more than 90%.

(C) Microcrystalline, nanocrystalline, and ultrananocrystalline diamond CVD: Experiment and modeling of the factors controlling growth rate, nucleation, and crystal size [30].

(D) (bottom) Plasma CVD deposition of diamond films (left) from the seeded substrates is shown schematically (right). The glow of plasma looks like a ball of fire interacting with diamond small particles (seeds) on the substrate. [32].

be deposited by (rf) plasma CVD method with the application of a sub-strate bias which controls the percentage of the sp^3 carbon in the films.

Ablation of graphite using a laser to produce diamond has of late been reported in research institutions with a couple of variations to the method. The principle used depends on the fact that laser-induced carbon vapor condenses as a sp^3-bonded matrix. The variations in the method are on the carrier gases, which are sometimes incorporated (Figure 2.16).

b. **Chemical vapor deposition:** We have already discussed the CVD method for the deposition of graphene and nanotubes, and it is the most popular method for single as well as polycrystalline diamond deposition. It requires diamond substrates or diamond seeds. In a low-pressure CVD where the synthesis was done, the deposition is achieved in non-equilibrium conditions because under equilibrium conditions, graphite is more stable than diamond. Furthermore, making the chamber rich with carbon favors the deposition of the graphitic phase. The thin-film CVD process begins with the genera-tion of growth species that is, radicals and ions from the temperature of the plasma. The diffusion of the gas phase is determined by the chamber pres-sure and the rate at which the gases are activated into reacting radicals. The homogeneity of gas-phase chemistry is determined by the gas composition, flow rates, and chamber pressure. It determines the sustenance of the growth species and hence the growth rate. There are a few types of CVD techniques.

  i. **Microwave plasma-enhanced CVD (MWCVD):** It uses microwave energy (2.45 GHz frequency) to activate the gases. It is perhaps one of the most widely used methods but is expensive to set up. Using a plasma CVD technique, high-density plasma and a very high temperature of elec-trons can be produced, which can produce diamonds of a high-crystalline order. Most importantly, the gas-phase doping of diamond can be made in a microwave CVD system with chemicals such as TMB that introduce boron into diamond films (Figure 2.17b). Microwave CVD is a clean tech-nique that does not introduce any metal impurities to diamonds.

  ii. **RF plasma CVD:** It uses radio frequencies (13.56 MHz) for activation purposes. It is often incorporated in other CVDs to improve the excita-tion of the feeder gases.

  iii. **DC plasma CVD:** Makes use of direct current (in milliamperes) (volt-age ~kV) electrical energy and is often incorporated in other CVD just like RF plasma.

  iv. **Hot filament CVD (HFCVD):** This is perhaps the simplest, less expen-sive to run and maintain, and allows large-area deposition for nanodia-mond films using multiple filaments. It makes use of a heated transition metal element such as tungsten/tantalum or molybdenum for the activa-tion of the precursor gases. A tungsten filament is known to catalyze the dissociation of hydrogen and nitrogen. This helps to improve the efficiency of the system. Overall, the filament determines the film mor-phology and grains. The synthesis of high-crystalline diamond films requires high energy (temperature) that can be provided by a hot fila-ment CVD technique; however, it can introduce contamination in the films from the metal electrodes.

58   Carbon Superstructures

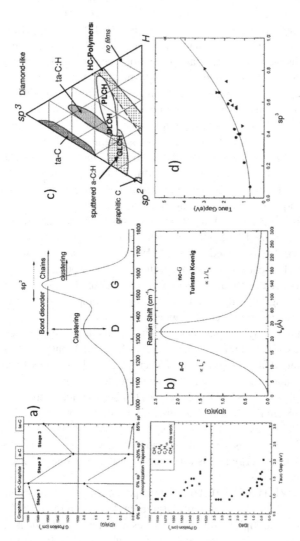

FIGURE 2.16 (a) Amorphization trajectory, displaying the schematic evolution of the *G* location for excitation energies of 244 and 514.5 nm. Amorphization trajectory for 244 and 514.5 nm excitation energies, indicating the potential for non-uniqueness in stages 2 and 3. (bottom) *I(D)/I(G)* ratio variation and the *G* location with Tauc gap. (b) Interpretation of Raman spectra of disordered and amorphous carbon. *I(D)/I(G)* ratio variation with L$_a$ indicates the general shift between the two regimes. Diagrammatic representation of impacts on Raman spectra. The indirect impact of the sp^3 content on the rising *G* position is indicated by a dotted arrow [32]. (c) Ternary phase diagram for amorphous carbons (consisting of sp^{2}C, sp^{3}C, and **H**) that are hydrogenated and carbon-free. (d) As-deposited a-C:H *G* location and *I(D)/I(G)* ratio vs. optical Tauc gap. Optical Tauc gap in relation to the sp^3 concentration of a-C:H as deposited. 100% sp^3 is determined to be the optimal point at 5 eV. The line fits the data in a quadratic way.

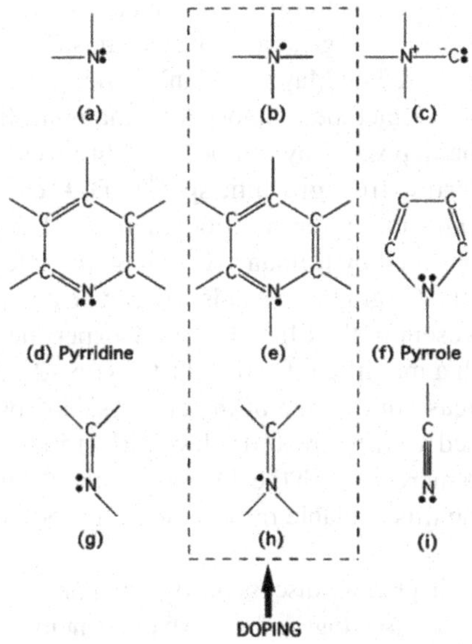

FIGURE 2.17 Possible configurations of nitrogen in a-C:H:N. The lines represent bonds, dots are unpaired electrons, and dot pairs are lone pairs. Possible configurations of nitrogen in a-C:H:N show nitrogen atoms in different configurations in a linear form as well as ring structures (hexagon and pentagon) suggesting doping and nondoping configurations. (a–c) In the top layer, three diagrams with nitrogen acting as a lone pair and nitrogen sitting in the middle of a tetragonal bond. (d–e) The middle layer contains two hexagons and one pentagon where nitrogen acts as a lone pair. (g–i) The bottom layer shows C and N are connected by a double line (bonds) or triple lines. The middle column of the figure is claimed to have a doping configuration in the tetragonal or a trigonal form since the nitrogen atom is donating one electron to the carbon system [34].

Understanding these processes will help in achieving faster growth rates of high-quality carbon structures including diamond films. At this point, we will look briefly at some of the growth models put forward by other workers on how nanodiamond films grow.

2.7.1 GROWTH MODEL OF NANODIAMOND FILMS

In the study by May et al., diamond growth is based upon the competition between –H, –CH$_3$, and C$_1$ radicals. They proposed that the –CH$_3$ radical is the main reactive radical that adds to the diamond lattice with continual abstraction of hydrogen by the –H radical. The diagrams below show some of the possible routes suggested by these researchers. The flow chart below shows the model put forward by May et al., for determining which diamond type will form. Diamond growth requires certain ratios of [CH$_3$], [C], and [H] radicals. For example, when [H]/[CH$_3$] > 5 instead of the diamond graphitic phase of carbon is produced. However, for [CH$_3$]/[C] + [CH] < 2000 microcrystalline diamond can be grown. Experiment and modeling of the deposition of ultrananocrystalline diamond films using hot filament CVD and Ar/CH$_4$/H$_2$ gas mixtures. In the CVD process, the ratio of CH$_4$ and H$_2$ as a function of Ar concentration determines the quality of diamonds ranging from nanocrystals to microcrystals. A generalized mechanism for ultrananocrystalline diamond growth was presented by May and Mankelevich.

In addition to the CVD methods, amorphous diamond-like films (instead of crystalline films) can be deposited by ion-beam techniques and laser ablation. A wide range of carbon starting from graphitic sp^2 C to DLC can be formed as a function of hydrogen concentration in the reaction chamber. The disorder level of the carbon films can be monitored by Raman spectra e.g. the integrated intensity ratio of Raman D (disorder) and G (graphite) bands and can be compared with the size of the graphitic crystals present in the films. From the experimental observation, this ratio was found to reach a maximum for the cluster size ~ 20 A. The disorder level decreases with the increase in the size of the clusters. The position of the Raman G-band can be correlated with the disorder level. Most importantly, the Tauc gap (band gap) was found to increase with the decrease of the disorder level and shift of the G peak. We shall use this valuable information for modeling carbon structures in Chapter 5.

In the last part of this chapter, we discuss nitrogen doping of diamond-like (tetrahedral amorphous carbon) films (see Figure 2.17). An efficient nitrogen doping of crystalline diamond can revolutionize the electronic industry, but nitrogen forms a deep ~1.4 eV center. Earlier, various researchers claimed that synthesis of the possible compound C_3N_4 is harder than diamond, but this local structure does not increase the conductivity of the associated carbon systems. In a-C:H weak n- and p-type doing by phosphorus and boron incorporation was demonstrated due to the creation of a large density of defects. Indeed, nitrogen incorporation to carbon creates defect centers that increase the conductivity of carbon films. Two decades ago, a new form of DLC Tetrahedra amorphous carbon (ta-C) was claimed to have been p-type with a band gap of 2.5 eV. It was also claimed that nitrogen incorporation reduces conductivity followed by an increase. These results were explained based on the movement of the Fermi level in the band gap away from near the valence band to close to the conduction band as if nitrogen acts as an efficient substitutional donor in this material. Based on the hypothesis, a variety of different bonding configurations with single and double bonds, lone pairs, and nonbonding configurations, were suggested in Figure 2.17. In a trivalent configuration, N forms three s bonds with the remaining two electrons in a lone pair which is a nondoping configuration. In a fourfold coordinated system, N occupies a "substitutional" site by using four electrons in s bonds and the fifth is unpaired and is available for doping. In the sp^2 configuration, N can also form pi bonds like in a benzene (doping) ring, pyridine, or pyrrole (nondoping) configuration. In addition to that, an olefinic variant of p bonding as well as "chain terminating" cyano grouping of nitrogen with a triple bond and a lone pair was suggested (nondoping). It seems most of the molecular structures of CN system are nondoping. In low-dimensional carbon structures, N can form other molecular forms; however, we do not have a very clear idea of these complex structures which will be discussed in Chapters 5–8.

Our experiments show that at low nitrogen concentration, a shallow donor level near the CB seems to be formed. But, as the nitrogen concentration reaches >19%, the properties of nitrogenated carbon can be explained as compensated semiconductors rather than as truly substitutional doped. In this case, excess neutral nitrogen atoms (N_3^0) introduce strain in the structure forming N_4^+ and D^+ and C_3^- and D^-, and finally, an $N_4^+C_3^-$ configuration. Comparing a-C with compensated a-Si:H, we say that the shift of E_F at high nitrogen concentration seems to appear not only by dopant incorporation but mostly by the formation of new dangling bonds ~D^+ or D^-. For high doping, excess electrons are trapped in the localized orbital of the cluster, which may lead to a bond breaking, seriously affecting the electronic structure by creating new dangling bonds (D^-) close to the VB.

Q. Based on the gas-phase chemistry we show how the ion and neutral species can be produced in the plasma of an organic precursor and how they can recombine to form carbon molecular structures.

BIBLIOGRAPHY

1. P. M. Ajayan, and S. Iijima, Capillarity-induced filling of carbon nanotubes, *Nature* 361, 333 (1993).
2. R. Sen, A. Govindaraj, and C. N. R. Rao, Metal-filled and hollow carbon nanotubes obtained by the decomposition of metal-containing free precursor molecules, *Chemistry of Materials* 9, 2078 (1997).
3. C. N. R. Rao, A. Govindaraj, R. Sen, and B. C. Satishkumar, Synthesis of multi-walled and single-walled nanotubes, aligned-nanotube bundles and nanorods by employing organometallic precursors, *Material Research Innovations* 2, 128 (1998).
4. S. Bhattacharyya, A. Granier, and G. Turban, Investigation on the electronic properties, microstructure and growth of amorphous carbon nitride films, *Journal of Applied Physics* 86, 4668 (1999).
5. F. Ding, A. R. Harutyunyan, and B. I. Yakobson, Dislocation theory of chirality-controlled nanotube growth, *Proceedings of the National Academy of Sciences* 106, 2506–2509 (2009).
6. H.-B. Li, A. J. Page, S. Irle, and K. Morokuma, Theoretical insights into chirality-controlled SWCNT growth from a cycloparaphenylene template, *ChemPhysChem* 13, 1479–1485 (2012).
7. Q. Liu, X. Shi, Q. Jiang, R. Li, S. Zhong, and R. Zhang, Growth mechanism and kinetics of vertically aligned carbon nanotube arrays, *EcoMat* 3, e12118 (2021).
8. J. Tomada, T. Dienel, F. Hampel, R. Fasel, and K. Amsharov, Combinatorial design of molecular seeds for chirality-controlled synthesis of single-walled carbon nanotubes, *Nature Communications* 10, 3278 (2019).
9. C. Coleman et al., Quantum linear magnetoresistance and shubnikov de-Haas oscillation in suspended wrinkled and smooth multilayer graphene, *Journal of Nanoscience and Nanotechnology* 17, 5408 (2017).
10. Q. Wu, S. Jeon, and Y. J. Song, Growth phase diagram of graphene grown through chemical vapor deposition on copper, *Nano: Brief Reports and Reviews* 15, 10, 2050137 (2020).
11. D. Chen, Y. Zheng, L. Liu, and H. Zhuang, Stone–Wales defects preserve hyperuniformity in amorphous two-dimensional networks, *Proceedings of the National Academy of Sciences* 118, e2016862118 (2021).
12. I. S. Mosse (unpublished).
13. C. Maddi, F. Bourquard, V. Barnier, et al., Nano-architecture of nitrogen-doped graphene films synthesized from a solid CN source, *Scientific Reports* 8, 3247 (2018).

14. R. Lv, Q. Li, A. R. Botello-Méndez, T. Hayashi, B. Wang, A. Berkdemir, … Terrones, M., Nitrogen-doped graphene: Beyond single substitution and enhanced molecular sensing, *Scientific Reports* 2(1), 1–8, (2012).

15. L. P. Ding, B. McLean, Z. Xu, X. Kong, D. Hedman, L. Qiu, A. J. Page, and F. Ding, Why carbon nanotubes grow, *Journal of the American Chemical Society* 144, 5606 (2022).

16. S. Shaik, D. Danovich, W. Wu, P. Su, H. S. Rzepa, and P. C. Hiberty, Quadruple bonding in C2 and analogous eigh t-valence electron species, *Nature Chemistry* 4, 195 (2012).

17. J.-Y. Raty, F. Gygi and G. Galli, Growth of carbon nanotubes on metal nanoparticles: A microscopic mechanism from ab initio molecular dynamics simulations, *Physical Review Letters* 95, 096103 (2005).

18. S. Hofmann et al., In situ observations of catalyst dynamics during surface-bound carbon nanotube nucleation, *Nano Letters* 7, 602–608 (2007).

19. S. Hofmann, G. Csanyi, A. C. Ferrari, M. C. Payne, and J. Robertson, Surface diffusion: The low activation energy path for nanotube growth, *Physical Review Letters* **95**, 3 (2005).

20. Y. Y. Wang, B. Li, P. S. Ho, Z. Yao, and L. Shi, Effect of supporting layer on growth of carbon nanotubes by thermal chemical vapor deposition, *Applied Physics Letters* 89, 183113 (2006).

21. S. Iijima, Helical microtubules of graphitic carbon, *Nature* 354, 56–58 (1991).

22. N. Papasimakis, V. Fedotov, V. Savinov, et al., Electromagnetic toroidal excitations in matter and free space, *Nature Materials* 15, 263 (2016).

23. A. Banerjee, A. Saha, and B. K. Saha, Understanding the behavior of π–π interactions in crystal structures in light of geometry corrected statistical analysis: Similarities and differences with the theoretical models, *Crystal Growth & Design* 19, 4, 2245–2252 (2019).

24. M. He, et al., Growth kinetics of single-walled carbon nanotubes with a $(2n, n)$ chirality selection, *Science Advances* 5, eaav9668 (2019).

25. V. Artyukhov, E. Penev, and B. Yakobson, Why nanotubes grow chiral, *Nature Communications* **5**, 4892 (2014). https://doi.org/10.1038/ncomms5892.

26. E. S. Penev, V. I. Artyukhov, and B. I. Yakobson, Extensive energy landscape sampling of nanotube end-caps reveals no chiral-angle bias for their nucleation, *ACS Nano*, 8, 2, 1899–1906 (2014).

27. A. Ghedjatti, Y. Magnin, F. Fossard, G. Wang, H. Amara, E. Flahaut, J.-S. Lauret, and A. Loiseau, Structural properties of double-walled carbon nanotubes driven by mechanical interlayer coupling, *ACS Nano* 11, 4840–4847 (2017).

28. T. Lin, G. Y. Yu, A. T. S. Wee, Z. X. Shen, and K.P. Loh, Compositional mapping of the Ar- Ch4-H2 system for polycrystalline to nanocrystalline diamond film growth in an HFCVD system, *Applied Physics Letters* 77, 17 (2000).

29. P. W. May, M. N. R. Ashfold, and Y. A. Mankelevich, Microcrystalline, nanocrystalline, and ultrananocrystalline diamond chemical vapor deposition: Experiment and modeling of the factors controlling growth rate, nucleation, and crystal size, *Journal of Applied Physics* 101, 053115 (2007).

30. P. W. May and Y. A. Mankelevich, Experiment and modeling of the deposition of ultrananocrystalline diamond films using hot filament chemical vapor deposition and Ar/CH4/H2 gas mixtures: A generalized mechanism for ultrananocrystalline diamond growth. *Journal of Applied Physics* 100, 024301 (2006).

31. S. Bhattacharyya, Microstructure and anisotropic order parameter of boron-doped nanocrystalline diamond films, *Crystals*, 12, 1031 (2022).

32. A. C. Ferrari and J. Robertson, Interpretation of Raman spectra of disordered and amorphous carbon, *Physical Review B* 61, 14095 (2000).

33. S. R. P. Silva, J. Robertson, G. A. J. Amaratunga, B. Rafferty, L. M. Brown, J. Schwan, D. F. Franceschini, and G. M. Mariotto, Nitrogen modification of hydrogenated amorphous carbon films, *Journal of Applied Physics* 81, 2626 (1997).

34. S. Iijima, and T. Ichihashi, Single-shell carbon nanotubes of 1-nm diameter, *Nature* 363, 603–605 (1993); D. S. Bethune, C. H. Kiang, M. S. De Vries, G. Gorman, R. Savoy, J. Vazquez, and R. Beyers, Cobalt-catalysed growth of carbon nanotubes with single-atomic-layer walls, *Nature* 363, 605–607 (1993).

35. H. Naser, M. A. Alghoul, M. K. Hossain, et al., The role of laser ablation technique parameters in synthesis of nanoparticles from different target types, *Journal of Nanoparticle Research* 21, 249 (2019).

36. I. S. Mosse, et al., Tuning magnetic properties of a carbon nanotube-lanthanide hybrid molecular complex through controlled functionalization, *Molecules* 26, 563 (2021).

37. A. Dager, T. Uchida, T. Maekawa, et al., Synthesis and characterization of mono-disperse carbon quantum dots from fennel seeds: Photoluminescence analysis using machine learning, *Scientific Reports* 9, 14004 (2019).

38. A. Banerjee, A. Saha, and B. K. Saha, Understanding the behavior of π–π interactions in crystal structures in light of geometry corrected statistical analysis: Similarities and differences with the theoretical models, *Crystal Growth & Design* 19, 4, 2245–2252 (2019). https://doi.org/10.1021/acs.cgd.8b01857.

3 Confined Low-Dimensional Structures

3.1 QUANTUM DOTS IN CARBON NANOTUBES

In this chapter, we discuss electronic transport in carbon nanostructures, particularly, in carbon nanotubes (CNTs) and graphene, based on localized states, that is, defects and quantum dots. In the previous chapters, we have explained quantum dots. They can be considered as semiconducting or metallic particles of a few nanometers in size. Due to the quantum confinement of wave functions, they have different optical and physical properties than those of larger particles. An electron in a quantum dot (QD) can be excited by photons to a higher energy state. For a semiconducting QD, an electron transition occurs from the valence band to the conducting band. The electronic wave function in QDs is similar to that of atoms. Artificial molecules can be created by coupling two or more QDs. Such systems exhibit hybridization at room temperature. QDs have properties of both atoms and bulk semiconductors. Optoelectronic properties of the quantum dot like the wavelength of the emitted light depend on the size and shape or the geometry of the quantum dot. The electronic properties of nanostructures attached to metallic QDs depend on the depth (size) and geometrical arrangements (interdot coupling).

3.1.1 STRONG LOCALIZATION

Anderson localization (or strong localization) is a phenomenon (named after P.W. Anderson) where diffusion of waves in a disordered medium is absent. It was suggested that the localization of electrons could occur in a lattice potential in the presence of a sufficiently large degree of disorder in the lattice. It applies to the transport of certain types of waves like electromagnetic waves, acoustic waves, and spin waves which arise from the interference of waves between several scattering paths; however, it is different from weak localization (which will be discussed in the next chapters). It can be distinguished from Mott localization since the transition from conducting (metallic) to insulating behavior is caused by Coulomb interactions between the incoming wave and ions/electrons in the medium rather than disorder. In the strong limit waves inside the scattering medium are halted due to interference. For non-interacting electrons, a transition known as the metal-insulator transition (MIT) occurs in the absence of SOC and a magnetic field. Although MIT cannot be observed in either the 1D or 2D systems, MIT can exist

DOI: 10.1201/9781003316411-3

in two dimensions since states are only marginally localized which results in SOC. Since the localization lengths of a 2D system with the potential disorder are large, a localization-delocalization transition can be found regardless of the magnitude of the disorder and system size in numerical simulations. Numerical approaches typically use the tight-binding Anderson Hamiltonian (see Chapter 5). Participation numbers which are obtained by diagonalization can be used to determine the properties of the electronic eigenstates. Another useful method that can be used to compute localization lengths is referred to as the transfer matrix method. This method provides numerical proof that a one-parameter scaling function exists. This will be discussed in Chapter 5 in detail. In this chapter, we show that localized states can be created by the application of gate voltage as well as at the junction of the electrodes. In the first case, the depth of the potential and the degree of localization can be controlled by the applied bias which will create ON/OFF state of a transistor, that is, from a delocalized to a localized state.

Localization arises from the disorder in the system which drives the electronic conduction from a diffusive to a ballistic regime. For a device, this can be controlled by incorporating the quality of the conduction channels. The localized wave function moves in a CNT as a surface state, like a doughnut. Since it is delocalized it can follow a helical structure which can induce SOIs to the electron transport and the system will be topologically protected. In the presence of a bound state (created by a strong Coulombic potential) which will produce a zero-bias conductance peak (ZBCP) with satellite peaks and oscillatory features. Similarly, if we consider the edge state of a piece of graphene, then it will be topologically protected and also support the SO interaction. This effect would produce oscillatory features in conductance.

3.1.1.1 Coulomb Blockade

In general, a tunnel junction is thought of as a very thin (a few nanometers thick) insulating barrier between two conducting materials that permits tunnel current to flow when a bias voltage is applied. A single electron should ideally charge the circuit so that it behaves like a capacitor. The ratio of the elementary charge to the junction's capacitance determines the voltage that develops in the capacitor. Due to the very low capacitance of a tunnel junction, the voltage buildup may be sufficient to block the tunneling of another electron. At low bias voltages, the electric current is thus reduced, and the device's differential resistance rises close to zero bias. The phenomenon is regarded as Coulomb blockade which shows a reduction in conductance of an electrical device associated with a tunnel junction at low bias voltages. Due to the coulomb blockade, the conductance is not always constant at low bias and may even vanish if the bias falls below a specific threshold. The best places to see coulomb blockades are tiny gadgets like QDs. In such systems, electrons will strongly repel one another, blocking the flow of other electrons. The connection between current and voltage will resemble a step function. Even while Coulomb blockades can be used to demonstrate the quantization of charge, it is a classical effect. We shall see this effect in CNT junctions and as a function of gate voltages.

3.2   CARBON NANOTUBES OVERVIEW: QUANTUM WIRES IN CARBON SYSTEMS

The speed of carbon (device) is determined by the QDs and ballistic motion in quantum wires.

CNTs find a wide variety of applications as single-walled nanotubes are semiconducting and can be used as a thin-film material. They are easily fabricated through printing while being flexible and stretchable. This allows them to be used in bendable as well as sturdy electronic devices and has the potential to replace current glass- and silicon-based devices which are mostly fragile. Since they are transparent, they can be used in mobile devices in the processing and memory units that are see-through and thus act as a screen. Printed TFTs can be used in OLED televisions.

The semiconducting and electronic properties of CNTs lend themselves to making transistors as well as buses. This is due to a very high carrier mobility of greater than $10\,cm^2/Vs$ and an on/off ratio of 10^6. With such properties, CNTs can be used as buses to transport data while the processing units are studded on these nanotubes themselves. Furthermore, carbon being compatible with biological systems makes ideal choices for ingestible nanorobots which have gained interest for medicinal and therapeutic purposes. They can be used to make nanotube-based random access memory (NRAM) which is much faster and uses lower power than conventional RAM.

3.2.1   ELECTRONIC PROPERTIES OF SWNTS

In CNT the p-orbital electrons unbonded and free to couple with corresponding electrons to form π-orbitals. These electrons due to their weaker bonding are free to move about and are responsible for the electrical conductivity in CNTs. The electronic band structure transport properties depend on how the CNT has been rolled up from the 2D graphene sheet. There are three basic ways to roll up, namely. armchair, zigzag, and chiral (Figure 3.1). These different ways to roll up can be described by a "roll-up" vector (n,m) which represents the integer components of the circumference vector relative to the lattice with the nanotube axis perpendicular to it. In terms of the translation vector connecting two crystallographic equivalent sites, these will be, $C_h = na_1 + ma_2$, where a_1 and a_2 are lattice vectors. The conditions that are imposed on (m,n) give rise to the above-mentioned "roll ups" and hence determine the electronic properties of the CNT. For example, $n = m$ corresponds to armchair CNT which is metallic, $m = 0$ corresponds to zigzag CNT, and all other conditions on (n,m) result in chiral CNTs. Both zigzag and chiral CNTs can be either semiconducting or metallic. The roll-up vector along with length can solely specify both the physical and fundamental properties of the CNT.

These roll-up vectors (n,m) are linearly independent components that represent the contribution of two distinct states to the final state. They can be used to describe the probability of obtaining the two states and can be used as a quantum bit that is of the standard form $\alpha|1\rangle + \beta|2\rangle$ with α replaced by 'n' and β replaced by m. The quantum bit so obtained can be read and written by detecting the physical properties and manipulating the roll-up, respectively.

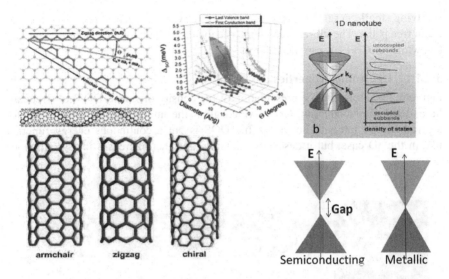

FIGURE 3.1 Top Left and Middle: The variation of the spin–orbit coupling (energy) is plotted as a function of the tube diameter and the chiral angle. A plot between these parameters shows a decrease in the coupling with the increase in tube diameter. Top Left: A square covered by hexagons. The sides of the square act as vectors a_1 and a_2. Right: Left-handed and right-handed cylinders made of a hexagon-filled surface. (a) The graphene band structure is impacted by nanotube curvature. (a) The Dirac points move away from the Brillouin zone's corners due to the curvature in a (4,1) nanotube. In terms of visibility, the transformation has been overestimated by 15 times. The top inset shows how the displacement vector c_v is divided into components that are parallel to and perpendicular to the nanotube axis. The shift happens with respect to the nanotube's circumference at a 3° angle. Nanotube shift for a chair, bottom inset. Since the shift in these structures occurs along the quantization lines, curves do not cause gaps in them. (b) Dirac cones at K and K_0 valleys, showing how horizontal shifts by c_v open a band gap in a presumably metallic tube with and without curvature effects (solid and dotted) [16]. (b) Figure 3.1a displays a diagrammatic depiction of the band structure and DOS for carbon nanotubes. Tube for semiconductors (Armchair) (zigzag). (c) An illustration of the band structure of a metallic and semiconducting carbon nanotube. The propagation direction is determined by the tube axis. The tube axis plane's intersection with the 2D graphene dispersion is where the 1D dispersion of CNTs is identified [17].

Other important properties of a CNT are the diameter and the mean free path. Generally, there are more subbands in CNTs with large diameters than smaller ones. This dependence of the number of subbands on the diameter is due to the quantization of the transverse wave vector which determines the allowed k-spacing of the subbands. This quantization arises from preserving the periodicity around the circumference of a CNT during the roll-up process.

The mean free path in CNTs is unusually large. This is due to the limited number of available electron states in CNTs which causes a large difference in the momentum of forward and backward-moving states. As such charge carriers cannot be scattered by room temperature phonons which have lower momentum, the scattering due to electron-phonon interaction is significantly less. Contrast this with the case in metals, even in good conductors such as gold or copper, the electron-phonon scattering involving small energies and momentum occurs very frequently resulting in small mean free paths. Therefore, even gold has a mean free path of a few tens of nanometers while for CNTs, it's in the order of ~ 1 μm.

3.2.1.1 Electronic Properties of CNTs

Figure 3.2 is a schematic representation of the density of states (DOS) of (i) metallic and (ii) semiconducting CNTs which gives the number of available states for a given energy interval. For a 1D, the DOS is not a continuous energy function like in the 3D case, but it descends gradually and then increases in discontinuous

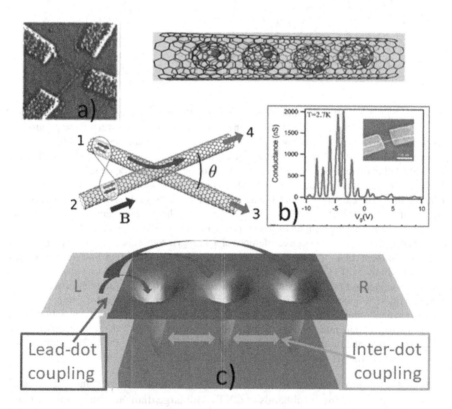

FIGURE 3.2 Tube axis defines propagation direction; intersection of plane defined by tube axis with 2D dispersion of graphene defines the 1D dispersion of CNTs. (a) Energy division $E_{ii} \Delta E$ as a function of CNT diameter d_t, with 0.7 dt 3.0 nm, for all values of n and m. The peaks of semiconducting and metallic tubes are indicated by crosses and open circles, respectively. For zigzag tubes, solid squares represent the $E_{ii} d$ values. Trigonal warping effect of carbon nanotubes [18]. (b) shows different regimes of conduction in nanotubes. Diffusion coefficient in (10,10) metallic nanotubes for three typical conduction regimes. On a long time scale, the ballistic, diffusive, and localized regimes are visible. Kubo conductance for the ballistic case is at the bottom. Band structure for the 5,5 armchair nanotube is on the left, density of states is in the middle, and conductance is on the right. (c) An array of quantum dots can be created between electrodes to study quantum transport as a function of the coupling between dots and with the electrodes.

spikes known as van Hove singularities (VHSs) as shown in Figure 3.2. It is given by $(E)_{1D} = \dfrac{1}{\hbar\pi}\sqrt{\dfrac{m^*}{2(E-E_c)}}$, where the electron mass is given by m^* and E_c denotes the kinetic energy and E is the energy related to the system. The energy difference between the peaks in the DOS of the VHS can be shown to first order to be related to the tube diameter based on the band folding approach. In addition to the individual nanotubes the properties of the nanotube junctions produce strong peak features in the electronic spectra which can be modeled based on QDs (Figure 3.3).

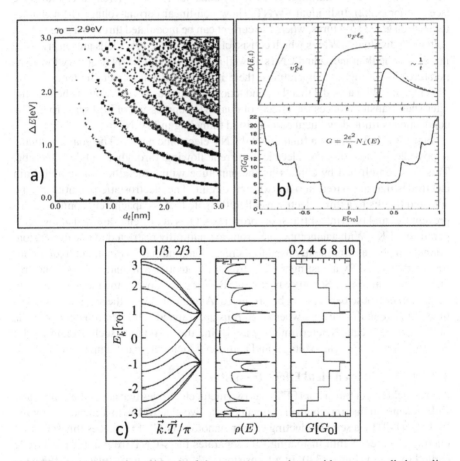

FIGURE 3.3 (a) 1D string of islands of charge – quantum dots – with gate-controlled coupling between them. The spins of the dots are used to form qubits of quantum computers. No dynamic g-factor modulation, no electron spin resonance, and no local magnetic fields are required to tune the qubits. For this operation, tuning the coupling between neighboring quantum dots along the nanotube is necessary. Operations only require nearest neighbor coupling in 1D, scaling does not lead to a proliferation of interconnect elements. (b) Spin filtering and entanglement detection due to spin–orbit interaction in carbon nanotube cross-junctions [19]. (c) Quantum information can be stored in quantum wells and Quantum dots as resonant states and accessed by field and frequency through resonance. Effect of structure and disorder is crucial [20].

3.2.2   ELECTRONIC TRANSPORT PROPERTIES OF CARBON NANOTUBES

The electronic properties of individual SWNTs depend on which type is being characterized as they display both semiconducting and electronic transport conditions from the synthesis of both types. Short SWNTs having low defect density behave like molecular wires which show ballistic conductance. At very low temperatures, they exhibit Coulomb blockade (CB) effects characterized by quantized conductance features which can be readily seen in gated structures (see Section 3.2.1). At higher temperatures, the CB effects are overcome by thermal effects and the CNTs display power law dependencies like the Luttinger liquid (LL) behavior in the differential conductance (see Section 3.2.2). Individual SWNTs show non-linear current-voltage (I-V) characteristics on gated structures, where the current can be modulated through a gate-source voltage. Individual MWNTs which consist of several concentric shells are surprisingly more metallic than individual CNTs as well as a network of SWNTs at very low temperatures but at higher temperatures, there is no conversion to metallic temperature. This matches with a model that the conduction of electrons is via the outer shells as the inter-shell transfer reduces the effects of quasi-1D-suppression of backscattering. The conduction is limited by fluctuation-assisted tunneling in MWNTs.

Figures 3.4–3.6 depict a fundamental CNT electronic device. The goal is to make it possible to quantify the electrical current flowing through a single nanotube. This is accomplished by contacting the nanotube with a metallic source and drain electrodes that are wired into an external circuit. The electrostatic potential can be adjusted thanks to a third electrode called the gate that is capacitively connected. To prevent thermal blurring of transport properties, QDs are often examined at low temperatures (1 K). With a nanotube, electrons are naturally restricted to one dimension. Tunnel barriers are frequently used to increase longitudinal confinement to quantum transport studies. By adjusting the electrostatic potential with gate voltages and frequently making use of Schottky barriers (SBs) induced close to the metal contacts in the nanotube, these barriers can be produced. A quantum dot is the section of a nanotube in between the barriers where electrons are trapped. The electron energy levels in the nanotube can be determined by analyzing the current via such a quantum dot as a function of bias, gate voltage, and additional factors like a magnetic field.

3.2.2.1   CNT-Based Field Effect Transistor

A basic field effect transistor (FET) has two metal electrodes that act as the source and sink. A semiconducting channel connects these two electrodes. In a carbon nanotube FET (CNTFET), a semiconducting carbon nanotube (s-CNT) replaces this semiconducting channel. A thin insulating film separates the s-CNT from a third electrode (called the gate) (Figure 3.4). In a transistor, if there is no charge placed at the gate, then there is no flow of charge in the channel. For an n-type FET, if a positive charge is placed at the gate, an electron current will flow through the channel. Similarly, if a negative charge is placed at the gate of a p-type FET then a hole current will flow. Due to the reduced phase space in metallic CNTs, there will be no scattering of charge carriers and the nature of transport will be ballistic in MWCNTs. However, in s-CNTs scattering of charge carriers by short-range potentials is more efficient. Since the energy states in s-CNTs are quite close to each other in terms of energy, inter-band scattering needs to be considered since it implies that the transport may

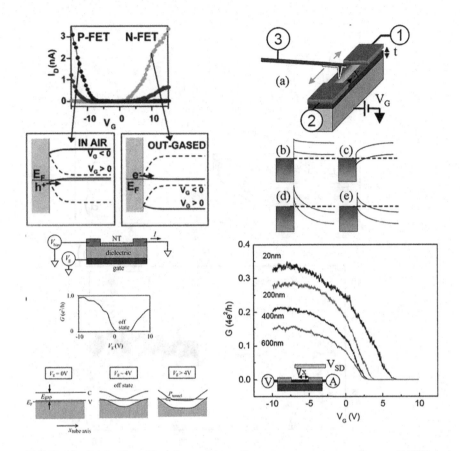

FIGURE 3.4 Right Top: The plot consists of two curves corresponding to P-FET and N-FET, both have a valley in the middle and rise on either side of 0 V_g; Bottom: The picture without-gassed has an arrow from the N-FET curve that rises on the right side of 0 and the one with in air has an arrow linking to P-FET. The lower layer is Si (back gate); above it is SiO$_2$. Above that on the left is Au(source) and a little gap separates it from Au(drain). Over this Au is an MWNT or SWNT connecting Au(Source) to Au(drain) through the gap. Left Top: Conductance is ON when 1D channel is partially occupied. Conductance is OFF when 1D channel is fully occupied (energy above energy gap) or empty (energy below energy gap). In ON state, same conductance can be found in metallic CNT [21]. The plot of conductance G vs. Voltage shows a skewed graph with a well in the middle and the curve rising on both sides. While it rises to 1 on the left, it rises less on the right. The well is the off state, and the right side is conduction states partially occupied. Three images at the bottom show $V = 0$ V, $V = 4$ V, and $V > 4$ V which refer to parts on the graph, where 0 and 4 V is the off state, and > 4 V is in the on state. In the latter two cases, the energy band is curved. Qualitative diagram showing the lineup of the valence and conduction bands of a CNT with the metal Fermi level at the source–CNT junction first in air and after annealing in vacuum [22]. (top right). A three-probe measurement setup consisting of two electrodes and the AFM tip which can be used as a voltage source, current probe, or voltage probe (a). (middle right) [b–e] Band diagrams are shown for the p-region and the n-region for Ohmic and Schottky contacts, respectively [22]. (bottom right). Conductance curves as a function of gate voltage are plotted for distances x between tip and drain ($x = 20$, 200, 400 and 600 nm). Inset: The source-drain bias VSD is applied to the tip, while the drain (right) electrode is connected to current amplifier, and the left electrode to a voltmeter [22].

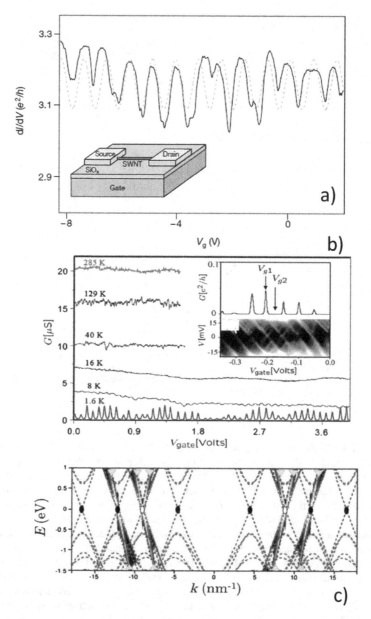

FIGURE 3.5 (a) Single CNT device conductance variations: Metallic CNT, 4K. $R > 7\,k\Omega$, 200 nm long. Zero-bias ($V_g = 0$) differential conductance (dI/dV). Oscillations of differential conductance < 10 K [23]. (b) As a function of the external gate voltage applied to the single-walled nanotube in contact with the noble metal, conductance spectrum for various temperatures was calculated. G(V_g) for a device at 4.2 K is shown in the inset, with a recurring pattern of Coulomb conductance peaks [24]. (c) Density of states (dI/dV) map after being Fourier-transformed as a function of electron momentum (k) and sample bias (V). The corners of the Brillouin zone closest (elongated black dots) or second closest (elongated white dots) to the Γ point are indicated by ovals close to the Fermi levels. Curves give the energy dispersions of the (19,7) tube by tight binding, whereas superimposed indicate the Luttinger liquid result.

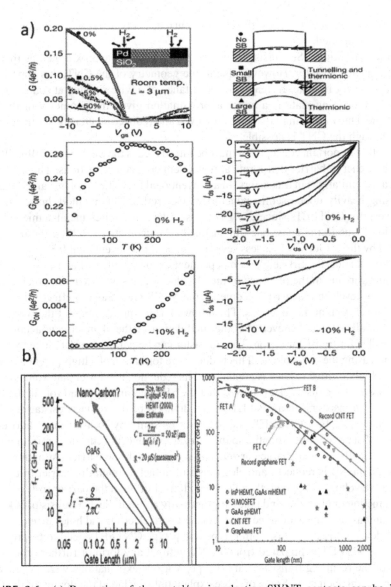

FIGURE 3.6 (a) Properties of the metal/semiconducting-SWNT contacts can be influenced by in situ metal work-function modification. (top left) Linear G - V_{gs} for a SWNT-FET (L = 3 m) before and after the pure Pd contacts were exposed to various concentrations of H_2. (top left), Band diagrams for the device show ON state (for p-channel) shows the development of SBs to the valence band for higher [H_2]. G_{ON} vs. T for the device with pure Pd contacts before exposure to H_2. I_{ds} - V_{ds} curves for the device under various V_{gs} at room temperature. G_{ON} versus T after exposure of the Pd electrodes to 10% H_2. I_{ds} - V_{ds} curves under various V_{gs} at room temperature exhibiting typical SB-FET behavior with inflection points in the curves [21]. (b) (left) The cut-off frequency versus gate length for some common FET transistors in comparison to CNT FETs [25]. (right) Predicted dynamic impedance for Ohmic contacts, for two different values with two different LL parameter (g) values. (Adapted from ref. [26].) The line of FET B which is Si MOSFET starts at 1000 GHz and goes a linear downward slope to 20 GHz at 2000 nm. The FET A, which is InP HEMT, GaAs mHEMT increases first and then takes a downslope, the GaAs pHEMT starts in the middle, increases first, and then decreases.

be diffusive. At low voltages, the channel current is observed to be proportional to the voltage applied.

In a conventional Si transistor, the gate capacitance is approximated as that of a parallel plate capacitor. However, due to the geometry of the CNT, this approximation is not very reliable. Instead, the capacitance of a wire situated at some distance from a conducting plate is a better approximation given the geometry of the gate electrode. Due to the small distances involved and reduced scattering, ballistic transport through the CNT is possible.

Another important concept that must be introduced is the SB (Figure 3.4b). SBs are a type of energy barrier that forms at metal-semiconductor junctions. These barriers can cause limitations in transistors. In a conventional FET, the problem can be avoided by using heavily doped semiconductors as electrodes. This method, however, does not work for CNTFETs since any CNT-based device requires metal-semiconductor junctions. Due to the quasi-one-dimensional (1D) geometry of CNTs, the limitation caused by the SB for CNTs is less severe compared to conventional FETs. In a planar metal conductor, the SB height is independent of the metal used and is a fraction of the band gap due to the presence of metal-induced gap states (MIGS) which can pin the Fermi level. In a quasi-1D geometry, the MIGS are unable to effectively pin the Fermi level deep at the junctions. This allows for the modification of the work function. Initially, it was believed that the modulation of the channel conductance was responsible for transistor action. The transistor can be put in an on/off state by reducing/increasing the gate voltage. This is due to the formation of a high-potential barrier when a high positive voltage is applied and its reduction at a lower gate voltage.

However, later discoveries showed that the SB plays an important role in the functioning of transistors. In a SB-FET the modulation of the contact resistance affects the conduction. In a CNTFET, the SB behaves similarly to a potential barrier for charges that are coming from the source. At zero voltage, the current is minimal since only thermally excited charge carriers can pass through the barrier. At high positive voltages, the SB is reduced, and electrons can tunnel through the barrier (n-type). If a negative voltage is applied, then holes can tunnel through the barrier (p-type). Both the tunneling of electrons and holes are thermally assisted. Such a transistor is known as an ambipolar transistor since it can conduct both electron and hole currents.

To develop logic circuits, both n-type and p-type CNTFETs need to be prepared. When a CNTFET is prepared from a CNT without any form of further processing, most of the resulting transistors are p-type. Various explanations were put forward to explain this behavior. Some of these explanations included the doping of the CNT by unknown dopants, charge transfer from the metal electrodes, and so on. There are two ways of converting a p-type transistor to an n-type transistor. The first method is to dope the p-type transistor using electron-donating dopants such as alkali metals.

The second method involves the heating of p-type transistors in a vacuum that causes any adsorbed gas such as oxygen to be desorbed which converts the p-type transistors into the n-type. An advantage of this method is that the process is reversible, that is, the n-type transistor can be reconverted to a p-type transistor through adsorption of oxygen. Initial experiments suggested that oxygen could add holes to CNTs. However, later observations did not support the claim. This is because if oxygen was indeed the responsible dopant for p-type characteristics then its removal should make the CNT intrinsic. An explanation was put forward which suggests that oxygen affects the contact SB. The Fermi level is more deeply pinned in the presence of oxygen which results in p-type characteristics. The desorption of oxygen causes the Fermi level to change its position due to the bending of the electron band structure near the contacts. Another important thing to note is that ambipolar behavior occurs only if the device is passivated at low temperatures by a silicon dioxide film. If the temperature is high, then oxygen can diffuse through the film and adsorb to the device (thus converting the device into a purely p-type transistor).

3.2.2.2 Electron Transport in Mesoscopic Systems

Electron transport in mesoscopic systems is determined by four characteristic lengths:

1. The mean free path is the average distance that an electron travels before it is scattered.
2. The phase relaxation length is the distance an electron travels before a significant change in phase. Elastic scattering is the main scattering phenomenon if the phases are coherent in a region.
3. Localization length is the average distance in which an electron wave function is trapped.
4. System dimension or length is the size of the system in question.

The way electrons are transported in nanotubes is dependent on these four lengths relative to each other. These give rise to three regimes of electron transport:

1. **Ballistic regime:** When the mean free path is larger than the phase relaxation length which in turn is larger than the system size, the electron can move through the entire system without losing phase or scattering. When that happens the electron is said to travel ballistically with no resistance at all. Ballistic transport has been reported in some CNTs.
2. **Localized regime:** When the mean free path is smaller than the phase relaxation length and both are smaller than system dimensions, the electron gets trapped in potential wells or defects and thus becomes localized. If the phase relaxation length is smaller than the localization length, the transport is diffusive, and it is called weak localization. During diffusive transport, electron wave interference, spin interaction, and magnetic impurities affect the transport. On the other hand, if the phase relaxation length is larger than the localization length, the transport takes place by hopping from one localized state to another. This is a strong localization.
3. **Classical regime:** This is the regime that is usually observed. In this case, the phase relaxation length is smaller than the mean free path and both are smaller than the size of the system. Ohm's law and other classical laws are followed in this regime. The factors affecting transport in this regime are material imperfections, lattice vibrations, and impurity scatters.
4. A localized state produces periodic oscillations of the current or the differential conductance with a ZBCP (Figure 3.5). This ZBCP is a signature of a bound state which is useful for quantum computation.

3.2.3 CARBON NANOTUBES AS HF DEVICES

There are some fundamental aspects of transport and potential applications of low-dimensional carbon and CNTs in the high-frequency regime but ballistic devices as an integral part of the circuit have yet to be realized (Figure 3.6). In this work, we were able to demonstrate for the first time a carbon-based device comparable to a conventional HEMT device on a much smaller scale, which is difficult to achieve in silicon using today's technology. A current major limitation to their application is that the measured performance is far below the predicted intrinsic capabilities, e.g., the highest measured cut-off frequency is about 40–100 GHz compared to the predicted 1 THz theoretical value. The cut-off frequency versus gate length for some CNT-based FET is compared to other standard semiconductor devices. The performance of carbon materials in high-frequency regimes can motivate the research community in this field.

The LL model is based on collective excitation of charge carriers due to the sufficient Coulomb interactions at 1D/plasmon. LL theoretical predictions made in 1963. The generation of these plasmons leads to the evolution of high-frequency performance and applications of SWNTs. Elusive LL phenomena can be observed experimentally at low temperatures through power law dependence of conductance, scaling behavior, and non-Fermi differential conductance in CNTs. The conditions that should be satisfied are (1) low defect conduction path, (2) length of the conductor much greater than the electron mean free path, and (3) low contact resistance. However, observation of related phenomena such as spin charge separation through frequency dependence has been remained challenging.

In 2D (3D) materials strong SOC can be observed which becomes weak in the 1D system due to the absence of orbit-like structures. In 1D strongly correlated fermions are characterized by bosonic (quasi-particles that can carry the same quantum number) excitations of collective charge and spin fluctuations moving with different velocities. The inhomogeneous electron interactions in 1D vary with the velocity and charge of the charge-wave excitations. This may be regarded as charge-wave reflection resulting in charge-spin separation which can be described as the properties of plasmonic oscillations or LL. This property can be observed in a system with dynamical correlations achieved by applying time-dependent signals in AC transport (Figure 3.7).

3.2.3.1 Diffusive to Ballistic Transport

LL behavior is known to be observable in very few systems, namely 1D electron-phonon, metals with impurities, and edge states in the quantum Hall effect. LL response is easily observable in a ballistic transport regime where the scattering rate is minimized. In DC transport, this can be achieved by reducing the dimensions of the conduction channel so that it is comparable to the mean free path. Since the correlations between the excitations are anomalous, they show up as interaction-dependent non-universal power laws in many physical quantities observed in Figure 3.7a and b

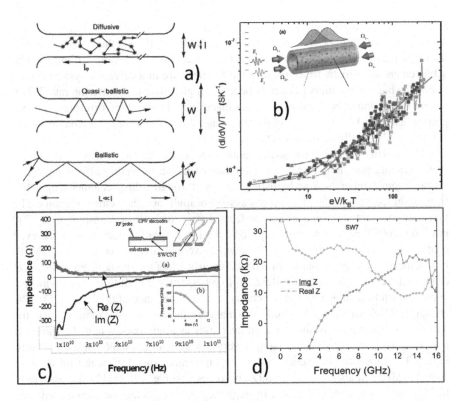

FIGURE 3.7 (a) An Illustration of transition from diffusive to ballistic conduction (a) when the separation between contacts (W) is greater than the mean free path (I), and the conduction is diffusive [11]. (b) Three transport regimes are shown. The first one has a chaotic zigzag line which is labeled diffusive, the second one has a simple zigzag line which is called quasi-ballistic, and the third one has a line that passes through without much zigzag and is labeled ballistic. Scaled differential conductance of a SWNT rope. The curves in universal scaling coordinates, when scaled using the power law exponent the curves at different temperatures roughly collapse to one as predicted by Luttinger liquid theory [27]. (c) Real and imaginary impedance from 10 to 110 GHz for a 700 nm long SWCNT rope. Top inset: Design of the coplanar electrodes. Bottom inset: Variation of the crossover frequency with bias [28]. (d) The real and imaginary impedance of sample SW7, showing a crossover of the two at around 10.3 GHz. Temperature-induced diffusive to ballistic transport in double-walled carbon nanotubes [29].

in the DC transport. This differentiates them from ordinary metals that are characterized by universal interaction independent powers. It enables experimentalists to infer LL behavior from both AC and DC measurements. However, previous DC measurements (Figure 3.7a) have shown indirect LL through tunneling experiments but there is still some ambiguity in these interpretations. In the case of AC transport, ballistic transport can be achieved in two ways, (i) by reducing the sample dimensions to the range of the mean free path or (ii) by using a time-dependent signal whose frequency is greater than the scattering rate in the material. The Drude model for AC transport expresses the conductivity made of two parts, e.g., the real part and imaginary part of Drude's formula $\sigma(\omega)=\sigma_0/(1+i\omega\tau)$, where τ is the momentum scattering time and ω is the frequency. The imaginary part of the conductivity is given by $i\omega\tau$, and it changes sign depending on whether it is capacitive or inductive. This aspect can, therefore, be used to determine experimentally whether the transport is the diffusive or ballistic regime if the real and imaginary impedance is known. The achievement of ballistic transport at room temperature in a defective system can be realized when the stimulus frequency becomes higher than the scattering rate. When the real impedance is equal to the imaginary impedance, this corresponds to the point when $\omega\tau = 1$. This point will therefore signal the crossover point from diffusive where $\omega\tau < 1$ to ballistic transport where $\omega\tau > 1$.

CNTs having 1D electronic structure are expected to develop a new class of integrated circuits, not limited by the electron transit time based on the ballistic transport. So far, room temperature ac ballistic transport through a long channel length has been observed only in the electron gas system of high mobility semiconductors. The first measurement showed a crossover from diffuse to ballistic transport in a ~700 nm long SWNT rope at a frequency of 92.5 GHz at room temperature is observed (Figure 3.7c). The transmitted power shows an average attenuation of around 28 dB, indicating a total rope resistance of 2.65 kΩ (nearly 1/5 of the fundamental resistance) close to the calculated value for Re(Z) at low frequency. Re(Z) then drops with frequency down to a value close to the characteristic impedance of the line (50 Ω). The change of Re(Z) with frequency indicates that the value measured is mainly due to the nanotube's quantum resistance (since it decreases as it approaches ballistic transport with increasing frequency) suggesting a very low contact resistance of a few Ohms. Figure 3.7c shows the frequency variation of the impedance Re(Z) and Im(Z) from 10 to 110 GHz. At about 70 GHz, Im(Z) becomes positive, followed by a crossover of Re(Z) and Im(Z) at ~90 GHz, where frequency, τ_m^{-1} (τ_m being the momentum scattering time), is satisfied and we get the value $\tau_m > 1$ ps. This crossover point can be downshifted by applying a bias suggesting a further increase of τ_m up to several ps and the mobility exceeding 10^5 cm^2/(Vs) in the tube (Figure 3.7c bottom inset).

Substituting the value of m^* (~$0.003m_e$), the effective ballistic mobility can reach the value of more than 10^6 cm^2/(Vs). The corresponding value for the Fermi velocity can also be estimated to be $> 10^6$ m/s, which is higher than graphite and any carbon system. It is interesting to note that instead of using 2DEG underneath the GaAs, the intrinsic property of ballistic conduction through the surface of CNT can show the crossover from diffusive to ballistic regime. From the measured value of the resistance of the tube ($R = m^*/n_e\tau^2$), the value of the kinetic inductance is found to be ~80 pH.

The characteristic length (l_{mfp}) and time (τ) of SWCNT have been determined from the DC resistance measured at low frequency and using the relationship $R_{dc} = (h/4e^2)$ (L/l_{mfp}), where the value of l_{mfp} of 1.1 μm is suitable to support a ballistic type conductance process in the present system. Assuming the Fermi velocity as ~10^6 m/s, an estimated value of the relaxation time of several ps will allow for the cut-off frequency of the SWCNTs to go beyond a limit of 100 GHz. The crossover point depends on the quality of the CNTs as well as the contact resistance which can be improved through synthesis and engineering. The real and imaginary impedance of a sample shows a crossover of the two at around 10.3 GHz in Figure 3.7d. The correlation parameter g, commonly known as the LL parameter, was determined, which is a ratio of single electron charging energy to the single particle energy spacing. The standing voltage waves in a transmission line are equivalent to exciting 1D plasmon waves and measuring the plasmon velocity can be a measure of 'g'.

In the measurement system by applying a voltage, low-lying energy excitation can be achieved. The strength of interactions between the electrons is measured by the correlation parameter g defined in terms of the resonance frequency as $g = v_f/4L\,F_{res}$, where F_{res} is the frequency for the first resonant peak. Using $v_f = 10^4$ m/s as the effective Fermi velocity (for reasons explained earlier) and the 13.6 GHz as the first resonance frequency (in the ballistic regime) from this experimental data for SWNTs, we obtain g to be $\sim 0.18 \pm 0.02$, which is close to the theoretically predicted value of 0.25.

3.3   ELECTRONS IN GRAPHENE

ELECTRONIC PROPERTIES OF GRAPHENE

- Electrons can propagate in graphene plane (2D). The propagation speed is not uniform and depends on the propagation direction: this defines a 2D dispersion $E(k)$ relation. For certain directions, there is no energy gap: Electrons can gain energy continuously making graphene visible.
- Use state-of-the-art microfabrication tools direct observation of isolated monolayers, direct measurement of a single nanostructure, and measure nano properties at micrometer scale can be achieved. However, the performance of graphene devices depends on the quality of contacts (electrodes). Contact with metals generates barriers which affects charge transfer and local band bending. Contact resistance (100–1000 Ω.cm) can be comparable to resistance of graphene channel itself.
- The gate layer is very important for the performance of graphene devices. Si_2N_4, thin Al layer oxidized in situ, DLC, and h-BN are used as gate dielectric insulator for graphene. Graphene grown on thermally annealed SiC layer at 1450°C produce carrier mobility 900–1500 cm²/Vs and carrier density 3×10^{12}/cm².
- High-quality isolated graphene sheet exhibits a strong ambipolar electric field effect with a room temperature mobility of 10,000 cm²/Vs at a concentration of 10^{13} cm⁻³. It shows ambipolar electric field effect where gate bias controls the number of charge carriers (n proportional to Fermi energy squared). $N < 10^{13}$ cm⁻², $\mu < 300,000$ cm²/V s. The characteristics of graphene devices remain unchanged with increasing temperature.
- Waver scale RF graphene FETS can be operated at 100 GHz which shows mobility 1500 cm²/Vs for a gate length of 240 nm. Cut-off frequency was claimed as 100 GHz for graphene FETs having a gate length of 240 nm, and drain bias 2.5 V which is much higher than Si MOSFET (40 GHz for gate length 240 nm). However, the performance of the devices is continuously improving with other technological developments. Hence, graphene FETs are suitable for fast analog electronics.

- In a good quality graphene device charge carrier mobility can be as high as 300,000 cm²/Vs. Electrons propagate in graphene like photons. Ballistic transport in graphene devices can be seen at sub-micrometer range (for channel length less than 300 nm) and at room temperature. Ballistic transport in short channel lengths can show quantum interference effects. High mobility can be seen even at high field-induced concentrations.
- It can achieve high current-carrying capacity: $< 10^9$ A/cm² as well as high thermal conductivity: 5000 W/m K. Graphene FET does not turn off at charge neutrality point (minimum conductivity; On/Off ratio: ~10). It seems performance of graphene devices remains constant over a wide range of temperatures (4–300 K).

For a material to function effectively in electronic devices like transistors, it needs a band gap like in silicon. As we know that, graphene is classified as a semimetal in its inherent state since it does not have an energy band gap. So, the material cannot be switched "on" and "off" through an electric field which is available in silicon-based electronics.

According to recent claims, a research team at Georgia Tech under the direction of Walter de Heer produced the first graphene-based semiconductor device in history using silicon carbide substrates, called epigraphene (SEG). It is generally known that graphene multilayers are produced when carbon-rich silicon carbide crystal surfaces solidify and evaporate. The first graphitic layer to develop on the silicon-terminated face of SiC is an insulating epigraphene layer that is partly covalently bound to the SiC surface. A functional graphene semiconductor with a band gap may be produced by "doping" the graphene with atoms that "donate" electrons to the system.

SEG grown on single-crystal silicon carbide substrates has a band gap of 0.6 eV. The room temperature mobilities can be 5000 cm² $(Vs)^{-1}$, which is ten times greater than that of silicon and twenty times bigger than that of conventional two-dimensional semiconductors. A quasi-equilibrium annealing technique is created that yields SEG on terraces that are macroscopic and atomically flat. The manufacture of SEG includes a confinement-controlled sublimation furnace where a semi-insulating SiC chip is annealed in a graphite crucible under an argon environment. The rate at which silicon escapes from the crucible is a crucial factor in determining the temperature and graphene creation rate, which are carefully regulated.

It is possible to go from producing silicon wafers to the silicon carbide wafers required for epitaxial graphene. Additionally, this technique may be incorporated into already-in-use production procedures. Future applications for graphene-based semiconductors include quantum computing. Electrons in graphene, like light, may be accessible in devices thanks to their quantum mechanical wave-like qualities, especially at very low temperatures. We do not know whether graphene-based semiconductors can outperform the most sophisticated quantum computers' superconducting technology. This will be discussed in Chapter 10 [35].

Graphene is a promising candidate for future electronic devices as it has high crystalline quality and high mobility along with electronic properties that can be tuned by applying gate voltage or by using multilayer structures. Moreover, both spintronic and valleytronic devices can be made as they show both spin and valley degrees of freedom. Unlike most 2D materials, graphene is thermodynamically stable. Moreover, it has a continuous crystalline structure with very high crystal quality which allows the transport of charge carriers to be ballistic as they can travel large distances without scattering. The electronic properties of graphene and their applications in FETs are described in Figures 3.8a,b and 3.9a,b.

Understanding the geometry of graphene lattice is a must for understanding its properties and how they may be used. It contains two non-equivalent sublattices A(black) and B as shown in Figure 3.10a with lattice separation $a = 1.42$ Å. Graphene having a honeycomb lattice is not expressible as a Bravais lattice, that is, one cannot build the lattice by considering one atom as the basis. This gives rise to the choice of having two atoms as the basis and reducing the lattice to a hexagonal lattice. Therefore, in the reciprocal space, the Brillouin zone (BZ) has the geometry as shown in the right panel of Figure 3.10 with four high symmetry points Γ, M, K, and K', where K and K' are non-equivalent to each other. Another property of graphene that has a bearing on its usability and understanding is the band structure. The first studies of the band structures of graphene were by P.R. Wallace and showed its semi-metallic behavior. He used these ideas to understand the properties of graphite that were used in nuclear reactors without thinking about graphene. The work of Mcclurie in 1957 and Slonczewski and Weiss in 1958 led to the SWM model named after their initials that were successful in explaining experimental data. For studying multilayered graphene, the model was modified recently by incorporating the van der Waals-like interaction of many body effects.

In addition to the semi-metallic behavior of graphene, other interesting properties of graphene have been revealed by the tight-binding model. As shown in Figure 3.10a and b, the dispersion relation is linear around the non-equivalent points in the BZ. This is the result of low energy excitations, that is, electrons and holes behaving as massless chiral Dirac fermions. Hence, these points where the surface of the plot touches the non-equivalent points are called Dirac points. The lattice of graphene has two sublattices with a cosine-like energy band associated with each. These bands intersect near the edges of the BZ. This means that electronic states which lie close to each other have states that belong to each of the sublattices and their contribution in terms of quasi-particles must be considered. This requires the introduction of an index called the pseudospin since it is like the spin index (which consists of an up spin and a down spin).

The Hall effect is the change in the path of charge carriers and hence the creation of a potential difference in a conductor when a magnetic field is applied perpendicular to the current (Figure 3.10b). This causes an increase in resistivity in the material linearly proportional to the magnetic field applied and inversely to the carrier density. At low temperatures, the electrons are present in the lowest possible energy levels which are quantized as the Landau energy levels. When n Landau levels are filled, a proportional number of electrons are available to conduct. The carrier density is thus dependent on the energy available, the required energy to jump up a level increases

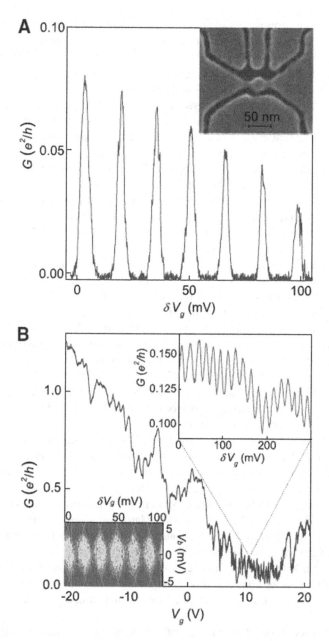

FIGURE 3.8 (A) Graphene single-electron transistor: Low-temperature (0.3 K, 4 K) conductance through quantum dot (50–250 nm) as a function of gate voltage Coulomb blockade confinement gap 0.5 V. (B) To create transistor activity, graphene can be carved using nanometer ribbons and quantum dots. At low T, comparably big quantum dots (diameter 0.25 m) show a Coulomb blockade. The conductance of such devices may be influenced by either the back gate or a side electrode made of graphene. Graphene constrictions with low T resistance appreciably greater than 100 K are employed as quantum barriers [30].

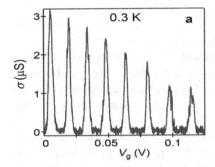
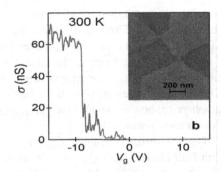

FIGURE 3.9 (a) Graphene single-electron transistor: Low-temperature (0.3 K, 4 K) conductance shows oscillations. (b) 10-nm-scale graphene structures are very resilient and can endure temperature changes to liquid helium temperature under normal conditions. A broad variety of gate voltages at the neutrality point may completely cut off the conductance of such devices, which display outstanding transistor behavior even at ambient temperature. (Scanning micrograph of two graphene dots of ~40 nm in diameter with narrower (< 10 nm) constrictions) [30].

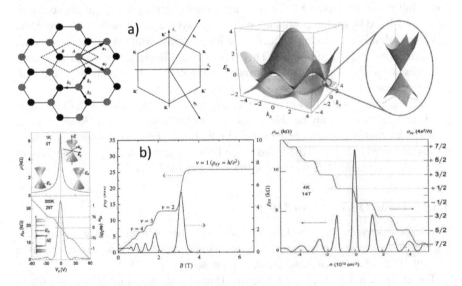

FIGURE 3.10 (a) (left) The two inequivalent atoms A and B, as well as the lattice vectors and bond lengths, are what make up the hexagonal lattice of graphene. (middle) The Brillouin zone of graphene is similarly hexagonal and has alternating inequal vertices known as the K and K′ points. At the Dirac points, the graphene dispersion relationship is linear; (Right) the sites where the valence and conduction bands overlap produce Dirac cones with varying chirality for the various valleys [31]. (b) The half-integer quantum Hall effect, which results from the Dirac nature of fermions, is seen in both the longitudinal and transverse resistivity. The steps in the transverse conductivity and the longitudinal and transverse resistivity as functions of the carrier concentration both clearly demonstrate the half-integer dependency [32].

with increasing magnetic field. Thus, the resistivity increases in steps as the magnetic field is increased. This is called the integer quantum Hall effect. Graphene shows this kind of effect even at room temperatures which are of interest.

Massless Dirac fermions in quantum electrodynamics (QED) show a similar situation. However, the electrons in graphene move at 1/300 of the speed of light. The speed of these electrons gives rise to very high cyclotron energy which allows graphene to show anomalous integer quantum Hall effect in the presence of magnetic field at room temperature.

One interesting phenomenon that was observed is known as the chiral quantum Hall effect. Other QED effects include the Gedanken Klein paradox and the zitterbewegung since these two effects cannot be observed in particle physics. The Klein paradox refers to a process where electrons can perfectly tunnel through barriers of arbitrary height and width. Zitterbewegung describes movements in electrons caused by interference between the electron and positron states that are part of the electron's wave packet. Graphene has some unique opportunities since the electronic states are not pinned deep and can be accessed via probing or tunneling. The chirality and pseudospin can be used to explain most of the electronic processes in graphene. Therefore, a lot of research interest is focused on determining these properties.

One major problem with graphene is that it remains metallic at the neutrality point. But it can be solved by engineering semiconductor gaps in graphene which may be achieved by spatial confinement in single-layered graphene. Another problem is the lack of anisotropic etching technique for graphene which leads to the conductive channel having irregular faces. This in turn results in electronic states associated with these edges having a sample-dependent conductance. A possible way is to consider graphene as a conductive sheet instead of a channel material.

Due to graphene's stability at nm sizes, single-electron transistors (SET) based on graphene may theoretically be able to operate at room temperatures. However, there are still two challenges that can be solved in the future. First, high-quality graphene wafers still need to be developed, and second, the properties of the graphene devices need to be independently controlled to ensure reproducibility.

3.3.1 DIRAC DISPERSION OF ELECTRONS

The linear dispersion predicted by Wallace led to the realization of graphene having exceptional quantum transport features. Graphene has been a source of a large amount of research activity due to these features.

The charge carriers obey the Dirac-like Hamiltonian given by $Hk = v_f \sigma \cdot p$, where v_f is the Fermi velocity of electrons in graphene ($v_f \sim 10^6$ m/s), and σ and p are the pseudospins and momentum operators, respectively. Some of the first investigations of graphene were made on its quantum Hall effect as it is an ideal 2D material. As shown in Figure 3.10, quantum Hall plateaus can be observed; however, the sequence of these plateaus is not the same as seen in traditional 2D materials. The Hall plateaus occur at half-integer values. The first one occurs at $2e^2/h$ and the subsequent ones at $(4e^2/h)$ $(N + \frac{1}{2})$, where N is an integer. Thus, the plot looks like a ladder of equidistant steps in the Hall resistivity that is continuous even at zero. This is the result of Landau quantization which is within ~0.2% experimental accuracy in graphene. Theoretical investigation of the half-integer QHE in graphene by several groups has related it to properties of massless Dirac fermions such as the existence of both electron and hole-like Landau states at exactly zero energy. These are direct consequences of the Atiyah–Singer index theorem as present in quantum field theory and superstring theory. Moreover, QHE signatures have been observed at room temperature at very low magnetic fields which can be understood because of the Dirac nature of charge carriers and hence large cyclotron gaps. In addition, other phenomena such as very high carrier concentration (up to 10^{13} cm^{-2}), high mobility at a wide range of temperatures (liquid helium to room temperature), and zero field conductivity are also responsible. Very high carrier concentration in graphene is reached with only a single 2D subband occupation which is important to populate the lowest Landau level. Other 2D systems like GaAs heterostructures occupy multiple subbands or can be depopulated in moderate magnetic fields which causes a reduction in the effective energy gap. The zero field conductivity in single-layer graphene is a noteworthy sign of the Dirac nature of fermions. Between 4 and 100 K, the conductivity never goes below a well-defined value.

A minimum resistance of 6.5 K has been observed in single-layer graphene devices, with an experimental error of 15%. It was discovered that this minimum resistivity was unaffected by the fact that their mobility might vary by a factor of 10. This lowest value has been connected to $h/fe^2 = 6.45$ K, where f is the quadruple degeneracy factor derived from Hall effect measurements and h/e^2 is the resistance quantum. The phenomenon, which is an inherent characteristic of electronic systems represented by the Dirac equation, differs from the conductance quantization seen in previous quantum transport studies. This characteristic results from the substantial suppression of localization effects for such relativistic charge carriers.

The topology of the band structure is also non-trivial, and the states close to the Dirac cones have a Berry phase (γ), which indicates that they change their sign when forming a closed circuit in the reciprocal space (from K to K' points). Several intriguing charge carrier phenomena in graphene are caused by this linear dispersion relation.

3.4 DEFORMATION THROUGH NANOMANIPULATION IN MULTILAYERED GRAPHENE

3.4.1 NANOMANIPULATION DEVICE FABRICATION TECHNIQUE

Even though there is a lot of research on using graphene for novel quantum electronic applications, making functional devices with the material still presents several difficulties. These difficulties include isolating single- or few-layered sheets as well as moving, suspending, and manipulating the substance to fabricate devices while ensuring high-quality reproducible conduction channels.

Magnetic field-dependent transport studies are one of the most extensively researched techniques for evaluating the quality of graphene devices. As previously indicated, graphene has shown extraordinary characteristics and is now thought to make an excellent platform for studying quantum linear magnetoresistance (QLMR) brought on by inhomogeneities or disorders. The Kapitsa linear law, which was first discovered by Abrikosov and then solved by Parish and Littlewood, was used to explain the magnetoresistance (MR) characteristics of elemental bismuth. QLMR is one of the historical riddles of quantum transport. In clean, homogenous single, bilayer, and trilayer graphene samples, the quantum LMR phenomenon is difficult to detect because it is overshadowed by powerful Shubnikov de Haas (SdH) oscillations. Nevertheless, multilayer graphene produced on silicon carbide, as well as bilayer graphene, provides evidence of QLMR.

Since many characteristics of the charge carriers may be gleaned from the influence of interlayer effects on SdH oscillations and temperature-dependent transport parameters in multilayer graphene devices, it is a topic of great interest. In comparison to single-layer graphene values, temperature-damped SdH tests on single-layer, bilayer, and trilayer graphene revealed reduced hole and electron-effective masses (m^*). This is typically attributed to the band structure's renormalization brought on by electron–electron interaction and the increase in effective mass with more layers. Investigating if this trend persists as the number of layers is further raised to few-layered samples is one area of investigation.

The frequency of the SdH oscillations could be varied by changing the gate voltage of the device as it has been established that the electric field has a prominent effect. Even though multilayer graphene systems are quite complex, the transport physics observed in bi- and trilayer graphene is a strong motivator to understand multilayered graphene. Hence, the utilization of a novel fabrication technique, nanomanipulation (see in Chapter 2), is presented in this chapter. The method allows for the creation of suspended multilayered graphene devices. Furthermore, deformations can be easily incorporated into devices, thus allowing tuning of transport properties between the ideal 2DEG to one where QLMR can be observed.

Conventional techniques, such as electron beam lithography combined with reactive plasma etching and techniques using scanning probe microscopy, particularly AFM-lithography and STM imaging, and probing, are used in the fabrication of graphene devices. These delicate and complex procedures are known to cause defects in the graphene layers. In addition, many reported devices contain a layer of graphene on a silicon oxide substrate rather than being suspended from support, which restricts electron mobility due to impurities trapped between the SiO_2 and graphene and can result in substrate doping.

As demonstrated in Figure 2.1a–e, this production method produces both smooth graphene and wrinkled graphene (in Chapter 2). The characteristics of the multilayer graphene samples are then ascertained using Raman spectroscopy. A comparison of the Raman spectra of a multilayered graphene device and a single-layer sample is provided in to demonstrate this. The 2D band's decomposition reveals a peak structure that is distinct from graphite stock but not identical to the particular peak structure seen in bi- or trilayer graphene samples. This suggests that there are more than four layers in the sample. A minor D peak at 1350 cm^{-1}, can be used to track disorder brought on by point defects in the graphene layers as well as folding and rippling. This Raman peak is a sign of material disorder and may take the form of folding, rippling, or point defects in the graphene layers. In wrinkled multilayer graphene, the MR properties observed are drastically different from smooth layered devices (Figure 3.11a). The MR shows QLMR (QLMR) at 300 mK in the magnetic

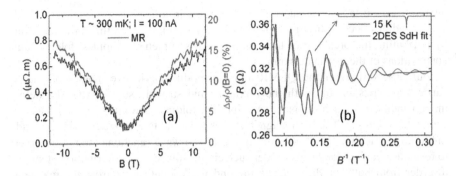

FIGURE 3.11 (a) Resistance temperature dependence of four-probe smooth flat, thin, and transparent multilayer graphene device at zero magnetic field (black curve) and at a perpendicular applied magnetic field of 8 T. (b) SdH oscillations in the field dependence of the longitudinal resistance of a multilayered graphene device at 15 K [33].

field range from −12 to 12 T, whereas thin, flat, and homogeneous graphene devices showed irregular SdH oscillation which starts at intermediate fields of above 4 T and occurs only at temperatures below 30 K.

3.4.2 Quantum Linear Magnetoresistance and Hall Effect

The measurements on QLMR (Figure 3.11a) and Hall effect (Figure 3.11b) are useful to determine the transport parameters of quantum transport physics in a wrinkled graphene system. The description of linear quantum MR used is: $\rho_{xx} = \dfrac{N_i H}{\pi n_s^2 ec} \propto H$ and $\rho_{xy} = R_H H = \dfrac{H}{n_s ec}$ where ρ_{xx} and ρ_{xy} are the longitudinal and transverse components of the MR, respectively, N_i is the concentration of static scattering centers, and n_s is the carrier density. This description allows us to find the concentration of static scattering centers and the carrier density which are in the range of 2.24×10^{-2} to 1.02×10^{14} and 5.2×10^9 to 3.5×10^{13} cm^{-2}, respectively.

The highest measured electron density in the heavily deformed sample is up to five times higher than the previously reported QLMR density for polycrystalline mosaic bilayer graphene ($p \sim 7 \times 10^{12}$ cm^{-2}). However, it is challenging to compare this figure with other findings on QLMR in multilayered graphene devices because those papers did not include information on electron density. The lack of observation of the QLMR effect is possibly the reason that the concentration of static impurity centers has not been determined in the reports. In our sample, the ratio of Ni/ns is roughly three at magnetic fields up to 12 T. At zero applied field and 12 T, the effective mass of charge carriers in the wrinkled graphene was found to be 0.0014 m_e and 0.121 m_e, respectively (where m_e is the electron rest mass with a value of 9.109×10^{-31} kg). This suggests that there is a spectrum in wrinkled graphene that permits both massive and massless fermionic excitations, similar to findings on the coexistence of massive and massless Dirac fermions seen in bilayer graphene samples with broken symmetry.

The mixing of Landau levels obliterates the normal plateaus in Hall conductivity, which causes the observed occurrences, according to theoretical works on the quantum Hall effect of massless Dirac fermions in a diminishing magnetic field. These findings once again highlight the importance of strain in this Dirac material by attributing the breaking of the time-reversal symmetry to ripples, folding, and corrugations of the graphene samples.

Our observations of the QLMR and the modified effective masses of charge carriers are thus in line with the literature and strongly suggest that the deformation induced in such devices (i.e., rippling, folding, and corrugations) leads to symmetry-breaking. The high magnetic fields will tend to localize the carriers leading to large effective mass; however, the localized wave functions are not observed to form discrete Landau levels. Under such circumstances, the contribution of strong disorder from both interlayer scattering and weak localization from the magnetic field will tend to make the Dirac fermions more massive (in this case $m^* = 0.121m_e$), which is 2.42 times more massive for ABA-stacked trilayer graphene and 3.46 times $m_{AB} \approx \frac{\gamma_1}{2v^2} \approx 0.035m_e$ more massive for AB stacked bilayer graphene. As a result, our observations of the QLMR and the changing effective masses of charge carriers are consistent with previous research and strongly imply that the symmetry-breaking caused by the deformation induced in such devices (i.e., rippling, folding, and corrugations) is real. In spite of the fact that the localized wave functions are not seen to form distinct Landau levels, the strong magnetic fields will tend to localize the carriers, resulting in a large effective mass. Under these conditions, the contribution of strong disorder from interlayer scattering as well as weak localization from the magnetic field will tend to make the Dirac fermions more massive (in this case, $m^* = 0.121m_e$), which is 2.42 times more massive for ABA-stacked trilayer graphene and 3.46 times $m_{AB} \approx \frac{\gamma_1}{2v^2} \approx 0.035m_e$ more massive for AB stacked bilayer graphene.

This novel nanomanipulation device fabrication to produce multilayered graphene devices thus establishes novel ways of investigating several rich and rare physical phenomena, most notably the elusive QLMR effect. This is strongly in opposition to conventional graphene devices where the graphene behaves as an ideal 2DEG, as demonstrated in the next section.

3.4.3 SHUBNIKOV DE HAAS OSCILLATIONS

When a magnetic field is applied to a two-dimensional electron gas, the resistivity parallel to the current in the edge states starts to oscillate with an inverse dependence on the magnetic field strength. This happens as the magnetic field forces the electrons to move in circles. While the electrons in the bulk do so, the ones at the edge get scattered from the interface that are then propelled forward by the field. The energy Eigenvalues for electrons in circular paths are the same as harmonic oscillators with the same frequency. This energy is quantized in terms of Landau levels. If the Landau energy levels are close to the Fermi energy, then scattering takes place, and peaks in SdH oscillations are seen, the SdH resistivity is zero otherwise. Furthermore, these peaks correspond to the change in resistivity of the quantum Hall effect.

The effect of magnetic field on resistance for a smooth multilayer graphene device was studied throughout a wide temperature range of 2–50 K with magnetic fields of up to 12 T applied perpendicular to the devices (Figure 3.11a). It was found that the device first displays a decrease in resistance as the temperature is raised to about 25 K and that as the temperature is elevated further, the resistance increases. Electron–electron interactions are responsible for the weakly field-dependent upturn of the temperature-dependent resistance, which happens at somewhat lower temperatures with zero applied field. Resistance fluctuations at strong magnetic fields (8 T) for temperatures below 35 K have also been reported.

The device's MR was measured to examine this oscillating behavior in the temperature-dependent resistance. The longitudinal MR was observed to oscillate with increasing amplitude when the field strength was raised at temperatures below 25 K, after initially being flat up to fields of 0–4 T. This exhibits the distinctive oscillating behavior known as the SdH effect, which is a hallmark of Landau quantization. Weak localization, which has been previously noted in multilayer graphene, is the cause of negative MR below 25 K. As the temperature rises, these oscillations are seen to be damped.

By determining the oscillatory frequency through Fourier transformation, the Landau level filling index can be determined, $v = \dfrac{n_s h}{2eB}$, where h and e are Planck's constant and the fundamental charge, respectively. The difference between two consecutive Landau levels is thus determined by $v_1 - v_2 = \dfrac{n_s h}{2eB_1} - \dfrac{n_s h}{2eB_2} = 1$. Thus, it is possible to evaluate the carrier density n_s via the equation: $n_s = \dfrac{2e}{h}\left(\dfrac{1}{\dfrac{1}{B_1} - \dfrac{1}{B_2}}\right)$ The difference between the reciprocal of the field for two peaks can be obtained from the plot. Our analysis considers only major peaks while neglecting the higher harmonics which is expected to originate from interlayer effects. The carrier density seems to be relatively independent of the applied magnetic field in the range of 1.39×10^{12} to 2.85×10^{12} cm^{-2}. The values of carrier concentration are comparable with those reported in a single layer (1.47×10^{12}–3.43×10^{12} cm^{-2}), bilayer graphene (3.19×10^{12}–7.12×10^{12} cm^{-2}), and trilayer graphene samples (4.25×10^{12}–8.04×10^{12} cm^{-2}). The effective mass of charge carriers obtained from carrier density is in the range of $0.22m_e$–$0.032m_e$, which is comparable to single-layer, bilayer, and trilayer graphene which have effective masses in the range of $0.33m_e$, $0.25m_e$–$0.55m_e$, $0.35m_e$–$0.72m_e$, respectively. The ABA-stacked trilayer graphene also has a comparable effective mass of $\sim 0.05m_e$. We can also extract other transport parameters from the SdH oscillations, namely the phase coherence length, the quantum scattering time, and the carrier lifetime. The carrier lifetime is determined by following the formula for the general form of MR in a two-dimensional electron system (2DES) within the SdH oscillation regime: $R = R_0\left[1 + \xi \sum\limits_1^{\infty} D(X)\exp\left(\dfrac{s\pi}{w_c\tau}\right)\cos\left(s\dfrac{\hbar s_F}{eB} - s\pi + s\varphi_0\right)\right]$ from which the first order oscillation amplitude be expressed as $A = \xi D(X)\exp\left(-\dfrac{\pi}{w_c\tau}\right)$, where $w_c = eB/m^*$. $w_c = eB/m^*$ is the cyclotron frequency, $D(X) = \dfrac{2\pi k_B T / \hbar w_c}{\sinh(2\pi k_B T / \hbar w_c)}$

is the damping factor, k_B is the Boltzmann constant, T is temperature, and ξ is a constant. The carrier lifetime is thus found to be in the range of 11–36 fs for a mean effective mass 0.026 m_e (corresponding to a mean carrier density of 1.95×10^{12} cm^{-2}) by graphing the extrema amplitudes (A) vs. $1/B$ of the oscillation. The lifetime of 54 fs observed in single-layer graphene is lower than this value. The phase coherence length (L_φ) is estimated using the formula: $\Delta B = \dfrac{\varphi_0}{L_\varphi^2}$. And determined to be 31 nm at 15 K at 10 T. Using the inverse period of SdH oscillations to determine the cross-sectional area (A) of the Fermi surface pocket perpendicular to the magnetic field using the formula: $\Delta\left(\dfrac{1}{B}\right) = \dfrac{2\pi e}{A\hbar}$. We get a value of ~$4.5 \times 10^{17}$ m^{-2} for the Fermi surface area at 10 T field and 15 K temperature. Also, it has been found that the amplitude of oscillations in the high field domain decreases when the temperature rises above 20 K. The oscillations are also erratic, which can be attributed to interlayer interference effects or the modulation of the SdH oscillations by other types of oscillations known as magneto-intersubband oscillations (MISO) and phonon-induced resistance oscillations. The origin of these oscillations can be found in the scattering-assisted coupling of carrier states in various Landau levels, which can be investigated using Raman spectroscopy.

3.5 APPLICATIONS OF CNT HYBRID CIRCUITS AND GRAPHENE QDS AT HIGH FREQUENCIES

In the previous part of this chapter (Sections 3.2–3.2.2), we show the high-frequency behavior of CNTs associated with one-dimensional transport. Here we present an example of how a circuit of CNTs can be constructed combined with diamonds (nitrogen-vacancy centers).

Figure 3.12a is a high-resolution transmission microscopy image showing a single MINT. In Figure 3.12b, a current-carrying nanotube is suspended above a diamond sample in which individual optically resolvable NV centers are implanted 5–10 nm below its surface. Figure 3.12c is a diagram of a quantum hybrid system: a two-dimensional NV centers array in a thin diamond film located under a DC current-carrying CNT. The array of microwave strip lines, which do not cross each other, is used to address the centers independently (see Figure 3.12). We shall extend discussions on CNT-based qubits in Chapter 10.

3.5.1 GRAPHENE-BASED QUBITS

In the previous part of this chapter (Sections 3.1.2 and 3.3–3.5), we discussed the quantum dot (Coulomb blockade) behavior of graphene. Electronic properties of defective graphene films are demonstrated which can extend to bilayer and even multilayer films. Bi-layers of graphene films show very important electronic properties in manipulating the carriers in the form of QDs which can act as qubits (see Figure 3.13). The electron and hole Bloch states in bilayer graphene have topologically opposing orbital magnetic moments, which enables variable valley polarization in an out-of-plane magnetic field. Bilayer graphene's electron and hole QDs are attractive for valley and spin valley qubits due to this characteristic. Researchers measured the magnetic moments connected to the Berry curvature at the electron-hole crossover in a bilayer graphene QD. The tunneling barriers were regulated separately

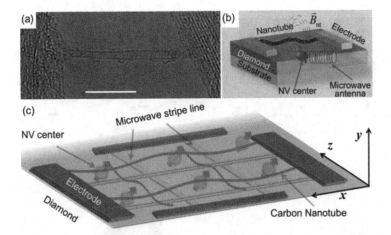

FIGURE 3.12 A high-resolution transmission microscopy picture (a) shows a single MINT. (b) A current-carrying nanotube is hung above a diamond sample that has individual optically resolvable NV centers implanted 5–10 nm below its surface. (c) A representation of a quantum hybrid system with a DC-current-carrying CNT on top of an array of two-dimensional NV centers in a thin diamond layer. Utilizing a variety of microwave strip lines that do not cross one another, the centers are individually addressed [34].

using three layers of top gates, changing the occupation from the few-hole domain to the few-electron regime, and bridging the displacement-field controlled band gap. The electron and hole dots charge at energies between 3 and 5 meV, respectively, while the band gap is approximately 25 meV. With modest B-fields, the extracted valley g-factor, which is roughly 17, causes opposite valley polarization for electrons and holes. These measurements are found to correlate with precise calculations. Discussions on graphene-based qubits will be extended in Chapter 10.

Q. Show the energy band diagram for nanotube transistors and calculate the h parameters.

Main problem discussed in this chapter: Strong localization due to bending of space(time) in a nanocarbon device.

What has been achieved?

Spacetime distortion will be discussed in Chapter 3 in the form of the gate bias, QDs, and quantum wells. The wells are separated by a barrier that resembles a spin-triplet state or the vortices are connected through a tunnel barrier. It is a four-level system of qubits.

What has not been achieved?

Understanding dimensionality in carbon is challenging due to the folding and bending of bonds that introduce an extra dimension to the system. For nanotubes, we have one dimension and the chirality of the tubes. We do not know how this extra dimension can control the electronic properties of nanotubes. Quantum simulations of this complicated structure should be performed.

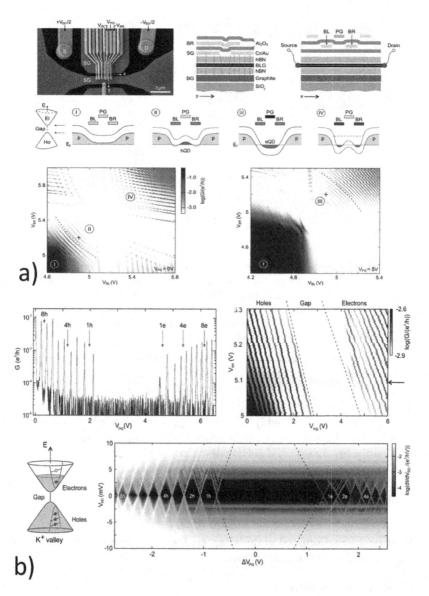

FIGURE 3.13 (a) The split gates (SGs) in the QD device's voltage-modulated conducting channel are visible in the SEM picture. The rear gate is made of graphite flakes. The four regimes (I, II, III, and IV) established by the barrier gate and plunger gate voltages are illustrated by the valence and conduction band edge profiles along the p-doped channel. The conductance is displayed in relation to the barrier gate voltages V_{BL} and V_{BR} at $V_{PG} = 0$ V and $V_{SD} = 200$ V in the charge stability diagram.

(b) Coulomb resonances with fixed $V_{BR} = 5.09$ V, $V_{BL} = 4.9$ V, and a V_{PG} ($V_{SD} = 200$ V) plunger gate voltage. The positions held by the band gap-divided hole and electron dot states. An illustration of the possible electron and hole QD states in the band structure around the K^+ valley. The gap originates in the transverse displacement field. Since the Fermi level is in the conduction band, the QD has one electron. The electron-hole crossover and the first Coulomb diamonds of the few-electron and hole QD are shown along the arrow in panel (b) of a finite bias spectroscopic observation.

BIBLIOGRAPHY

1. C. Kittel, *Introduction to Solid State Physics*, Wiley, Hoboken, NJ, 1976; N. W. Ashcroft and D. Mermin, *Solid State Physics*. Holt, Rinehart, and Winston, New York, 1976.
2. D. D. Ferry, and S. M. Goodnick, *Transport in Nanostructures*, Cambridge University Press, Cambridge, 1997.
3. Y. Imry, *Introduction to Mesoscopic Physics*, Oxford University Press, Oxford, 1997.
4. Y. V. Nazarov, and B. M. Yaroslav, *Quantum Transport Introduction to Nanoscience*, Cambridge University Press, Cambridge, 2009.
5. S. M. Sze, and K. K. Ng, *Physics of Semiconductor Devices*, Wiley, Hoboken, NJ, 1981.
6. M. J. Kelly, *Low Dimensional Semiconductors*, Oxford University Press, Oxford, 1995.
7. S. Datta, *Quantum Transport Atom to Transistor*, Cambridge University Press, Cambridge, 2005.
8. M. S. Lundstrom, *Fundamentals of Carrier Transport*, Cambridge University Press, Cambridge, 2000.
9. G. D. Mahan, *Many-Particle Physics*, Plenum, Berlin, 1991.
10. S. Ncube, C. Coleman, A. Strydom, E. Flahaut, A. de Sousa, and S. Bhattacharyya, Enhanced magnetic properties and spin valve effects in gadolinium carbon nanotube supramolecular complex, *Scientific Reports* 8, 8057 (2018).
11. S. Datta, *Electronic Transport in Mesoscopic System*, Cambridge University Press, Cambridge, 1995.
12. R. Schmechel, Gaussian disorder model for high carrier densities: Theoretical aspects and applications to experiments, *Physical Review B* 66, 235206 (2002).
13. A. Bachtold, M. Henny, C. Terrier, C. Strunk, C. Schonenberger, J. P. Salvetat, J. M. Bonard, L. Forro, Contacting carbon nanotubes selectively with low Ohmic contacts for four probe electric measurements, *Applied Physics Letters* 73, 274 (1998).
14. G. Bergmann, Weak localization in tunnel junctions, *Physical Review B* 39, 11280 (1989).
15. L. Banszerus, A. Rothstein, T. Fabian, S. Möller, E. Icking, S. Trellenkamp, F. Lentz, D. Neumaier, K. Watanabe, T. Taniguchi, F. Libisch, C. Volk, and C. Stampfer, Electron-hole crossover in gate-controlled bilayer graphene quantum dots, *Nano Letters* 20, 7709–7715 (2020).
16. V. V. Maslyuk, R. Gutierrez, and G. Cuniberti, Spin–orbit coupling in nearly metallic chiral carbon nanotubes: A density-functional based study, *Physical Chemistry Chemical Physics* 19, 8848 (2017).
17. M. Burghard, Carbon-based field-effect transistors for nanoelectronics, *Advanced Materials* 21, 2586 (2009).
18. R. Saito, G. Dresselhaus, and M. S. Dresselhaus, Trigonal warping effect of carbon nanotubes, *Physical Review B* 61, 2981 (2000).
19. F. Mazza, B. Braunecker, P. Recher, and A. L. Yeyati, Spin filtering and entanglement detection due to spin-orbit interaction in carbon nanotube cross-junctions, *Physical Review B* 88, 195403 (2013).
20. D. Churochkin, R. McIntosh, and S. Bhattacharyya, Spin filtering and entanglement detection due to spin-orbit interaction in carbon nanotube cross-junctions, *Journal of Applied Physics* 113, 044305 (2013).
21. A. Javey, J. Guo, Q. Wang, et al., Ballistic carbon nanotube field-effect transistors, *Nature* 424, 654–657 (2003).
22. Y. Yaish, J.-Y. Park, S. Rosenblatt, V. Sazonova, M. Brink, and P. L. McEuen, Electrical nanoprobing of semiconducting carbon nanotubes using an atomic force microscope, *Physical Review Letters* 92, 046401 (2004). E.D. Minot, Tuning the band structure of carbon nanotubes, Ph.D. Thesis, Cornell University, (2004)
23. W. Liang, M. Bockrath, D. Bozovic, J. H. Hafner, M. Tinkham, and P. H. Fabry, Perot interference in a nanotube electron waveguide, *Nature* 7, 665–669 (2001).

24. J. Nygård, D. H. Cobden, M. Bockrath, P. L. McEuen, and P. E. Lindelof, Electrical transport measurements on single-walled carbon nanotubes, *Applied Physics A* 69, 297–304 (1999).

25. P. Burke, Luttinger liquid theory as a model of the gigahertz electrical properties of carbon nanotubes, *IEEE Transactions on Nanotechnology* 1, 129 (2002); D. F. a. J. Appenzeller, AC and DC characteristics of carbon nanotube field-effect transistors, *IEEE Electron Devices Letters* 25, 34 (2004).

26. Z. Yu, and P. J. Burke, Microwave transport in metallic single-walled carbon nanotubes, *Nano Letters* 5, 1403 (2005); P. J. Burke, Z. Yu, C. Rutherglen, Carbon nanotubes for RF and microwaves, *Proceedings European Microwave Week* (2005).

27. S. Ncube, G. Chimowa, Z. Chiguvare, and S. Bhattacharyya, Realizing one-dimensional quantum and high-frequency transport features in aligned single-walled carbon nanotube ropes, *Journal of Applied Physics* 116, 024306 (2014).

28. L. Gomez-Rojas, S. Bhattacharyya, and S. R. P. Silva (unpublished).

29. G. Chimowa, E. Flahaut, and S. Bhattacharyya, Temperature-dependent diffusive to ballistic transport transition in aligned double walled carbon nanotubes in the high frequency regime, *Applied Physics Letters* 105, 173511 (2014).

30. L. A. Ponomarenko, F. Schedin, M. I. Katsnelson, R. Yang, E. W. Hill, K. S. Novoselov, and A. K. Geim, Chaotic Dirac billiard in graphene quantum dots, *Science* 32, 356 (2008).

31. P. Adroguer, W. E. Liu, D. Culcer, E. M. Hankiewicz, Conductivity corrections for topological insulators with spin-orbit impurities: Hikami-Larkin-Nagaoka formula revisited, *Physical Review B* 92, 241402 (2015).

32. K. S. Novoselov, A. K. Geim, S. Morozov, D. Jiang, M. Katsnelson, I. Grigorieva, S. Dubonos, and A. A. Firsov, Two-dimensional gas of massless Dirac fermions in graphene, *Nature* 438, 197 (2005); K. S. Novoselov, E. McCann, S. V. Morozov, V. I. Fal'ko, M. I. Katsnelson, U. Zeitler, D. Jiang, F. Schedin, and A. K. Geim, Unconventional quantum Hall effect and Berry's phase of 2π in bilayer graphene, *Nature Physics* 2, 177 (2006).

33. C. Coleman, et al., Quantum linear magnetoresistance and shubnikov de-Haas oscillation in suspended wrinkled and smooth multilayer graphene, *Journal of Nanoscience and Nanotechnology* 17, 5408 (2017).

34. P.-B. Li, Z.-L. Xiang, P. Rabl, and F. Nori, Hybrid quantum device with nitrogen-vacancy centers in diamond coupled to carbon nanotubes, *Physical Review Letters* 117, 015502 (2016).

35. J. Zhao, P. Ji, Y. Li, R. Li, K. Zhang, H. Tian, K. Yu, B. Bian, L. Hao, X. Xiao, W. Griffin, N. Dudeck, R. Moro, L. Ma, and Walt A. de Heer, Ultrahigh-mobility semiconducting epitaxial graphene on silicon carbide, *Nature* 625, 60 (2024).

4 Quantum SPIN Tunneling in Carbon Nanostructures and Devices

4.1 SPINTRONICS OVERVIEW

Conventional electronic devices are made using silicon chips. The charge on electrons is used to store information in a semiconductor material as well as to transport it. The important point of spintronics is to provide low-power, high-speed, and high-density logic and memory electronic devices. Instead of just charging, like normal electronics, spintronics also considers the spin of an electron as well as the magnetic moment associated with it. As the spin of an electron is quantized, it can be influenced by an external non-uniform magnetic field that realigns the spin of the electrons. Based on sufficient interactions between transport electrons and local magnetic moments spintronic devices such as spin valve, spin current amplifier, spin capacitors, spin-based integrated circuits, spin LEDS, and spin lasers can be designed. These devices having two magnetic layers take advantage of mostly two effects, namely Giant Magnetoresistance (GMR) and Magnetic Tunnel Junction (MTJ) effect, are discussed in the following sub-sections.

This device is composed of two ferromagnetic materials and an insulating layer between them. The resistivity of the device depends on the alignment of spins of the two layers relative to each other. The spin valve uses the MTJ effect to manipulate this resistivity. A high resistivity would give the bit a value of 0 and low resistivity a value of 1. This is how quantum computations would be performed using spintronics.

Generally, spintronic devices operate in three steps: first, the information is written into spins. The orientation of these spins can be up or down. Second, they are attached to mobile electrons and carry the information along a wire. In the last step, the information is read at the terminal. A good spintronics device needs to have an efficient spin injection, a possibility of spin manipulation, and then spin detection. The spin injection is a process in which spin is injected from a ferromagnetic material to a nonmagnetic semiconductor, the spin needs to have a long enough spin diffusion to allow the possibility of efficient spin manipulation, and the minimum threshold is in the order of nanometers. There are a few different ways of spin detection, including Silsbee-Johnson spin-charge coupling and the more famous spin-valve effect. The spin orientation of conduction electrons lasts for a long time approximately 10^6 s. This makes spintronic devices very interesting compared to conventional electronic devices.

DOI: 10.1201/9781003316411-4

4.1.1 MAGNETIC TUNNEL JUNCTION EFFECT

Tunneling magnetoresistance (TMR) is characterized as an adjustment in the electrical resistance of a MTJ due to an outer magnetic field. M. Jullière (College of Rennes, France) initially discovered the MTJ effect in 1975 in the investigation of Fe/GeO/Co junctions at 4.2 K. Maekawa and Gäfvert first detailed room temperature tunneling effect in 1982 with an arrangement of Ni/NiO/FM system, where FM is Fe, Co, or Ni. The 2007 Physics Nobel prize winners, Albert Fert and Peter Grünberg, discovered GMR in 1988. This discovery is the one that paved the way for the ignition of spintronics as a new research field. They revolutionized data storage on computer drives as their science led to more compact drives with much higher capacity. Stuart Parkin and his colleagues made a breakthrough that transformed the magnetic data storage industry when they discovered the "spin valve," a device with the ability to alter the magnetic state of materials at the atomic level.

An MTJ comprises of two ferromagnetic metals which are terminal electrodes isolated by a nonmagnetic insulating layer namely a tunnel barrier. The electrons can tunnel through the tunnel barrier made of very thin layers (nm). This tunneling phenomenon of the electron is related to the wave nature of the electron. However, this ephemeral transmission of electrons through the tunnel barrier leads to an exponential dependence of the tunneling current on the barrier thickness.

The external magnetic field governs the direction of magnetization of the two ferromagnetic layers. The bottom ferromagnet will have a "fixed" spin, meaning that all the electrons have the same spin. While the top ferromagnet has free spin, the electrons can change their alignment due to external magnetic fields. When the magnetizations are in a parallel orientation, more flow of electrons will be tunneling through the insulating layer. There will be less tunneling current if they are in opposite (antiparallel) orientations. Consequently, this junction switches between two states of electrical resistance: one with low and the other with very high resistance. The reason for the change in resistance can be explained using the transfer of spin from the first layer to the second and by using John Bardeen's approach of tunneling as determined by the overlap of wave functions entering the tunnel barrier.

Spin transfer describes the process by which the orientation of the magnetic layer in an MTJ or spin valve can be modified using a spin-polarized current. Experiments have demonstrated that if a spin-polarized current passes from a thick fixed ferromagnetic layer, through a nonmagnetic layer, to another thin film "free" nanomagnet by spin-dependent scattering of the polarized current an angular momentum can be transferred to this layer which can be used to excite oscillations in the nanomagnet. In the absence of a strong external magnetic field, this spin-dependent scattering can also result in the reversal of the orientation of the magnetic moment of the free nanomagnet with the final orientation relative to the fixed layer being dependent on the direction of the current flow.

Figure 4.1 gives two current models for parallel and antiparallel orientation of magnetization, where arrows indicate the flow of the electron passing through the tunnel barrier. Tunnel magnetoresistance (TMR) has been explained by using Bardeen's approach

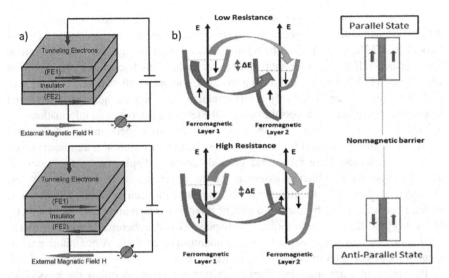

FIGURE 4.1 (a) Spin-valve device that can demonstrate the functionality of a device using spin transmission between ferromagnetic layers through a thin layer of an insulator. (b) Schematic diagram of tunneling magnetoresistance (from a high to a low resistance) effect due to spin tunneling between two FM layers. (right) The spins change from parallel to antiparallel state.

of tunneling, as being determined by the overlap of wave functions entering the tunnel barrier with the terminal electrodes with a little transmission probability. The TMR ratio is determined as the difference in terms of the conductance of the parallel and antiparallel orientation given by Jullière's model: $^{\mathrm{TMR}}\text{Ratio} = 2G_P G_{AP}/(1 - G_P G_{AP})$. The terms G_P and G_{AP} denote the conductance of both parallel and antiparallel states. The GMR effect originates from the spin-dependent electronic transport inherent to magnetic metal systems. It is understood by using the Mott model which was proposed to understand the increase in resistivity of ferromagnetic metal above their Curie temperature. It assumed the electrons in a magnetic metal traverse in two independent channels coinciding to spin-up and spin-down states, and hence the conduction occurs in parallel within two spin channels. The second assumption is that the electron scattering for the spin-up and spin-down is independent so the density of states (DoS) at the Fermi level is not the same. Hence, the resistivities are different for electrons with different spins as shown in Figure 4.1. Hence, by using Mott's argument, the GMR is explained by stating that electrons with a spin antiparallel to the external **B** field are strongly scattered and electrons with a spin parallel to the applied **B** field are weakly scattered, which results in a low and high resistance corresponding to the scattering rates, respectively. There are different types of GMR effects, namely multilayer, spin valve, pseudo-spin, and granular. The magnetic layers are separated by very thin (few nm) insulating layers in GMR devices. The GMR effect is used mainly for the development of hard disks and hard drives along with magnetoresistive random-access memory (MRAM).

This resulted in a significant increase in the storage capacity. It facilitated paving the way for some of the currently trending devices and online applications. We can store information in spin and hence do computations because we can change this information by inducing a magnetic field.

The way to do this in computers is that, instead of capacitors, we have magnetoresistive RAM that has tunneling magnetoresistive structures. When a computer sends a current through the TMR, if the spins on both layers are aligned, a low resistance path is formed which gives a 1. If the spins are antiparallel to each other, a high resistance path is formed which gives a 0. We have a 1 and if not a zero. The computer will read the information of a TMR in this manner. However, it is challenging to upscale these quantum effects, and hence, it is very difficult to access the states of the ferromagnet to encode information.

Quantum devices based on spin qubits rely on our understanding of the spin of magnetic molecules and the ability to manipulate this spin at a nanoscale level. Magnetic molecules have a bi-stable ground state and display magnetic hysteresis at low temperatures which is essential for the development of quantum devices. However, at these low temperatures, the magnetic molecules show decoherence, which is characterized by the rapid loss of quantum information and the magnetic bistability that the magnetic molecules display. This decoherence prevents an efficient application of magnetic molecules to quantum technology. A quantitative treatment of the effect of temperature on electrical resistance is given by the Kondo effect.

Possibly important material: This "spin-transfer" process opens the possibility of new nanoscale devices for memory and other spin electronics applications. One application, in addition to direct current addressable magnetic memory, might be the use of spin transfer to excite a uniform spin wave in a nanomagnet and then to use this nanomagnet as a processing spin filter to inject a coherent spin pulse into a semiconductor structure. These nanomagnets can be single-molecule magnets attached to carbon structures. Carbon intrinsically has a low SOC. However, ferromagnetic elements (quantum dots) can be added to (defective) carbon structures with high mobility, such as graphene and spin–spin interactions can be found.

4.1.1.1   Kondo Effect

Kondo resonance and SOC are the most observed phenomena in spin-dependent interactions in nonmagnetic semiconductors. Kondo resonance and SOC depend on the crystal symmetries of the material and the structural orientation of semiconductor-based hetero-structures and therefore exist in different functional forms.

The Kondo effect is a standard model in correlated electron physics that describes the scattering behavior of conduction electrons when a magnetic impurity is present. The effect is observed as a characteristic change in electrical resistance with temperature as shown in Figure 4.2. For normal metals, the scattering of conduction electrons reduces with temperature as the electron–phonon scattering is lesser at lower temperatures reducing backscattering. This phenomenon continues till absolute zero is reached. However, for nonmagnetic materials with magnetic impurities, a minimum in resistance is observed above absolute zero.

The effect is named after its discoverer Jun Kondo who used perturbation theory on dilute magnetic alloys and hypothesized that the scattering rate due to the magnetic impurity should diverge as the temperature approaches zero. This was the effect of conduction electrons aligning their spins opposite to that of a nearby impurity. The impurity deviates from the electron propagation at low temperatures and consequently flips its spin deflecting its pathway. These electrons then pair with the localized electrons to form a net zero spin state: a singlet that is responsible for the characteristic change in resistance as the temperature decreases. The point in

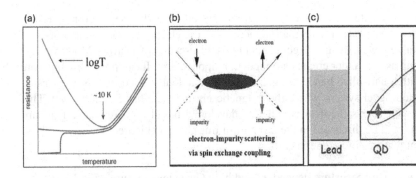

FIGURE 4.2 (a) Resistance-temperature variation for a typical magnetic impurity in a metal traditional Kondo system. At low temperatures, the spin scattering off the impurities enhances the resistance. (b) Demonstration of electron impurity-scattering via spin exchange. (c) Demonstration of the Kondo effect in a 0D quantum dot system connected to electrodes which arises when an unpaired spin from the quantum dot is paired to an itinerant electron to form a Kondo system.

temperature below which the Kondo scattering becomes significant (relative to the bifurcation temperature observed in susceptibility) is called the Kondo temperature (Figure 4.2a and b). Hence, in essence, a battle ensues when a magnetic impurity is placed in nonmagnetic metal.

The Kondo effect is influenced by the size of the sample and nonmagnetic random scattering events. It has no dependence upon the properties of magnetic impurity or nonmagnetic host.

It was demonstrated that Anderson's more microscopic model gave rise to Kondo's phenomenological one, which states that when electrons tunnel on and off the local site, delocalized electrons have the opposite spin to local electrons. Heavy fermions and intermetallic insulators containing rare earth elements have both shown the Kondo effect. In a quantum dot nanostructure with a specified gate, the change in resistance can be examined in more depth.

More generally, the Kondo effect has been observed when an even number of electrons occupy a quantum dot, at a singlet-triplet degeneracy, and need not include the screening of a localized spin, but rather the screening of any localized degeneracy. A magnetic impurity in a metal is equivalent to the classic Kondo system of the quantum dot with at least one unpaired electron. It behaves as a magnetic impurity and itinerant electrons can scatter off the dot when it is connected to a metallic conduction band. A topological Kondo effect has also been shown to exist in Majorana fermions, quantum simulations, and ultracold atoms.

Coming back to carbon, we think that the theory applied to heavy fermions and standard magnetic materials cannot be applied to carbon since it is a light element. Indeed, incorporation of metal clusters deforms the spacetime of a carbon structure like a vacancy center. The pi orbitals around such a center can overlap, and a closed (loop) structure is formed which is topologically protected from scattering. Such loops can be created from the edge states of graphene which is different from the ring structures of graphene or nanotubes where currents from the adjacent rings cancel each other. This topic will be discussed thoroughly in this chapter. The loops also create an imbalance between the minority and the majority carriers and spin-up or spin-down states, which creates spin-valve effect in any system. Finally, the Aharonov–Bohm (A-B) effect can be seen from the interactions of the electronic wave as it splits into two interacting waves which is equivalent to the bifurcation of a lattice. Moreover, the loop acting as a cylinder creates a (magnetic) dipole effect. It can produce a bipartite lattice with a half-filled band. Altogether, we have a double-degenerate state consisting of a four-level system where the magnetic ground state has $S = \frac{1}{2}(n_A - n_B)$. This matches with Lieb's theorem which was developed based on a repulsive Hubbard model and will be discussed in the last part of this chapter (Figure 4.3).

Salient features of magnetic carbon

- Magnetism: Structure dependent and high Curie temperature ~500 K.
- Exchange interaction in nanographene.
- The interplay between edge states and conduction pi electrons: Like s-d interaction in transition-metal magnets.
- Weak magnetic effect: With and without magnetic impurity.
- Magnetization can be tuned with doping, spin glass transition.
- pi-conjugated organic semiconductors offer promising alternative approaches to semiconductor spintronics.
- Strong e-p coupling and large spin coherence.
- Weak spin-orbit and hyperfine interaction: Long (~50 nm) spin diffusion length gives 40% magnetoresistance.

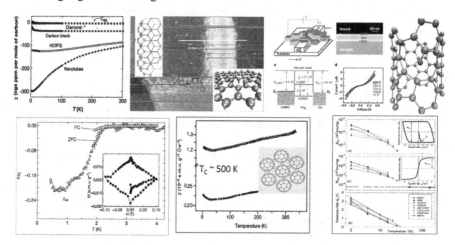

FIGURE 4.3 Examples of ferromagnetic Carbon: C_{60}, ferromagnetic graphite, molecular junction: switching. (top) Ferromagnetism in carbon nanostructures has been claimed in fullerenes, ion-irradiated graphite, molecular junctions, and nanotubes. (top left) Magnetic susceptibility was plotted as a function of temperature of carbon nanotubes, HOPG, carbon black, diamond, and C_{60}. (bottom left) Temperature-dependent susceptibility was plotted for boron-doped diamond and (middle) polymers of fullerenes. (right) Parameters obtained from the fits to the m(H) results of nitrogen-doped nanocrystalline diamond. The insets show the hysteresis loop models for the SC and SPM contributions with the meaning of the plotted parameters [J. Barzola-Quiquia, M. Stiller, P.D. Esquinazi, A. Molle, R. Wunderlich, S. Pezzagna, J. Meijer, W. Kossack, S. Buga, Unconventional Magnetization below 25 K in Nitrogen-doped Diamond provides hints for the existence of Superconductivity and Superparamagnetism, Sci. Rep. 9, 8743 (2019).].

4.2 MAGNETISM IN PURE CARBON?

The presence of magnetic properties and associated properties is reported in "pure" carbon systems. We will present a few examples of magnetic carbon nanostructures; however, it is difficult to identify a single effect that can explain the magnetic properties in a disordered carbon system. A major requisite for the occurrence of magnetism in an all-carbon structure is the presence and stability of carbon radicals. The occurrence of radicals, which can introduce an unpaired spin, is cut down by the strong ability to pair all valence electrons in covalent bonds.

What do we need for magnetism? First of all – unpaired spins, second: spins must interact! These can be achieved in carbon-disordered nanostructures in different ways, e.g., vacancies, edge states, curvatures, and addition of other atoms. Atoms of light elements like hydrogen, oxygen, boron, and nitrogen can also contribute to the creation of carbon-based magnetic materials.

4.2.1 FULLERENES

The paramagnetic properties of carbon materials have been scrutinized since the early 1950s, but the surface chemistry of carbon was not studied carefully at that time to explain the magnetic behavior. The initial idea of getting ferromagnetic carbon was to heat organic molecules in a vacuum, although the magnetic moments obtained are small ($\sim 10^{-3}$ Am^2/kg) enough to be attributed to naturally occurring impurities in our environment. Researchers attempted to achieve higher values of the magnetic moment in very different kinds of structures. Earlier there was a search for a spin carrier in a fullerene matrix. A polymeric network containing charged fullerenes has been created by doping. For example, in 2001, Makarova and others reported ferromagnetic ordering in pure carbon which was C_{60} of rhombohedral structure. They reported a value of $\sim 10^{-3}$ Am^2/kg. Doping creates stable molecular ions $C_{60}\pm$ or fullerene radical adducts $C_{60}R$. The interaction between charged fullerenes in the polymeric phase is ferromagnetic and long range. C_{60} becomes magnetically active due to the spin (and charge) transfer from dopants. Effective exchange interaction of the pair of $C_{60}\pm$ ions in a polymerized fullerene matrix is ferromagnetic ($J > 0$) and rather strong for high T_C. Energy difference $E\uparrow\downarrow - E\uparrow\uparrow$ is of the order of several tenths of eV. J is positive for the cases.

Interaction between fullerene radical adducts in a 2D layer of the tetragonal polymeric phase is ferromagnetic and does not fall within the two first coordination spheres (Figure 4.4). The researchers suggested that magnetism was caused by induced disorder with a partial change of hybridization from the sp^2 type to the sp^3 type. A mixture of sp^2–sp^3 carbon atoms may lead to a ferromagnetic state with a magnetization larger than pure Fe. The exchange is strong enough to account for the high-temperature ferromagnetism since there was a long-range ordering in the fullerene matrix. The possibility that nonfullerene (i.e., edge-site-containing) structures may be responsible for ferromagnetic behavior has also been mentioned from XRD evidence for the presence of "amorphous carbon" in their "polymerized" fullerenes obtained by treatment at high pressure and high temperature, although rhombohedral fullerenes exhibited an absence of ferromagnetism. Recent studies on carbon surface chemistry claimed that the spins responsible for the superparamagnetic or ferromagnetic behavior and for the broad electron spin

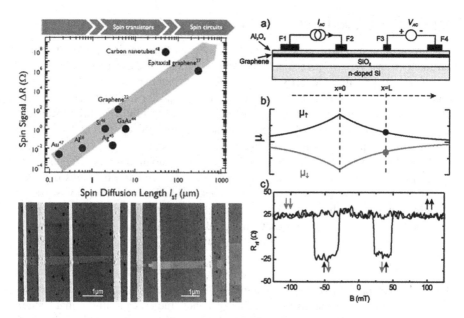

FIGURE 4.4 (top, left) Spin diffusion length in some carbon nanostructures can be very long compared to semiconductors. In graphene it can be 100 μm which would be very useful to make spin logic gates and low-power memory. (right) Graphene devices demonstrate anisotropic magnetoresistance [19]. (bottom, left) SEM images of two graphene spin valves (type III devices, fabricated by etching).

resonance (ESR) peak are of sp^3-type radicals in the product structure originating from the dangling bonds reflecting the partially remaining adamantane structure (after thermal CVD at 1000°C); they also noted that "pyrolytic carbon prepared from adamantane shows completely different magnetic features compared with other pyrolytic products prepared at 200–220°C that have been revealed to have a triplet state... on the average."

4.2.2 GRAPHITE AND GRAPHENE

In the previous chapters, we have reviewed the honeycomb structure of graphene consists of sp^2-hybridized bonds that can have exceptionally long electron scattering length (in the ballistic range) and long spin diffusion length so that spintronic device applications can be made possible. However, clean graphene shows intrinsic diamagnetism due to the absence of unpaired electrons, which limits its potential uses. Bulk graphite with a perfect structure is a diamagnetic material whose magnetic susceptibility is second only to that of superconductors. Graphite containing certain defects can exhibit spontaneous magnetization. These defects are adatoms, vacancies, zigzag edges, zigzag boundaries, and negatively curved graphitic surfaces. The techniques for imprinting magnetism into graphene lattices have recently attracted a lot of attention. Numerous research have demonstrated that graphene holds a lot of promise for applications in spintronic devices by further incorporating magnetism into graphene on the assumption of leveraging its special electron transport capabilities.

In 2003, Esquinazi and others found that graphite with a parallel magnetic field showed a magnetization value from 0.3×10^{-3} to 2.5×10^3 Am²/kg. This range is much more than that can be attributed to only iron impurities. But the problem with this report was that different samples showed different magnetization and hence the results were not reproducible. A similar behavior of carbon was also seen in a 50,000-year-old meteorite by Coey and others. The graphite-rich fragments from this meteorite showed strong magnetization of about 20×10^3 Am²/kg out. Although the asteroid can be rich in mineral content, the magnetization property was claimed due to carbon content itself.

It may also be attributed to the diamagnetic field created in carbon samples due to the magnetic field applied during the measurement by the equipment itself. The increased diamagnetic susceptibility in semi-metal graphite (or in nanotubes) can originate from the overlap of the valence with the conduction band. In fact, the susceptibility is large enough in the perpendicular direction to swamp away any ferromagnetic signal intrinsic to the sample. In the parallel direction, measurement is easier, but there's still some diamagnetic susceptibility. In non-conducting carbons, these problems are greatly reduced.

Nanoscale magnetism can be generated from the vacancy centers in the carbon lattice. The three carbon atoms bonded to this vacant site will have a modified configuration. Two of them adjust their bonds due to this vacancy which causes the third one to protrude out of the plane. This leaves away a dangling sp² orbital which is responsible for 1 μB of magnetization. By irradiating a graphite sample with protons local magnetic moments can be created, although irradiating by alpha-particles such effects were not found. Another way to create magnetism in carbon is to produce an adatom between two carbon atoms and form a sort of bridge. This adatom will have sp² hybridization out of which two will make covalent bonds with carbon on each side while the third orbital will be left dangling. The p-orbital will share an electron with this dangling orbital while lying parallel to the graphene surface and then contribute 0.5 μB of magnetization to the system.

A concrete theory of such magnetism in carbon is still lacking, but there are plausible explanations for the phenomenon. All of these ad-hoc theories depend on the concept of a local magnetic moment that interacts via a long-range exchange interaction which prefers ferromagnetic ordering. A theory was proposed by Lieb based on the splitting up of a flat energy band in a graphitic ribbon with a zigzag edge. The energy bands for spin-up and spin-down split up with the latter having higher energy and thus a ferromagnetic ordering is born. According to the theory, the electrons can hop from a lattice site A to B or vice versa but not from A to A or B to B. This gives us a total number of spin states which is half of the difference between these two sites. Now, if the edge states terminate differently, say with different numbers of hydrogen atoms on each side, then one of the lattice sites has an edge over the other and therefore a magnetic ordering is developed. The occurrence of such graphene sheets with many spins has been verified from electronic structure calculations but hasn't been observed experimentally. We shall present more description of graphene magnetism in the last part of this chapter (Figure 4.5).

In addition to the intrinsic magnetism, the graphene layers show very interesting properties in the spintronic application, including spin injection onto graphene, and defect-induced magnetism in graphene. Moreover, investigation of SOC and spin relaxation mechanism in graphene could give rise to high Curie temperature and can extend magnetic storage information for different applications. The theory predicts that the spin lifetime in pristine graphene is about 1 μs, while experimental values range from

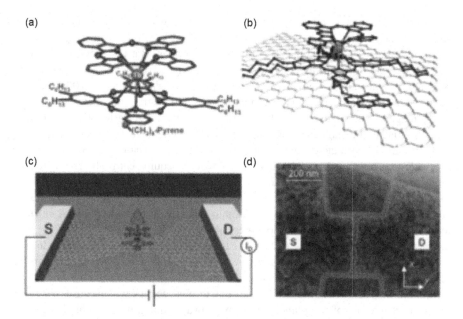

FIGURE 4.5 (a) Schematic diagram of the single-molecule magnet, (b) attachment of single-molecule magnet on the graphene sheet, (c) architecture of device with source and drain electrodes and, and (d) SEM micrograph of the device. SEM micrograph of the sample (a), device configuration (b), spin-valve feature of the device (c), and (d) trace measurements recorded at 20 mT/s [30].

10 ps to a few nanoseconds. However, modification of the graphene surface layer with molecular nanomagnets reveals a significant change in terms of the electrical and magnetic properties of the material. Candini et al. reported that in the low-temperature measurement, the magneto-conductivity signal is more than 20% found for the spin reversal, which reveals uniaxial magnetic anisotropy of the $TbPc_2$ quantum magnets (Figure 4.6).

4.2.2.1 Carbon Nanotubes

Like defective graphene and fullerenes, CNT (polymeric) networks can be formed to produce strong magnetic properties. This can be achieved by linking the CNTs with metal ligands which will be discussed thoroughly in this chapter. In the later part of this chapter, we extend this study by including intrasite and intersite tunneling properties which determine the speed of devices made of carbon nanostructures. Certain CNTs that have naturally little to no impurity allow for the insertion and use of electrons with polarized spin in them for computations. This makes them a good test tube for the study of spintronics. They have existed as macroscopic in one dimension and as nanoscopic in the other two dimensions therefore a good platform to perform coupling. CNTs have been employed as electrodes for molecular devices, which can exhibit better performance than traditional metal electrodes. That can be due to the low-density states of single-walled CNTs.

A monolayer was created of the magnetic organic molecule, from spin-polarized tunneling current, the results show that a higher current arises when the magnetization vector of the tip and molecules were parallel. In such a spin-valve system, the alignment of the molecule magnets can act as a valve for the tunneling of spin-polarized electrons.

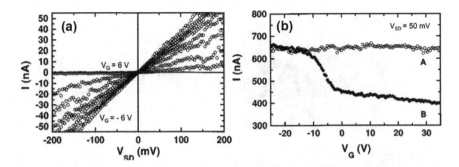

FIGURE 4.6 (a) $I\text{-}V_{sd}$ curve measured for SWCNTs device at gate voltage range from −6 V to 6 V. (b) I-V_G curve measured for MWCNTs device in comparison with that of a collapsed MWNT of similar cross section [31]. The y-axis goes from −50 to +50, and the x-axis goes from −200 to +200. Many lines emerging from the origin seem to spread across linearly in the first and third quadrants to end at the other side. Lines originate from the y-axis and move parallel to the x-axis for some time before they all coalesce on the x-axis. Transport holes dominated. Estimated hole mobility 20 cm²/(Vs) (diffusive transport), comparable to Si smaller than graphite [31].

However, functionalized CNTs have been modified, which can be used to enhance the magnetic and transport properties of the material for spintronic-based device fabrication. Recently, CNTs with side-attached single-molecule magnets TbPc₂ were used for this purpose. It shows GMR (spin valve effect) up to 1000% including the Coulomb blockade in the nanotubes with side-attached molecular magnets. Some researchers reported on the strong magnetic coupling between a metallic ion and a radical spin of TbPc₂. These are multi-terminal devices made by CNTs laterally coupled to the single-molecule magnet (SMM). This gives extra resonances in the hysteresis loop due to a strong coupling of Tb and its radical. The results of previous experiments on TbPc₂ coupled to a non-suspended CNT have shown that the spin relaxation was mainly enabled by bulk phonons in the environment of the individual single-molecule magnets. This can only be coupled to a one-dimensional phonon, associated with the nanomechanical motion of the CNTs. However, this work shows strong spin-phonon coupling between an SMM and CNTs. The results predict that a coupling of this magnitude induces strong non-linearities in the mechanical motion of the CNTs. That can be used to enhance the sensibility of CNTs-based magnetometers (Figures 4.7 and 4.8).

The modification of CNT or graphene can lead to a new class of molecule. This offers the spin degree that can be used to control the charge transport in the conducting system. Many works have been studied related to SWCNT, DWNT, and MWCNT. Most of the studies were focused on the conductivity and magnetic properties of CNTs. reported the superconductivity of double-walled CNTs, resistance was inversely proportional to temperature, and superconducting at the temperature below 6.8 K. Recently, a certain project was based on the MWCNT, showing the improvement on magnetic properties when MWCNTs is attached (chelate) with gadolinium which gives a large moment of 15.79 μB and non-super paramagnetic behavior governed by sufficient spin interaction. Spin valve behaviors of up to 8% were observed in aligned Gd-DTPA MWNT devices.

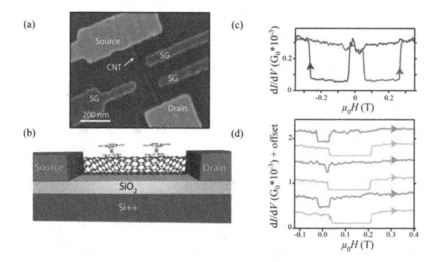

FIGURE 4.7 SEM micrograph of the sample (a), device configuration (b), spin-valve feature of the device (c), and (d) trace measurements recorded at 20 mT/s [32].

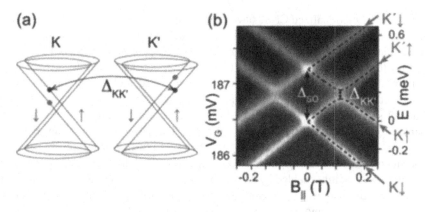

FIGURE 4.8 (a) Resistance-temperature variation for a typical magnetic impurity in a metal traditional Kondo system. At low temperatures, the spin scattering off the impurities enhances the resistance. (b) Demonstration of electron impurity-scattering via spin exchange. Demonstration of the Kondo effect in a 0D quantum dot system connected to electrodes which arises when an unpaired spin from the quantum dot is paired to an itinerant electron to form a Kondo system [33, 34].

4.2.2.2 Suspending Single Magnetic Molecules in CNTs

One can build these spin-nanochemical hybrid quantum devices. This results in a system that has profoundly unique superparamagnetic behavior under certain circumstances (Figure 4.9).

SMM has unique functional properties, including quantum tunneling and magnetic bistability which can be exploited for application in quantum technology. However, we cannot access these properties because there is limited interaction with the macroscopic

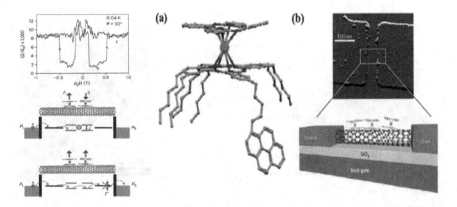

FIGURE 4.9 An investigation of this system also showed that non-covalent interactions result in highly efficient transport (a) and suspension of the SMM inside nanotubes as shown in the device structure (b). This allows the SMMs to have movement inside the tube and align themselves along an applied magnetic field. SMM exhibits slow relaxation while suspended in the CNTs [32].

world. CNTs seek to be the link to connecting the SMM to the macroscopic world without diminishing the functionality of the SMM. By encapsulating the SMM in CNTs, the molecule is also protected from external interactions that could impose strong decoherence and hence reduce the ability for quantum information processing.

Magnetic bistability is an important feature for data storage. It is a consequence of two phenomena. The super-exchange interactions inside the molecule lead to a ground state with high spin. The second phenomenon is due to the fact that the zero-field splitting parameter D is large and negative, and it results in a preferred direction of magnetization. A combination of these two factors results in a slow relaxation of the magnetization and magnetic bistability and therefore we have more accessible spin states. An example of how the suspension of SMM inside CNTs would look like using $Mn_{12}Ac$ SMM is given in this chapter (Figure 4.10).

HRTEM was used to investigate the morphology of the Gd-Fctn-MWNT. The functionalized MWNTs (Figure 4.11a–d) show that Gd^{3+} centers are accommodated by fibril and spherical-shaped nanostructures of approximately 2 nm in diameter, with a relatively uniform distribution in proximity but not continuous coverage of the outermost surface of the MWNTs. The atomic resolution of the Gd-DTPA aggregate can be seen on the HRTEM image (Figure 4.11d).

4.2.3 ELECTRONIC TRANSPORT

Figure 4.12 shows the *I-V* characteristics of the Gd-DTPA-MWNT network device at various temperatures ranging from 300 mK to 70 K, and the inset shows a typical device used in this work; the Gd-Fctn-MWNT bundle deposited around the electrodes. *I-V* characteristics change progressively over the temperature range and a large deviation from linearity is seen at 300 mK. The strong nonlinearity is an indication of transport in the Coulomb blockade regime. The charging energy for a

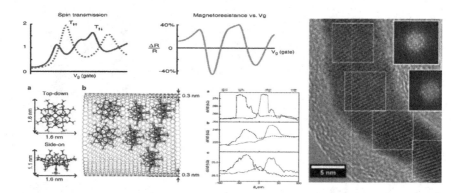

FIGURE 4.10 (a, b) Encapsulation of $Mn_{12}Ac$ SMMs inside a carbon nanotube [33, 34]. Top and middle panels show magnetoresistance and spin transmission in carbon nanotube based magnetic tunnel junctions by the application of a gate bias. (left) HRTEM image shows carbon nanotube-lanthanide hybrid molecular complex structure [21].

device with a channel length of 1 μm is calculated to be 340 meV, which is roughly ten times the value reported for a single MWNT device and substantially larger than the thermal energy of the system at 300 mK (0.026 meV). The resistance of the device at this temperature was recorded to be approximately 300 kΩ (i.e., larger than the fundamental resistance quantum; 25.9 kΩ), and thus the requirements for the Coulomb blockade regime are met. The conductance was measured as a function of temperature for the same range and shows a steady increase till approximately 4 K and then saturates below this temperature. Analysis of the resistance vs. temperature data indicates that the Gd-DTPA-MWNT networks do not follow an activated conduction (Eq. 4.1) or variable range hopping (Eq. 4.2) which are the expected mechanism for thin CNT networks (given below).

$$\frac{R(T)}{R(0)} = \left(exp\left(\frac{T_0}{T}\right)^1 \right) \tag{4.1}$$

$$\frac{R(T)}{R(0)} = \left(exp\left(\frac{T_0}{T}\right)^{0.25} \right) \tag{4.2}$$

This was concluded after failure to linearize the logarithmic normalized conductance as a function of T^β, where β is a critical exponent representing the dimension scale of the hopping. The devices do, however, display similar trends to those reported for thicker SWNT networks and conducting polymers that suggest interrupted metallic conduction mediated by fluctuation-induced tunneling (FIT). A nonlinear fit to the data set gives a relation below.

$$\sigma(T) = \sigma_1 T + \sigma_2 e^{-\frac{T_1}{T+T_0}} \tag{4.3}$$

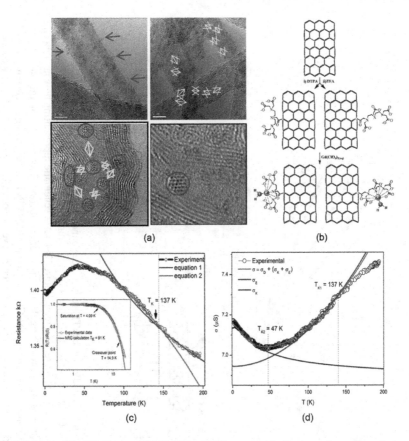

FIGURE 4.11   HRTEM image of MWNTs. (a) The arrows indicate regions along the nanotube that have strand-like filament structures, believed to be a result of the DTPA ligand attachment. These strands are generally in close proximity to the multiple Gd^{3+} centers. (b) The MWNTs show a fairly homogenous distribution of the Gd-DTPA attachment throughout the sample, forming a network-like cover of the outer wall; distances between centers are indicated by the arrows. Multiple Gd-DTPA centers separated by a distance greater than 10 nm are indicated by the arrows. Morphology of aggregated centers viewed under higher magnification circled [28]. (c) The resistance as a function of temperature are fitted to equations 1 and 2 which yields a Kondo temperature of 137 K. Inset shows the resistance-temperature curve for the functionalized material which shows clear saturation fitted by the NRG equation yielding a Kondo temperature of 90 K. (d) The conductance as a function of temperature for the Gd- filled MWNT sample: here, the conductances show a minimum at around 40 K and then increase sharply till finally saturating at 10 K [28].

In this model, the conductance is separated into two terms, the first term scales linearly with temperature, while the second term considers fluctuation-assisted tunneling. Here σ_1 and σ_2 are constants, and T_1 represents the activation energy required to tunnel through the barriers and T_0 is the temperature at which the crossover from the saturating to activated transport occurs. This model has been successfully utilized in a range of disordered carbon networks. It should also be noted that some

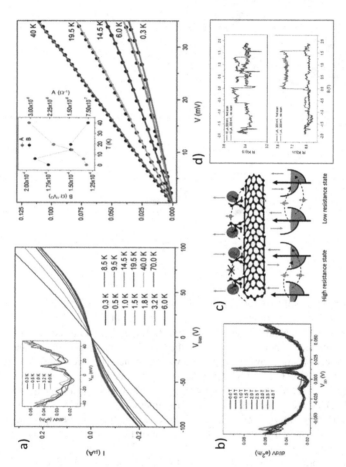

FIGURE 4.12 (a, b) The current-voltage characteristics of the filled bundle device. There is a clear band gap opening at temperatures below 40 K compared to linear behavior above that. Differential conductance (inset) shows the formation of temperature-dependent resonance peaks. The current-voltage plots are fitted. The inset shows the evolution of the coefficients with increasing temperature [28]. (c) A schematic representing how neighboring Gd-DTPA complexes can act as spin valve barriers effectively preventing or promoting the transport of conduction electrons depending on local spin densities coupled to the Gd-DTPA complex [28]. (d) Magnetoresistance switching for the Gd-functionalized MWNT device passing a current of 10 μA. The switching is asymmetric with respect to forward and reverse field sweeps and multiple switching events occur, like nonlocal spin valve behavior. (b) There is a large increase in the switching effect, roughly 7%, when the current is lowered to 1 μA.

studies have linked the saturation in the resistance at low temperatures to tunneling between the outer and secondary shells of the MWNT; however, due to the strong magnetic response and *I-V* analysis, it is believed in this system that the low-temperature behavior is due to electron and spin correlations.

This led to the probing of the saturating resistance considering the numerical renormalized group (NRG) calculation as presented in Figure 4.11c and d, which shows the normalized resistance (concerning the saturation resistance) as a function of temperature. A clear saturation is observed below approximately 4 K. The solid line fits the numerical renormalization group equation $\dfrac{R(T)}{R(0)} = \left(1 - c\left(\dfrac{T}{T_K}\right)^2\right)$, where $c = 6.088$ and T_K is the Kondo temperature.

From the fitting $T_K = 91$ K is extracted, although surprisingly high this value is very similar to what has recently been observed in disordered graphene using the same fitting. In addition, it is observed that the equation fits the data set best in the region below 10 K, signifying the crossover from thermally activated transport mechanism (FIT fitting) at higher temperatures. The crossover temperature is significant due to observations in the differential conductance as shown in Figure 4.11c and d.

The increase in resistance down to 40 K has been analyzed considering Kondo scattering utilizing the Fermi liquid description which has recently been successfully used for graphene devices.

$$\rho(T) = \rho_{c1} + \rho_0\left(1 - \left(\frac{T}{T_K}\right)^2\right) \text{ and } \rho(T) = \rho_{c2} + \frac{\rho_0}{2}\left(1 - 0.47\ln\left(\frac{1.2T}{T_k}\right)^2\right) \quad (4.4)$$

The equations are valid for temperatures between resistance maximums up to approximately 150 K. There is a change in the transport mechanism that is described by the other equation. Both equations give the best fit to the data with a Kondo temperature of 137 K. The values of ρ_{c1} and ρ_{c2} have a discrepancy of 12%, indicating self-consistency between the two equations. This model does not, however, explain the additional turning point in the *R-T* curve indicating that additional mechanisms are at play. To establish the nature of decreasing low-temperature resistance, the conductivity data is analyzed considering an interpolation scheme considering the competition between the Kondo effect and co-tunneling which has been successfully used to describe MTJ with superparamagnetic particles embedded within the dielectric layer. A combination of the two-interpolation formula is used to indicate the crossover from a co-tunneling dominated to Kondo dominated regime. This is given by $\sigma = a\sigma_D + b(\sigma_E + \sigma_K)$, with $a + b = 1$. Here, σ_D is the direct tunneling conductivity between the electrodes and has previously been neglected when using this model. For a more realistic fit, a conductance offset determined from the fitting to be $\sigma_D = 6.6 \times 10^{-6}$ μS has been related to σ_D, this effectively sets the coefficients a and b as 0.75 and 0.25, respectively. σ_K represents a conductivity term involving the Kondo effect and σ_E is the elastic conductivity not including spin-flip events. The conductivity term associated with the Kondo effect takes the form:

$$\sigma_K = \sigma_0 \left(\frac{T_{K2}^2}{T^2 + T_{K2}^2} \right)^S \text{ where } T_{K2}^2 = \frac{T_K}{\sqrt{2^{1/S} - 1}} \tag{4.5}$$

As usual, $S \sim 0.2$ and σ_0 are related to the occupancy of electron clusters near the magnetic particles within the tubes, for perfectly symmetric barriers it will have a value of $\sigma_0 \leq \left(2e^2\right)/h$. For these devices, a value of approximately 7×10^{-9} well within the expected range demonstrated by other MTJ devices is obtained. The electrical conductivity is given by a combination of sequential and co-tunneling conductance terms, and these take the form: $\sigma_{cot} = \frac{2h}{3e^2} \frac{1}{R_T^2} \left(\frac{k_B T}{E_C} \right)^2 ; \sigma_{seq} = \frac{1}{2R_T^2} \left(1 + \frac{E_C}{k_B T} \right)^{-1}$.

Here, R_T is the tunneling resistance between electrodes and the MWNT network, and this quantity is determined to be $R_T = 5.3$ kΩ which is slightly higher but of the same order of magnitude as what has been reported for double MTJs with embedded superparamagnetic NiFe nanoparticles. E_c is the charging energy which is related to the size and shape and hence the capacitance of the magnetic nanoparticles and can as a first approximation be expressed as $E_c = 2\pi\varepsilon\varepsilon_0 d$, where ε is the relative permittivity of the tunneling barriers and d is the diameter of the nanoparticle rod. The Kondo temperature from this fitting is 47 K, which agrees with the point of inflection in the conductivity temperature curve. Thus, the RT behavior is explained as a competition between tunneling events and the strengthening of the Kondo effect as the temperature is lowered, at some point the Kondo effect dominates over the tunneling (47 K) and an inflection is observed. To verify the co-existence of co-tunneling conductivity in the system, an analysis of the current-voltage characteristics is undertaken as shown in Figure 4.11. The slope of the I-V sweeps changes markedly upon lowering the temperature and observe the formation of a gap (strong deviation from linearity) at a temperature below 40 K is observed. As shown in the inset of Figure 4.11a and b, the differential conductance exhibits resonance peak features which are strongly temperature dependent. To probe the interaction of sequential and co-tunneling conductivity, the theory developed for Coulomb gap inelastic co-tunneling is relied on

$I(V) = \left(\frac{\hbar}{3\pi e^2 R_T^2 E_C^2} \right) \left[(2\pi k_B T)^2 V + e^2 V^3 \right]$, which describes the current as a sum of two terms, one linear and the other cubic in voltage. The I-V curves at different temperatures are fitted to the phenomenological expression: $I(V) = AV + BV^3$. Here the ratio between the coefficients A and B is related to the strength of contribution, either sequential or co-tunneling, to the current. The fit is shown in Figure 4.11b, and the trend of the coefficients A and B are shown in the inset as a function of temperature as can be seen there the range of increase of coefficient A is seen to increase at a faster rate than the change in coefficient B, this was taken to explain the dominance of co-tunneling over sequential tunneling in this low current regime.

As mentioned previously the differential conductance sweeps exhibit strong resonance peaks at around 10 mV. These peaks are temperature and magnetic field-dependent, and exponentially decrease in height when increasing either parameter. Upon examining the temperature-dependent dI/dV, an apparent shifting and broadening of the peaks is observed.

These features can be related to the competition between co-tunneling and the Kondo effect. As the temperature is increased, the Kondo effect is suppressed, and co-tunneling is favored, analogous to the conductance vs. temperature. This leads to the formation of co-tunneling peaks at a slightly lower voltage indicated as a shoulder and eventual spectral shift of the Kondo peak. Similar features are observed in the B field dependence. Following conventional analysis, the Kondo temperature can be in this case the zero-field 300 mK resonance peak gives a Kondo temperature of 40 K. This value is in excellent agreement with the value obtained from the conductivity fitting and is a significant temperature point as indicated by the susceptibility blocking temperature indicating that spin coherence becomes important for electron correlations.

The magnetic properties of MWNTs filled with $GdCl_3$ are substantially modified, which clearly shows the co-existence of Kondo and co-tunneling effects. The susceptibility measurements indicate an antiferromagnetic exchange interaction, with a bifurcation in susceptibility depending on whether the magnetic field is applied or not during cooling, a signature of superparamagnetism. Co-tunneling of conduction electrons interfering with a Kondo-type interaction has been verified from the exponential decay of the intensity of the zero-bias anomalies conductance peaks which also show strong resonant features which were observed only in $GdCl_3$ filled MWNT devices. This study raises a new possibility of tailoring magnetic interactions for spintronic applications in CNT systems.

The dI/dV vs V_{SD} show a gapped region between -0.05 and 0.05 eV. Near the zero-bias point, resonance-like features are observed which have not been reported in previous studies on pristine MWNTs. These resonance peaks are strongly temperature dependent and only appear in the temperature region below 15 K. The peaks are located at approximately ± 10 mV and are asymmetric and robust, and these peaks are observed at all temperatures below 15 K. In addition, the asymmetry in the resonant peaks has led to investigate the possibility of Fano-Kondo resonance in this system that can be fitted with the Fano formula:

$$\frac{dI}{dV} \sim \frac{(\varepsilon + q)^2}{\varepsilon^2 + 1} \tag{4.6}$$

In this formula (4.4), ε is related to the width and energy of the resonance, and q is the Fano parameter which is related to the asymmetry of the peaks. The Fano formula is a good fit for the 10 mV peaks and has tracked these peaks as a function of the applied field. The position and width of the peaks are fairly constant as the field is increased up to 3.5 T but notice a decrease in dI/dV peak height. From the width of the resonance $T_K = 81$–90 K is calculated, in close agreement with our value arrived at through the NRG fitting. Such features are most commonly attributed to Kondo resonances which arise here due to the interaction between localized spin atGd-DTPA]$^{n+}$ and the electrons of the surrounding host (MWNT). It was also observed that the peaks are strongly affected by the strength of the field, some of which increase in intensity, while others diminish with the increasing magnetic field.

At 2 K, the resonance modes are still observed but are, however, less distinct due to thermal effects. As the conduction is weakened at lower temperatures, probably due to the Coulomb blockade effect, the RKKY exchange interaction will be suppressed essentially freezing the spin on the Gd-DTPA molecule and thus lead to more stable magnetic tunnel barriers which may lead to the observed resonance states. It should be noted that such resonance peaks are commonly observed only with scanning tunneling microscope (STM) measurements over single magnetic impurities, or for quantum dot devices, their presence in this network device is thus one of the first demonstrations of a multi-body mesoscopic Kondo effect in a carbon system and are thus rather significant.

To further probe the magnetic properties of the composite material, the dependence of the resistance on the magnetic field is investigated. The spin-switching effect is observed at 300 mK showing asymmetry. Similar results have been reported for CNT devices fabricated with multiple nonlocal ferromagnetic contacts, where it was shown that the orientation of the different ferromagnetic contacts can change the switching fields and magnetoresistance difference quite drastically. To further probe the nature of the spin-switching effect, the influence of current strength was considered as shown in Figure 4.12c and d. There are a few noteworthy features such as multiple switching at various field strengths and asymmetry in the forward and reverse sweeps. As the Gd-functionalized MWNTs have many magnetic domains along the length of the tubes, it is not surprising that the multiple switching effect is observed. This indicates that interaction and alignment between adjacent magnetic domains are important when analyzing the observed jumps in the resistance. It was found that when passing a current of 10 µA, the largest magnetoresistance jump was around 3%. Also the step-like features in some of the jumps have been observed before and were attributed to sequential tunneling (Figure 4.12c and d). When the current is lowered to 1 µA, a large increase in the TMR to about 7.5% was observed. At the lower current, it is expected that the Coulomb blockade effect is stronger due to lower electron temperature. This ensures that the system is more resistive than at higher currents due to stronger Coulomb repulsion. In this state, it is expected that co-tunneling effects dominate over ground-state sequential tunneling; this second-order co-tunneling process takes place through a virtual intermediate state that enhances TMR (observed in the device).

The enhancement of TMR due to co-tunneling is a result of spin accumulation, whereby an imbalance in the densities of states in the antiferromagnetic configuration occurs due to the preferred tunneling of one spin state over the other. When a field is applied to the system, the densities of the state for the two regions will shift, and this results in spin flipping of states above the chemical potential as shown schematically in Figure 4.12c. However, nonlocal measurements must be performed on such spin systems to separate features relating to the Magneto-Coulomb effect and spin accumulation both manifest similar features. Such measurements are yet to be conducted on the samples studied in this work. Another seemingly anomalous observation in the magnetoresistance is the shifting of switching fields as indicated by the susceptibility analysis there is antiferromagnetic coupling in the system. It is well known that exchange bias can cause the shifting of switching fields in spin valves, and again these hints at inter-domain interaction, coupled with possible spin accumulation from the co-tunneling events. This leads to changes in the local magnetic moment and field of single domains. Further tunneling into or out of a domain can have a cascading effect and lead to spin flipping which can enhance or reduce the resistance (Figures 4.13 and 4.14).

4.2.3.1 Magnetism in Boron-Doped Diamond Films

In the beginning of this chapter, we reviewed the disagreement over the cause of the ferromagnetism found in metal-free carbon allotropes. In general, the inherent ferromagnetism found in other forms of carbon, such as fullerenes, graphene, CNTs, and graphite, has been attributed to hydrogenation and sp^2/sp^3 defects. It is claimed that hydrogenation of a graphene sheet will cause the localization of electrons at the unhydrogenated carbon sites, and the uncompensated spin polarizations at these sites will transform the system into a ferromagnet. Hydrogen plays an important role in changing the ratio of carbene-to-carbyne structures and ferromagnetism in carbon. Considering the effects of electron–electron interactions in the context

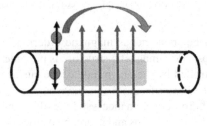

Sequential tunneling dominant regime:

TK1, kondo effect between GdCl3 nanoparticle and incident electron, signified by onset blocking temperature. (spin coherence of electron and magnetic particle match)

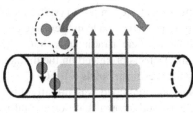

Co-tunneling regime:

As temperature is lowered co-tunnelling events through formation of virtual states dominates over sequential, related to collective quasiparticles with preferred combined spin state. Due to spin accumulation from longer spin coherence times of nanomagnet.

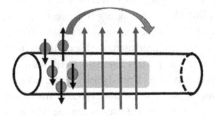

Electron-electron Kondo correlation regime:

As spin accumulation increases konod effect between localized electrons can lead to preferential spin configuaration for tunneling

FIGURE 4.13 Temperature dependence of the differential conductance. The peak heights decrease exponentially as temperature increases, and co-tunneling step features are observed along with a spectral shift as temperature increases. Magnetic field dependence of the dI/dV peaks: magnetic field is seen to exponentially suppress the peak intensity [28].

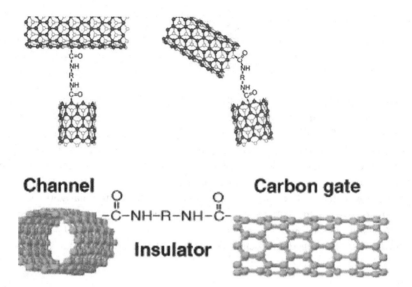

FIGURE 4.14 Nanotubes can be connected with each other by ligand molecules in various ways which can create circuits and memory.

of the Hubbard model shows that there can be spontaneous magnetic ordering in nanographite. In the case of zigzag-type strips, a ferrimagnetic structure is possible; the appearance of magnetic ordering is controlled by the magnitude of the surface deformation caused by electron–phonon interaction.

The same principle can be applied to boron-doped nanocrystalline diamond (BNCD) films since they accommodate both defective bonds and hydrogen on the surface and interface. The absence of ferromagnetism was tested in polycrystalline bulk diamond synthesized at high temperature and high pressure in a hydrogen-free environment. Bulk magnetization tests for diamond that has been irradiated with nitrogen or carbon ions have also revealed ferromagnetism. Defects produced on the diamond surface can rearrange the carbon atoms and produce resonant bonds (–C=C–C=C–). The ferromagnetism seen in bombarded graphite and diamond was initially attributed to bonding defects in sp^2 and sp^3 mixes. It is well known that polycrystalline diamond has abundant sp^2 and sp^3 mixes at the grain boundaries. The defects can be sp^2-like in diamonds or sp^3-like in graphene which produces resonant bonds, which is the key to magnetism. This result has been linked to the bonding imperfections and sp^2/sp^3 carbon mixture produced by the ion damage. In the resonant structure, orbital magnetism can be produced. We discuss some of the experimental observations from the boron-doped diamond films where magnetic properties seem to exist with the superconducting properties. Boron is used as a dopant to alter the electronic properties of diamonds. Diamond that has been doped with boron undergoes a transition from insulator to metal. If higher concentrations of boron are used, then the diamond becomes superconducting. Superconductivity in diamonds has resulted in research being performed on the Cooper pairing mechanism in doped materials. STM measurements have suggested that photon-mediated coupling occurs similarly to what is predicted by the Bardeen–Cooper–Schrieffer (BCS) theory. Due to the occurrence of the insulator–metal transition due to doping, the resonant valence band theory appears to be validated, and correlation-driven Cooper pairing occurs. Now boron-doped diamond also exhibits ferromagnetism. Theoretical models have proposed that hydrogen incorporation can turn a diamond into a ferromagnetic diamond. Hydrogen incorporation as well as defects in sp^2/sp^3 due to ion bombardment contributes to ferromagnetism. A cause of superconductivity in diamonds is the pairing of charge carriers with antiparallel spins at the domain wall. This is called domain wall superconductivity and it follows from the Anderson-Suhl theory. According to the Anderson-Suhl theory, a singlet superconducting state can co-exist with a ferromagnetic state under certain conditions. The resonant structure of carbon can also form a closed loop, like a Kagome lattice which can produce ferromagnetism and anti-ferromagnetism transitions. It is claimed that the arrangements of impurity spin at the diamond (111) surface [in addition to the rectangular spin-lattice for the (100) surface] are like a triangle that can form localized YSR states and YSR bands.

To verify this superconducting transitions, perpendicular and parallel applied magnetic fields were studied. Magnetic field dependence of the longitudinal resistivity showed giant PMR, where the anomalous $\rho_{xx}(T)$ dip. The $\rho_{xy}(H)$ demonstrates an anomalous Hall effect. The temperature dependencies of the critical field, $m_0H(T)$, measured in magnetic fields perpendicular and parallel to the sample. To build up the m_0H-T phase boundaries, a criterion was set at 95% of the normal-state resistivity to determine the onset critical temperature. The resistive superconducting transition was linearly extrapolated for the determination of the offset critical temperature in different magnetic fields (Figures 4.15 and 4.16).

4.2.3.2 GNRs

In general, the magnetism of GNRs is connected to localized zigzag states that may be controlled via structural engineering and the application of an external electric field. However, introducing two substitutional boron atoms into the carbon backbone chain provides a workable technique to imprint magnetism into the interior of GNRs. They demonstrated that the boron atom pair breaks the conjugation of topological bands, resulting in the local creation of two spin-polarized border states. On-surface synthesis (OSS, a "bottom-up" process) is typically used to place GNRs on the surface of a metal substrate. To measure the intrinsic magnetism of GNRs and prevent the influence of the eventual magnetic ground-state quenching caused by the strong interaction between boron atoms and the metal substrate, an STM tip can lift GNRs and remove substitutional 2B atoms from the metal substrate. The configuration in which only one B atom is isolated from the substrate is most likely where the observed Kondo resonance originates. This work shows that spin polarization may be tuned and provides a feasible method for integrating B atom pair chains into the GNR lattice.

Since there are one unit more carbon atoms in the majority sublattice (nA) than the minority sublattice (nB) in 7-AGNR, the insertion of the naphtho group disrupts the sublattice symmetry. A 1 B magnetic moment results from the translation of this sublattice imbalance into a spin imbalance. A significant split of 0.8 eV caused by Coulomb electron–electron contact may be seen in the dI/dV spectra of electrically decoupled edge extension states. Except for the appearance of a pair of in-gap spin-split states that are symmetrically situated around the Fermi level and separated in energy by 0.38 eV, the addition of the naphtho group mostly retains the electronic structure of 7-AGNR. The naphtho group's zigzag edge is the primary

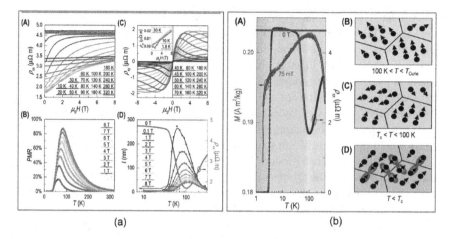

FIGURE 4.15 (a) Ferromagnetism in boron-doped superconducting nanocrystalline diamond films. (A) Magnetic field dependence of $\rho_{xx}(H)$, and (B) giant PMR is observed in the temperature window, where the anomalous $\rho_{xx}(T)$ dip is located. (C) $\rho_{xy}(H)$ demonstrates anomalous Hall effect. (D) The mean free path, deduced from the $\rho_{xx}(H)$ and $\rho_{xy}(H)$ measurements, is plotted together with the anomalous $\rho_{xx}(T)$ dip showing a strong variation of the mean free path.
(b) Correlation between $\rho_{xx}(T)$ and M(T), indicating the electronic entanglement of the ferromagnetic and superconducting states and the presence of a precursor phase, in which spin fluctuations intervene for the development of the domain wall superconductivity at low temperatures. (A) The $\rho_{xx}(T)$ and M(T) curves are roughly divided into three regions with respect to the temperature coefficient of ρ_{xx} and M. The corresponding spin configurations are schematically illustrated in (B–D). (B) In the temperature window of $100\,K < T < T_{Curie}$, the overall ferromagnetism of the system results from the ferromagnetic arrangement of the domains (domain walls represented by the black lines). (C) The precursor phase emerges in the temperature window of $T_c < T < 100\,K$, where spin fluctuations intervene in the system via antiferromagnetic arrangement of the domains. (D) When $T < T_c$, carriers with antiparallel-aligned spins can be an additional source for Cooper pairing at the domain walls, in addition to the boron doping-induced superconductivity in the HBD [36].

location of the magnetic moment that is contained in the singly occupied in-gap state, and it localizes on the majority sublattice close to the extension. In the region of the band edges, a spin polarization of around 60% is attained. Overall, our results point to possible spintronic uses for functionalized 7-AGNR, even at modest edge extension densities.

According to the Fujita/Klein argument, electron localization at zigzag edges results in spin polarization, which causes a "sharp peak in the density of states at the Fermi level," and this causes "lattice distortion due to the electron–phonon interaction and/or magnetic polarization due to the electron/electron interaction."

The parallel spins are postulated to be on the carbon instead of the heteroatoms (which are less common in common carbonaceous solids), and they are postulated to be on stabilized carbon zigzag edges instead of the periphery of highly reactive molecules like 2,6,10-tri-tert-butyltriangulene for which a triplet ground state is also anticipated. A saturation magnetization of several hundred emu/g is

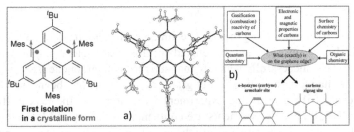

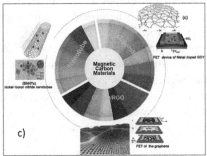

FIGURE 4.16 (a) To create a model of the chemical composition of the edges of graphene sheets in both flat and curved sp²-hybridized carbon materials. It is claimed that contrary to what is often believed, a sizeable portion of the oxygen-free edge sites under ambient circumstances are neither H-terminated nor oxygen free radicals. The triplet ground state is the most prevalent at the zigzag sites, which resemble carbenes. The most prevalent ground state at armchair locations is the singlet ground state, which is carbyne-like. This hypothesis not only agrees with the fundamental electrical characteristics and surface (re)activity behavior of carbons, but it can also account for some impurity-free carbon compounds' previously unknown yet perplexing ferromagnetic features [37].
(b) In materials science, attempts have been made to synthesize and isolate crystalline forms of hydrocarbons with a triplet ground state. One of the most well-known triplet ground-state benzenoid hydrocarbons is triangulene. Recently, the synthesis and isolation of a crystalline triangulene has been claimed. The application of bulky substituents to the reactive zigzag edges is essential for success. The achievement here will open the door for the synthesis and isolation of other hydrocarbons with higher spin multiplicity [38].
(c) Various carbon-based magnetic materials and applications e.g., hybrid nickel–boron nitride nanotubes (BNNTs), FET of the graphene, and FET device of metal-doped GDY [39]. All those research studies provide a theoretical basis to promote the application of 2D carbon-based magnetic materials in spintronic devices involving spin and charge modulation and lay a foundation for the preparation of high-performance carbon-based magnetic materials with specific magnetic properties.

projected for a short, open, zigzag single-wall nanotube, which is similar to 200 for R-Fe. A high-power laser was used to create an ultra-low-density carbon "foam" that was then examined for ferromagnetism earlier this year by John Giapintzakis of the University of Crete and colleagues. Giapintzakis et al. discovered the substance to be made up of randomly linked carbon clusters with an average diameter of between 6 and 9 nm using electron microscopy. At ambient temperature, the "nanofoam" possessed a magnetic moment of roughly 0.4 Amkg and a Curie temperature of 90 K, albeit this moment vanished shortly after the foam was created (Phys. World).

4.2.4 NITROGEN DOPING OF GRAPHENE

In addition to the edge engineering techniques discussed above, sp³ function-alization, light/transition-metal atom adsorption, and defect engineering tech-niques can also successfully generate magnetic into the graphene lattice. Inducing long-range magnetic coupling in 2D graphene networks has been the subject of a great deal of theoretical and experimental research, but the task is still quite difficult. Local magnetic moments are confirmed to be induced by point defects caused by ion irradiation in the graphene lattice. Another effective method to imprint magnetism onto graphene is heteroatom doping, which includes light and transition-metal atoms (Figure 4.17).

Nitrogen-doped graphene induces ferromagnetism that is highly dependent on the N concentration as well as bonding configurations, with a doping level up to 29.82 at%, one of the highest values ever reported in the heteroatom doping in low-dimensional graphene-based materials (LDGMs). In contrast to nonmagne-tism at doping concentrations lower than 5 at%, a ferromagnetic phase transition was discovered to occur at 69 K and the saturation magnetization can reach up to

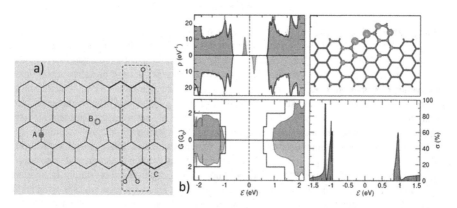

FIGURE 4.17   (a) Atomic flaws with a net magnetic moment might be the cause of the mag-netic order in carbon. A carbon "adatom" on the sheet's surface (A), a carbon vacancy (B), and an irregularly terminated zigzag edge are three different forms of defects that may exist in graphene sheets. The "unit cell" of a magnetic carbon ribbon, which has a spin angular momentum of 1/2, is enclosed by the dotted line, and the protrusions are hydrogen atoms. Although it is believed that each of these flaws can cause a magnetic moment, no mechanism has yet been discovered that explains how these moments interact to create a magnetically ordered state. https://physicsworld.com/a/the-magnetism-of-carbon/.

(b) To create a model of the chemical composition of the edges of graphene sheets in both flat and curved sp²-hybridized carbon materials, hitherto unconnected experimental results are linked with a theoretical analysis. It is claimed that contrary to what is often believed, a sizeable portion of the oxygen-free edge sites under ambient circumstances are neither H-terminated nor pure ó free radicals. The triplet ground state is the most prevalent at the zigzag sites, which resemble carbenes. The most prevalent ground state at armchair locations is the singlet ground state, which is carbyne-like. This hypothesis not only agrees with the fundamental electrical characteristics and surface (re)activity behavior of carbons, but it can also account for some impurity-free carbon compounds' previously unknown yet perplexing ferromagnetic features [40].

1.09 emu/g. Their calculations revealed that graphite N makes up the majority of the ferromagnetic state's contribution, whereas pyridinic N makes up a relatively small portion. Furthermore, it was determined through a theoretical calculation that the N-doped graphene system's delocalized magnetic moments are produced by itinerant p electrons of graphite N atoms. It is important to note that the enormous magnetic moments of N-doped graphene are influenced by nitrogen types in addition to being directly connected to the doping concentration. The ability of graphite N to improve the ferromagnetic coupling between the localized magnetic moments produced by faulty N was experimentally shown.

4.2.5 Magnetism Induced by In-Plane Manipulation in Graphene

4.2.5.1 Orbital Magnetism of the Moiré Graphene System

In general, shrinking graphene's lateral dimension to the nanoscale is advantageous for expanding the range of electrical applications and opening the bandgap caused by quantum confinement and edge effects (Figure 4.18). Graphene nanoflakes (GNFs), graphene quantum dots (GQDs), and graphene nanoribbons (GNRs) are examples of typical graphene nanostructures. The characteristics of these nanostructures are constantly influenced by their sizes, shapes, and edge structures. Edge engineering, defect engineering, and sp^3 functionalization are approaches to imprint magnetism in the case of large-size graphene sheets; these techniques have also been frequently reviewed in previous years. In addition, scientists have been paying close attention to the emerging phenomena of correlated stimulation states, superconducting states, and quantum Hall effect in the evolving moiré graphene system.

Earlier, a continuous model to describe the electrical structures of twisted bilayer graphene (TBG), in which the velocity at the Dirac point disappears at a particular angle of twist (the "magic angle") along with a flat moiré band. Recently, correlated insulating states at half-filling appear because of the flat bands at Fermi energy, successfully preparing the TBG with a twisted angle of about $1.11°$ and achieving the exciting idea made above. Strong-correlation physics, including superconducting phase, correlated insulating phase, and orbital magnets, would manifest because of the moiré flat band's variable filling. However, applying appropriate external pressure helps to reduce the interlayer of TBG and relax the twisted-angle criteria, at which the eigenvalue spectrum exhibits a flat band at the Fermi energy.

A diverse platform is made possible by the growth and success of TBG, which also sparks significant worry about the orbital ferromagnetism produced by electric current loops as opposed to intrinsic spin magnetic moments. First, a discussion of the magnetism in graphene nanostructures that depends on size, shape, and edge-topological states is presented. Such features may be further controlled by structural engineering, using an external electric field, and chemical doping. Second, we describe the magnetism caused by in-plane graphene modification, including defect engineering, sp^3 functionalization, and heteroatom adsorption. Finally, and especially, the topic of orbital magnetism in TBG, which broadens the idea of carbon magnetism, is briefly discussed.

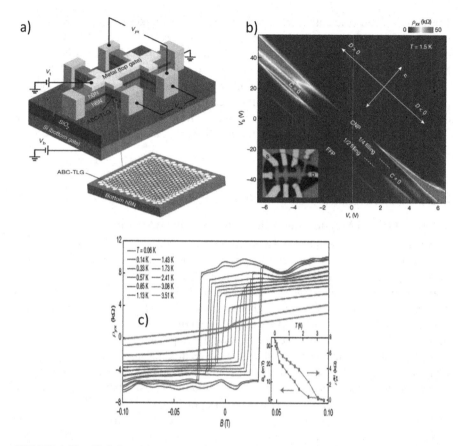

FIGURE 4.18 (a) Schematic of the dual-gated ABC-TLG/hBN moir superlattice Hall bar device and measurement configuration. The inset shows that the moiré pattern exists between ABC-TLG and bottom hBN. (b) Plot of the longitudinal resistivity ρ_{xx} as a function of V_t and V_b at $T = 1.5$ K. The arrows show the direction of changing doping n and displacement field D, respectively. (c) Magnetic field–dependent ρ_{yx} at 1/4 filling and $D = -0.5$ V/nm at different temperatures. The Hall resistivity displays a clear anomalous Hall signal with strong ferromagnetic hysteresis [41].

Q. Calculate the electronic scattering matrix and the tunneling probability in quantum wells.

MAIN PROBLEM DISCUSSED IN THIS CHAPTER

In Chapter 4, we introduce spins to the defect centers and study spin–spin (electron–electron) interactions. The spins associated with the magnetic impurity centers interact with the spins of the conduction electrons. Hence, we consider a spin-triplet state responsible for the transition of the conductivity (with temperatures).

WHAT HAS BEEN ACHIEVED?

In Chapter 4, we introduce spins to the defect centers and study spin–spin (electron–electron) interactions. The spins associated with the magnetic impurity centers interact with the spins of the conduction electrons. Hence, we consider a spin-triplet state responsible for the transition of the conductivity (with temperatures).

WHAT HAS NOT BEEN ACHIEVED?

In Chapter 4, one of the most important questions arises whether pure carbon structures can be ferromagnetic. Defects in carbon structures can change the properties of carbon, but we do not know the structure of the defective regions. We cannot control the defect structures as well as ferromagnetic doping.

BIBLIOGRAPHY

1. M. Bockrath, D. H. Cobden, J. Lu, A. G. Rinzler, R. E. Smalley, L. Balents and P. L. McEuen, Luttinger-liquid behaviour in carbon nanotubes, *Nature* 397, 598 (1999).
2. S. Ncube, G. Chimowa, Z. Chiguvare and S. Bhattacharyya, Realizing one-dimensional quantum and high-frequency transport features in aligned single-walled carbon nanotube ropes, *Journal of Applied Physics* 116, 024306 (2014).
3. G. Chimowa, S. Ncube and S. Bhattacharyya, Observation of one-dimensional plasmonic features in single walled carbon nanotube bundles excited by high frequency signals, *EPL* 111, 3 (2015).
4. R. Leturcq, C. Stampfer, K. Inderbitzin, L. Durrer, C. Hierold, E. Mariani, M. G. Schultz, F. von Oppen and K. Ensslin, Franck–Condon blockade in suspended carbon nanotube quantum dots, *Nature Physics* 5, 327–331 (2009).
5. H. W. Ch. Postma, T. Teepen, Z. Yao, M. Grifoni, and C. Dekker, Carbon nanotube single-electron transistors at room temperature, *Science* 6, 293 (2001).
6. Z. Yao, H. W. Ch. Postma, L. Balents, and C. Dekker, Carbon nanotube intramolecular junctions, *Nature* 402, 273–276 (1999).
7. P. Jarillo-Herrero, J. Kong, H. S. J. van der Zant, C. Dekker, L. P. Kouwenhoven and S. De Franceschi, *Nature* 434, 484–488 (2005).
8. J. Nygård, D. H. Cobden and P. E. Lindelof, Kondo physics in carbon nanotubes, *Nature* 408, 342–346 (2000).
9. J. Paaske, A. Rosch, P. Wölfle, N. Mason, C. M. Marcus, and J. Nygård, Non-equilibrium singlet-triplet Kondo effect in carbon nanotubes, *Nature Physics* 2, 460–464 (2006).
10. J.-D. Pillet, C. H. L. Quay, P. Morfin, C. Bena, A. L. Yeyati and P. Joyez, Revealing the electronic structure of a carbon nanotube carrying a supercurrent, *Nature Physics*. 6, 965–969 (2010).
11. J.-P. Cleuziou, W. Wernsdorfer, V. Bouchiat, T. Ondarçuhu, and M. Monthioux, Carbon nanotube superconducting quantum interference device, *Nature Nanotechnology* 1, 53–59 (2006).
12. R. M. Potok, I. G. Rau, H. Shtrikman, Y. Oreg, and D. Goldhaber-Gordon, Observation of the two-channel Kondo effect, *Nature* 446, 167–171 (2007).
13. V. Mourik, K. Zuo, S. M. Frolov, S. R Plissard, E. P. Bakkers, and L. P. Kouwenhoven, Signatures of Majorana fermions in hybrid superconductor-semiconductor nanowire devices, *Science* 336, 1003–1007 (2012).

14. C. Gómez-Navarro, P. J. De Pablo, J. Gómez-Herrero, B. Biel, F. J. Garcia-Vidal, A. Rubio, and F. Flores, Tuning the conductance of single-walled carbon nanotubes by ion irradiation in the Anderson localization regime, *Nature Materials* 4, 534–539 (2005); A. Bachtold, C. Strunk, J. -P. Salvetat, J. -M. Bonard, L. Forró, T. Nussbaumer, and C. Schönenberger, Aharonov–Bohm oscillations in carbon nanotubes, *Nature* 397, 673–675 (1999); V. Skalakova, A. B Kaiser, Y.-S. Woo, and S. Roth, Electronic transport in carbon nanotubes: From individual nanotubes to thin and thick networks, *Physical Review B* 74, 085403 (2006).

15. R. Schleser, T. Ihn, E. Ruh, K. Ensslin, M. Tews, D. Pfannkuche, D. C. Driscoll, and A. C. Gossard, Cotunneling-mediated transport through excited states in the coulomb-blockade regime, *Physical Review Letters*, 94, 206805–206807 (2005).

16. H. Imamura, S. Takahashi, and S. Maekawa, Spin-dependent Coulomb blockade in ferromagnet/normal-metal/ferromagnet double tunnel junctions, *Physical Review B* 59, 6017–6020 (1999).

17. S. Takahashi and S. Maekawa, Effect of Coulomb blockade on magnetoresistance in ferromagnetic tunnel junctions, *Physical Review B* 80, 1758–1760 (1998).

18. K. Kenmochi, K. Sato, A. Yanase, and H. Katayama-Yoshida, Materials design of ferromagnetic diamond, *Japanese Journal of Applied Physics* 44, L51–L53 (2005).

19. P. Seneor, B. Dlubak, M.B. Martin, et al. Spintronics with graphene. *MRS Bulletin* 37, 1245–1254 (2012).

20. Z. Remes, S.-J. Sun, M. Varga, H. Chou, H.-S. Hsu, A. Kromka, and P. Horak, Ferromagnetism appears in nitrogen implanted nanocrystalline diamond films, *Journal of Magnetism and Magnetic Materials* 394, 477 (2015).

21. T. L. Makarova, Magnetic properties of carbon structures, *Semiconductors* 38, 615–638 (2004).

22. Y. Wang, Y. Huang, Y. Song, X. Zhang, Y. Ma, J. Liang, and Y. Chen, Room-temperature ferromagnetism of graphene, *Nano Letters* 9, 220–224 (2009).

23. D.-C. Yan, S.-Y. Chen, M.-K. Wu, C.-C. Chi, J. H. Chao, and M. L. H. Green, Ferromagnetism of double-walled carbon nanotubes, *Applied Physics Letters* 96, 242503 (2010).

24. H. Ohldag, P. Esquinazi, E. Arenholz, D. Spemann, M. Rothermel, A. Setzer, and T. Butz, The role of hydrogen in room-temperature ferromagnetism at graphite surfaces, *New Journal of Physics* 12, 123012 (2010).

25. J. Zhou, Q. Wang, Q. Sun, X. S. Chen, Y. Kawazoe, and P. Jena, Ferromagnetism in semihydrogenated graphene sheet, *Nano Letters* 9, 3867–3870 (2009).

26. I. S. Mosse, V. R. Sodisetti, C. Coleman, S. Ncube, A. S. de Sousa, R. M. Erasmus, E. Flahaut, T. Blon, B. Lassagne, T. AamoAil, and S. Bhattacharyya, Tuning magnetic properties of a carbon nanotube-lanthanide hybrid molecular complex through controlled functionalization, *Molecules* 26, 563 (2021).

27. V. R. Sodisetti, S. Ncube, C. Coleman, R. M Erasmus, E. Flahaut, and S. Bhattacharyya. Strong spin-phonon coupling in gd-filled nanotubes, *Journal of Applied Physics* 130(21), 214301 (2021).

28. S. Ncube, C. Coleman, C. Nie, P. Lonchambon, A. Strydom, E. Flahaut, A. de Sousa, and S. Bhattacharyya, Modification of magnetic properties in gadolinium chloride filled multiwall carbon nanotubes, *Journal of Applied Physics* 123, 213901 (2018).

29. M. Popinciuc, C. Józsa, P. J. Zomer, N. Tombros, A. Veligura, H. T. Jonkman, and B. J. van Wees, Electronic spin transport in graphene field-effect transistors, *Physical Review B* 80, 214427 (2009).

30. A. Candini, S. Klyatskaya, M. Ruben, W. Wernsdorfer, and M. Affronte, Addressing a single molecular spin with graphene-based nanoarchitectures, *Journal of Nano Letters* 11, 2634–2639 (2011).

31. R. Martel, T. Schmidt, H. R. Shea, T. Hertel, and Ph. Avouris, Single- and multi-wall carbon nanotube field-effect transistors, *Applied Physics* 73, 17 (1998).
32. M. Urdampilleta, S. Klayatskaya, M. Ruben, and W. Wernsdorfer, Magnetic interaction between a radical spin and a single-molecule magnet in a molecular spin-valve, *ACS Nano* 9, 4458 (2015).
33. F. Kuemmeth, S. Ilani, D. Ralph, et al., Coupling of spin and orbital motion of electrons in carbon nanotubes, *Nature* 452, 448 (2008).
34. E. A. Laird, F. Kuemmeth, G. A. Steele, K. Grove-Rasmussen, J. Nygård, K. Flensberg, and L. P. Kouwenhoven, Quantum transport in carbon nanotubes, *Reviews of Modern Physics* 87, 703 (2015).
35. M. del Carmen Giménez-López, F. Moro, A. La Torre, et al., Encapsulation of single-molecule magnets in carbon nanotubes, *Nature Communications* 2, 407 (2011).
36. G. Zhang, Superconducting ferromagnetic nanodiamond, *ACS Nano* 11, 5358 (2017).
37. S. Arikawa, A. Shimizu, D. Shiomi, K. Sato, and R. Shintani, Synthesis and isolation of a kinetically stabilized crystalline triangulene, *Journal of the American Chemical Society* 143, 19599 (2001).
38. L. R. Radovic and B. Bockrath, On the chemical nature of graphene edges: Origin of stability and potential for magnetism in carbon materials, *Journal of the American Chemical Society* 127, 5917–5927 (2005).
39. R. Li, M. Zhang, X. Fu, J. Gao, C. Huang, and Y. Li, Research of low-dimensional carbon-based magnetic materials, *ACS Applied Electronic Materials* 4, 3263–3277 (2022); A. Lenin, P. Arumugam, and R. Shanmugham, A. Sonachalam, S. Paramasivam, A. P. Rao, G. Singaravelu, R. Venkatesan, Hybrid, Ni-Boron nitride nanotube magnetic semiconductor-a new material for spintronics, *ACS Omega* 5, 32, 20014–20020 (2020); M. Gao, X. Han, W. Liu, Z. Tian, Y. Mei, M. Zhang, K. C. Paul, E. Kan, T. Hu, Y. Du, S. Qiao, and Z. F. Di, Graphene-mediated ferromagnetic coupling in the nickel nano-islands/graphene hybrid, *Science Advances* 7, 7054 (2021); R. Li, X. Li, M. J. Zhang, Y. Li, Z. Yang, and C. S. Huang, A universal Fe/N incorporated graphdiyne for printing flexible ferromagnetic semiconducting electronics, *The Journal of Physical Chemistry Letters* 12, 204–210 (2021).
40. M. Pizzochero and E. Kaxiras, Imprinting tunable π-magnetism in graphene nanoribbons via edge extensions, *The Journal of Physical Chemistry Letters* 12, 1274 (2021).
41. G. Chen et al., Tunable correlated Chern insulator and ferromagnetism in a Moiré superlattice, *Nature* 579, 56 (2020).

5 Carbon Superstructures
Diamond Meets Graphene

5.1 DIAMOND TRANSISTORS

Although carbon structures possess the highest level of mechanical strength, sound wave velocity, and thermal conductivity, standard microelectronic device made of carbon has remained challenging. Nevertheless, carbon-based quantum device technology is attempting to supersede the current computing technology and meet the demand for handling massive amounts of information that can also be accessed at high speed. In the previous chapter we see the possibility for true one-dimensional transport in the high-frequency regime CNTs which has rarely been observed in other materials. Further high-speed switching capabilities of diamond (the hardest material) and graphene (the thinnest material) in the form of resonant tunnel devices operating in the gigahertz range are shown that find potential applications in deep space astronomy (in Chapters 7 and 8) (Figures 5.1 and 5.2). Ultimately, heavily boron-doped diamonds showing spin-triplet-like superconducting transition can offer a new class of quantum bits or spin qubits which is expected to play a key role in the development of quantum artificial intelligence.

- **Diamond:** Power transistor at high temperatures and high frequencies.
- High mobility (>4000 cm²/(Vs)), long carrier lifetime, and high breakdown voltage

Single-crystal diamond is known for their highest density. Polycrystalline and nanocrystalline diamond films containing large densities of sp^2 carbon layers can show very high hardness. Some phases of carbon mixed with nitrogen are claimed stronger than diamond. One of the most important aspects of 3D structure and thin films is the hardness (and stability) over a large area that cannot be utilized in 1D or 2D materials. The hardness of a structure is not directly related to transport properties, but it has been observed that an applied pressure could induce superconductivity in fullerenes, boron-doped diamonds, and graphene with shear stress. Carbon structures including polymers go to a metallic phase transition by applying pressure and into the superconducting phase. Pressure-induced transformation of graphitic carbon into diamond is well known. Application of pressure-induced electrical transport properties in 1D and 2D systems can be found very difficult due to sample handling problems. Therefore, we use a 3D structure where a zero bandgap carbon layer is pressed by two layers of gapped (diamond-like) carbon having large compressive force. Hence, a quantum well or a multilayer superlattice (SL) structure is formed based on a single element, carbon. Before we show the deep potentials of quantum well (QW) structures, examples

 DOI: 10.1201/9781003316411-5

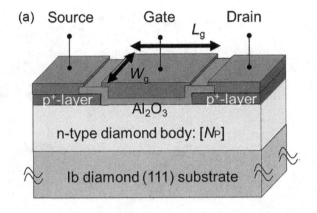

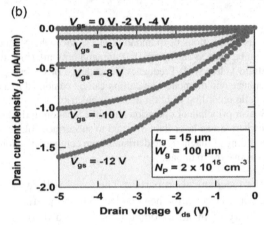

FIGURE 5.1 (a) Schematic of the cross-sectional structure of inversion channel diamond MOSFET on a high-pressure high-temperature (HPHT) synthetic Ib (111) semi-insulating single-crystal diamond substrate. (b) I_d–V_{ds} characteristics of the inversion channel diamond MOSFET with $N_P = 2 \times 10^{15} cm^{-3}$, $L_g = 15\,\mu m$, and $W_g = 100\,\mu m$ at RT. The applied V_{gs} and V_{ds} ranged from 0 to 12 V with a voltage step of 2 V and from 0 to 5 V with a voltage step of 0.1 V, respectively. Inversion channel mobility and interface state density of diamond MOSFET using N-type body with various phosphorus concentrations [56].

of shallow potentials such as doped/functionalized or corrugated graphene followed by graphene quantum dot clusters can be presented. Like a QW, these structures can show resonant tunneling features. An addition of a defect (or a vacancy) results in a potential step that can be considered an extra (2+1) dimension.

5.1.1 Defects in Graphene

The defects in real solids have different forms, and they vary from vacancies (due to the removal of ions), impurities (presence of extra charged ions), and atomic dislocations. The vacancies, as well as impurities, are responsible for most of the observed electrical as well as optical properties, while the dislocation of atoms has a more

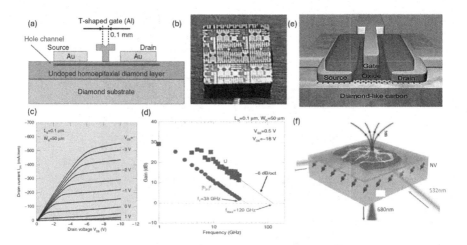

FIGURE 5.2 (a) Various applications of diamond in electronics where undoped homoepi-taxial diamond layers can be used as substrates (left: top and bottom). (b) A diamond device (c) Drain current vs. drain voltage. (d) Frequency dependence of current gain ($|h_{21}|^2$) and power gain (U). (e) Graphene and diamond-like carbon can be combined in a device [57]. (f) Microscopic-scale magnetic recording of brain neuronal electrical activity using a diamond quantum sensor (top) Sensor principle of operation and slice location in the brain. Schematic of the sensor operation where green laser light directed to subsurface colour centres (NV) in the diamond enables recording of magnetic field arising from compound action potentials in a brain tissue slice placed above the diamond.

prominent effect on the mechanical properties. The focus will be on vacancies and impurities as they have a stronger influence on electronic properties.

Given the impossibility of avoiding defects in solids, a theory by Hohenberg, Mermin, and Wagner involving some simple calculations of the Gibbs free energy of a system shows that as few defects as possible should be present in any solid to minimize the free energy. Moreover, the interplay between the dimensionality of the system and the thermodynamic equilibrium has a very important role in form-ing surface defects, where the energy required to form one of those defects at the thermal equilibrium is proportional to the linear dimensions of the sample. In the case of graphene, surface exposure is another factor that gives rise to the forma-tion of defects. In addition to that, the presence of phonons plays a similar role in creating surface defects as well as topological defects such as edge states and Stone–Wales type.

The realization of the importance of defects in graphene particularly and solids in general has led scientists to investigate the effect of defects, both experimentally and theoretically (Figures 5.3–5.5). For instance, techniques such as irradiation were used to create artificial defects in samples of graphene. Intense theoretical studies were carried out by different groups around the world to model defects and develop a theoretical understanding of the experimental results (see the references therein). Appreciable progress has been made since the first isolation of this noble material.

However, in previous work, situations where a carbon atom is removed, a vacancy is created, or a charged impurity is substituted were considered. Numerical calculations were performed using the tight-binding model and the recursive Green's function technique which are also employed in this work to study the electronic properties of defective 2D carbon systems. In the presence of vacancies, it was found that new states are formed at the Fermi level. Analysis of the inverse participation ratio (IPR) has revealed that those states are quasi-localized. Consequently, a gap opens near the Dirac points introducing the possibility of controlling the electronic properties using an applied voltage.

From a theoretical point of view, wave functions with large amplitudes at zero eigenenergies, called zero modes, arise when considering the propagation of the wave in a medium with unusual topological defects, which are vacancies in this study (Figure 5.4). Such a situation shows up in different fields of physics for example field theories (where applications in the chiral symmetry breaking in (1+1)-space-time QED have been shown), anti-phase boundaries in narrow-gap semiconductors, edge states in nano-graphite ribbon junctions, and many other examples in many different fields.

In the case of graphene, an analytical form for the wave function associated with zero eigenvalues was found, where the wave function decays as the electrons move away from the vacancy site. However, the position of the newly created states is exactly at the Fermi level, which has a very interesting consequence in

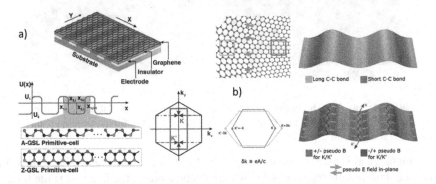

FIGURE 5.3 (a) (top) The diagram shows graphene superlattice structure. The electrostatic potential profile is changed by an array of back gate electrodes (bottom left). The corresponding atomic pattern in a primitive cell of the A-GSL and Z-GSL configurations is presented. The first Brillouin zone of the A-GSL configuration has two K points (bottom right) [58].
(b) Strained graphene is created by periodic pseudo electric and magnetic fields. High(low) density of carbon atoms and hence electrons are created in regions due to a strain gradient (upper left). This inhomogeneous charge distribution results in an electric field. Stretching of bonds cause the Dirac cones at K and K' points and shift symmetrically from their original unstrained positions in the reciprocal space (lower left). This creates pseudo-magnetic fields with opposite signs at the two valleys. The strain associated with rippling creates rare and dense regions in the graphene, effectively acting as two different materials in a superlattice (upper right). Pseudo-fields form near the interfaces of these "materials," both electric and magnetic are created (lower left). The up and down magnetic fields are separated by only few nanometers, same order as the magnetic length, making the individual Landau levels to interact. LDoS peaks are maximized at the ripple crests and troughs, where valley polarized snake states (curved lines) are also expected to form due to the reversal of the pseudospin dependent pseudo-magnetic fields across these lines [58].

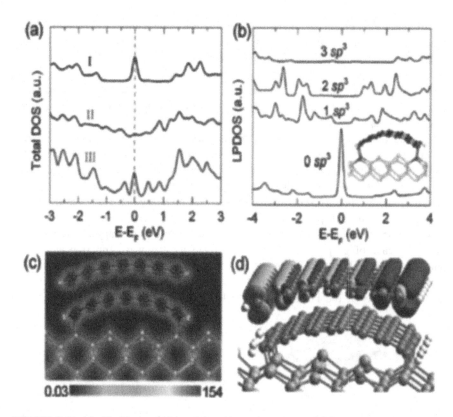

FIGURE 5.4   (a)–(d): The variation of the electronic structure of flat and curved graphene incorporated with sp³-bonded carbon atoms shows the possibility for the opening of an energy band gap at the Fermi level [59].

the topological sense; the states have to be localized at one lattice site. When the particle-hole symmetry is broken, two effects take place: the first one is the broadening of the states created at the Fermi level, and the second one is shifting the peak position by an amount that is approximately equal to the value of the second nearest hopping energy (denoted as t).

The second kind of defect, as mentioned previously, is impurities or charged particles (Figures 5.3 and 5.5). Boron, as well as nitrogen, is the common impurities in carbon allotropes. The difference between an impurity and a vacancy in terms of the tight-binding method is that the vacancy does not couple to the nearest neighbors. On the other hand, the impurity couples to its nearest neighbor with a coupling (hopping) energy which is different from the one between carbon sites due to the difference in the radii of the foreign atoms, and it is equal to t_p which is considered to be a free parameter in this work. In comparison to the work done, the on-site energy of the impurity is taken to be a non-zero value depending on the atom substituted, where it was zero. It is measured relative to the on-site energy of the carbon atoms in graphene (0 eV); for boron, it is taken to be 2.5 eV, while it is approximately −2.5 eV for nitrogen, which can be reliable in understanding some of the properties of nitrogenated graphene.

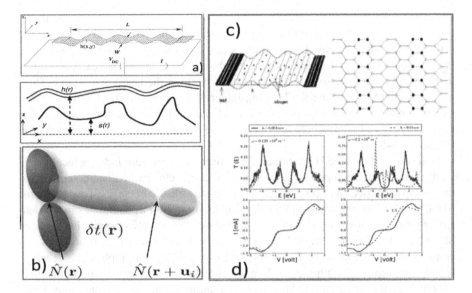

FIGURE 5.5 (a) Structure of rippled graphene is shown where the potential distribution is regular and irregular. (b) The orbitals are overlapping in the disordered structures. (c) Rippled graphene nanoribbons decorated with nitrogen atoms. (d) Transmission coefficient $T(E)$ and the corresponding currents for nitrogen strips incorporated in rippled graphene were calculated with C-N coupling $0.9t$ and N-N coupling $0.1t$, where t represents the hopping term for C-C bonds [60].

Another important feature of graphene is that it shares the properties of thin membranes, which are a third kind of defect. Curvatures in graphene are induced by many factors; for instance, graphene layers are usually placed on a SiO_2 substrate which leads to observations of ripples through scanning tunneling microscopy (STM). Moreover, the ripples were also observed using transmission electron microscopy (TEM) in a suspended layer of graphene. Regarding the effect of the substrate, it was also reported that imaging graphene using a single electron transistor (SET) showed electron-hole inhomogeneity limiting the transport. Such interpretation of this charge inhomogeneity is due to the remote effect of charged impurities on the substrate, which is also considered a key factor in affecting the transport properties of graphene. However, recent microscopic calculations showed that the curvature on the graphene surface may act as an additional source of charge inhomogeneity. In addition to this, the problem of Dirac fermions moving on a curved surface is analogous to problems in quantum gravity (Figures 5.3 and 5.4).

More interestingly, soft membranes gain energy due to bending, which might come from suspension on a solvent, but since graphene is an electronic membrane, extra energies have to be taken into account such as the elastic free energy that accounts for energies due to van der Waals' forces. This may appear between the sample and the substrate as attractive or repulsive forces, where the energy may arise from surface tension. Therefore, all these properties make graphene the first electronic membrane to be attached to leads and characterized. To understand the effect of corrugation on the electronic structure of graphene, one can think of an ideal layer of graphene where sp^2 orbitals are aligned to the xy-plane, while p_z orbitals are perpendicular to this plane. Then, curvature introduced to the system spoils the ideal picture of the orthogonal orbitals locally. For instance, when considering the first ideal scenario, the sp^2 orbital is not hybridized with the p_z orbitals, but in the second case, one infers that p_z is no longer perpendicular to the sp^2 orbitals of the nearest p_z orbital. Thus, it is convenient to redefine the hopping integrals (energies); instead of having one constant value t for the hopping between all sites in the cell, it is replaced with a hopping energy $\tilde{t} = t + \delta t$ where δt is a correction that depends on the local curvature (Figures 5.3–5.5).

In terms of the Dirac Hamiltonian formalism, the curvature induces local gauge fields that have two components; one scalar called the electrochemical potential, and the other one called the vector potential. The gauge fields in carbon-based systems like CNTs lead to suppression of weak localization as well as anomalies in the density of states. Details of the computational method of $\tilde{t}$ as well as the relation between Slater–Koster formalism (tight-binding formalism), and how gauge fields with two components arise. Furthermore, another interesting feature of graphene is the vibrational modes (phonons) Figure 5.5. In graphene, there are off-plane modes, and they are considered another cause of bending, accounting for the lack of long-range order which results in the formation of wrinkles as well as a crumbling effect.

5.1.2 GRAPHENE SUPERSTRUCTURES: TRANSPORT IN DEFECTIVE GRAPHENE

Heterostructures of graphene (either on an insulating substrate or forming a multilayer arrangement aligned vertically) have appeared to be important for practical devices aspects particularly, in fast switching devices based on the negative differential resistance (NDR) features. Although a large modulation of the bandgap in pure graphene seems to be difficult to produce several attempts have been made to open up the bandgap in graphene by creating nanostructures (e.g., vacancy clusters, quantum dots) or by artificially creating high barriers in the structure (e.g., curvatures or ripples) (Figure 5.5a and b). An array of artificial (periodic) barriers has been created in graphene by nitrogen incorporation in addition to creating vacancy centers. It was found that the NDR features could be controlled more effectively due to the interplay between the potential barrier created by the ripple and nitrogen strips. Prominent NDR features are seen for small heights and frequencies of ripples (Figure 5.5c) are depleted with the increase in height and frequency. The NDR features may be intensified by increasing the N–N coupling in the nitrogen strips and reducing the coupling strength of the C–N bonds. For the low frequency of the ripples, transmission coefficient ($T(E)$) and NDR features were approximately unaffected by the variation of the height, nevertheless, by increasing the frequency of the ripples the resonant states were found to be shifted to the lower energy regime with strong asymmetric features around the Fermi level (E_F). The calculated I–V curves (Figure 5.5c) show the weak NDR signature at relatively low biases which are different from a flat layer or rippled graphene with a distribution of nitrogen centers. A backbone of carbon can be constructed with the help of flexible graphene ribbons incorporated with sp^3 bonds which can change the electronic as well as the transport properties of the graphene ribbons significantly (Figure 5.5a–c).

5.1.3   MODEL: TRANSPORT THROUGH CARBON CLUSTERS

The incorporation of impurities in semiconductors, for example, carbon, silicon, and GaAs, can form clusters with a wide distribution in size and distance depending on the diffusion of the dopants at different energies. Resonant tunneling features due to impurity clusters embedded in an insulating matrix have been examined through the interplay between the size of the clusters and the intercluster distance (Figure 5.5a). Constructive interference phenomena were tuned through a systematic study of different geometrical configurations, thereby controlling confinement in quasi-bound states. The defects in a vertical row can be formed by high-energy ion implantation, which creates channels in carbon films. Gaussian trap potentials have been used to simulate the imperfect barrier-well interface associated with the disordered materials (Figure 5.7a). The horizontal distribution of the defects can be formed by diffusion of impurity (N) with controlled energy. The single-channel horizontal configuration is conducive to constructive quantum interference and consequently shows a strong NDR peak (Figure 5.7b). Constructive interference can also manifest in the vertical configuration, although this occurs at smaller wave vectors than in the horizontal configuration. This means the NDR features can only be achieved in ordered materials or very weakly disordered materials. In the horizontal configuration, breaking the symmetry by shifting one of the quantum clusters in the vertical direction decreases the NDR peak. As the asymmetry of the array increases, the constructive interference decreases so the strongly localized states become weakly localized. However, for both cases, either equal or different sizes of clusters can be formed where equal distributions do not produce NDR features for devices at least, similar to the previously observed SL structures.

Apart from the idealized horizontal configuration, arbitrary cluster arrangements were investigated in the form of triangular cluster configurations for implementation

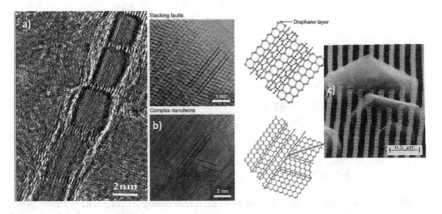

FIGURE 5.6   Various interfaces of low-dimensional structures of nanocrystalline diamonds. (a) HRTEM image shows microstructure of nanodiamond films which includes segments composing these nanowires [61]. (b) nanostructural elements found in a variety of naturally affected and laboratory-shocked diamonds, together with accompanying schematics of atomic structures including complicated patterns of novel nanotwin kinds and cubic–hexagonal sp^3–bonded stacking faults [61]. (c) Diamond nanocrystals are grown in combination with silicon nanowires.

in materials where the arrangement of impurity clusters may not be precisely controlled. Strongly localized states can be formed successfully despite weak the disorder by breaking the symmetry in the horizontal configuration. Since it is difficult to make either a perfectly horizontal or vertical configuration, in most cases, a triangular distribution of impurities can be expected (Figure 5.7c and d). Triangular configuration can be treated as a kind of very non-linear structure that is different from the 1D polymeric chains in which the nearest hopping can be considered. This

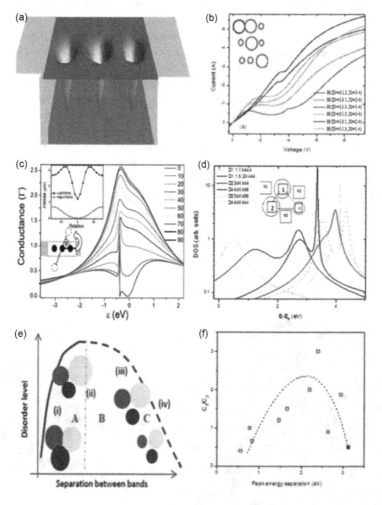

FIGURE 5.7 (a) Designing quantum dots of different configurations which shows how to control tunnel currents in graphitic carbon clusters. (b) Variation of current with voltage for horizontal configuration of different sizes and distances shows negative differential resistance features. (c) Variation of conductance with energy when dots in the horizontal configuration are rotated into vertical configuration (d) Variation of DOS with energy for asymmetric dot configuration. (e) The proposed growth model of the films through clusters of different sizes and configurations. (f) The disorder level described by the ratio of configuration parameters C_1/C_2 varies non-linearly with the energy separation of the resonant peaks [62].

analysis suggests that even in the non-ideal case, useful devices can still be engineered based on the short trapping time of the disordered configuration. The representation of the potential in a Gaussian form applies to weakly disordered systems where sharp discontinuities might not be possible, so this study can be extended to disordered quantum dot systems. The triangular configuration was studied as it represents a disordered system, albeit symmetric. The triangular configuration does not show prominent NDR features, although it can show step-like features when the correct parameters are chosen. Quantum structures of graphene can be made to develop a waveguide structure. The dots can be linked through barriers whose length is smaller than the mean free path of the electrons. NDR features can be tuned through varying configurations. A periodic potential array can show NDR features. By breaking the symmetry some (weak) NDR features can also be produced. This theoretical model can be correlated with the real growth model of carbon films and graphene (Figure 5.7e and f). The tuneable transport properties including resonant tunneling in amorphous carbon can be reproduced by adjusting the coupling between sp^2-bonded, finite-sized clusters of carbon and defect centers therein. A wide range of effective bandgaps can be created through the proper choice of cluster size and intermediate distance in the direction of wave propagation. The distribution of the interdot coupling constants identifies the optimum configurations for controlling the lifetime of carriers in these devices and realizes a carbon waveguide structure.

Several configurations of clusters having different sizes and separation forming triangles are considered systematically (Figure 5.7e). On the left part of the diagram, three large clusters are touching each other which means the intercluster coupling is maximum and the disorder effect is minimum (labeled as A and (i) in Figure 5.7e). In the middle part (labeled as B and (ii) in Figure 5.7e), two clusters are smaller than the third one which is represented as weak intercluster coupling. Due to the difference in the size clusters, the disorder level increases. In parts B–C or (iii)–(iv) of Figure 5.7e, this coupling becomes weaker, and for the small size clusters (in part C) which are well separated, the coupling becomes very weak. However, the disorder level can decrease due to the equal size of clusters. The coupling between clusters 1–2 and 2–3 denoted by C_1 and C_2, and their ratio determines the effect of the disorder level of the system. The parameter C_1/C_2 is plotted as a function of energy separation between the sharp and the broad peaks of the DOS spectra (Figure 5.7f). From the analysis of the spectra, a non-linear trend is realized which shows the increase of the disorder level (C_1/C_2) with the peak separation reaching a maximum at about 2 eV followed by a decrease. This model can explain the experimental observation of the disorder vs. optical gap in many amorphous carbon systems.

Similar configurations of QDs can be designed by a set of three qubits and can be simulated by a quantum computer (see Chapter 10). Interactions between quantum dots (or qubits) are created by changing the coupling between the qubits to produce resonant effects in the probability of electron transmission. This effect can be seen from the structure (the height and the width of the resonant peaks and their separations). Features of resonant tunneling are described in Sections 5.3 and 5.4.

5.1.4 QUANTUM WELL

A quantum well is a type of potential well that consists of discrete energy levels. The classical method used for the demonstration of a quantum well involves the confinement of particles that could move from three dimensions to two dimensions by forcing the particles to occupy a planar region. Quantum confinement becomes more noticeable as the thickness of the quantum well becomes like the de Broglie wavelength of the carriers which causes the formation of energy subbands which implies that the carriers will have discrete energy values.

Quantum well devices have been developed using the theory of quantum wells. Such devices are much faster, cheaper to operate, and are extremely important both for technological and telecommunications industries. Due to these reasons quantum well, devices are replacing conventional devices in many electrical devices. A simple quantum well system can be constructed by inserting a thin layer of a narrow gap semiconductor between two layers of wider band gap semiconductor. An example of this is having two layers of AlGaAs which have a large band gap and a thin layer of GaAs which has a small band gap. It is assumed that the potential well is along the z direction with no confinement. A quantum well will be created in the GaAs layer since the band gap of the GaAs layer is smaller than that of the AlGaAs layers. This difference can be interpreted as the change in potential felt by a carrier. Carriers can only exist in discrete states in a quantum well. An electron with low energy can become trapped in a quantum well. Holes on the other hand are trapped within the top of potential wells in the valence band.

5.1.5 MOTIVATION FOR CARBON SUPERLATTICE

Information science depends on the (i) availability of quantum states (density), (ii) connectivity between states, and (iii) lifetime of states (speed). The information stored in the potential wells should be neither too deep nor too shallow. The material should have high mobility of carriers for fast communication. Carbon as a rigid and strong material in general has several advantages in handling or manipulation. Rather than combining with other light elements to form a stable solid object carbon atoms find their counterpart within other carbon atoms hence superstructures of carbon can be formed. One can imagine that carbon atoms can change their configuration(s) from sp^2C to sp^3C and back to sp^2C by utilizing the available energy differentiating these two phases. One structure can complement its counterpart easily by controlling the disorder level. However, the main disadvantage is the structural disorder that arises from the flexibility of carbon-carbon bonds as well as their ability to accept many forms of the disorder. This material (burnt off into gaseous CO_2) may not be compared with other semiconductors (e.g., Si) since the oxide form of silicon is compatible with Si and can be utilized as a gate dielectric layer. Tunnel structures made from organic molecules have some real applications by overcoming the effect of the disorder. Can it be possible to make resonant tunnel devices using thin layers of carbon which are intermediate to molecules and solid-state devices? It seems one has to find the ultimate or finest elementary quasi-1D structure of (pure) carbon which can act as a backbone for the hybrid structure and carry information over a long path.

Stimulated by the high magnetic field and high frequency, it is possible to carry information at a very high speed unlike in bio-organic molecular structures and also can couple to the environment. We would like to search for the microstructural unit of carbon by comparing carbon SL structures of different forms; artificially created or in natural systems which might be extended in other related hybrid/organic/ molecular and even in biological systems.

To develop the proposed carbon SL structures in a conventional way, researchers have been searching for the solutions to the following problems: (i) opening an energy bandgap (even very narrow) in graphene or related materials, (ii) reducing the bandgap of diamond effectively by introducing delocalized states instead of localized states, and (iii) controlling the interface between the low bandgap and high bandgap carbon layers to form a stable hybrid structure. A few examples of two-dimensional graphene-diamond heterostructures (Figure 5.4a–d) and one-dimensional structures (Figure 5.6) may be used in describing the experimentally observed microstructures of nanocrystalline diamond films. At present, a comprehensive theoretical analysis of the structure theoretically appears to be challenging. To build up this structure one must understand different steps, such as a quasi-one-dimensional structure and how to modulate the bandgap by combining sp^3 and sp^2 structures of carbon. We reckon that the sp^2C requires intensive study particularly, in the context of controlling the disorder (and opening a gap) within it. In the next subsection, the electronic properties of q-1D sp^2C structures mixed with sp^3C structures and nitrogen atoms will be reviewed.

5.1.6 MULTILAYERED SUPERLATTICE STRUCTURES OF THE DISORDERED CARBON

Multilayered heterostructures of amorphous carbon (a-C) films have been studied in different directions like photodiode effect, electroluminescence, for optical communications, infrared detectors, hard protective coating, and other tribological applications that are sensitive to the period thickness of carbon multi-layers. Photoconductivity of multilayered a-CN_x films was also found to be very interesting and may open future applications in making heavy particle ion detectors. SL structures fabricated using high (2.8 eV) and low (1.5 eV) bandgap-modulated DLC thin-film layers exhibited some evidence of quantum size effects. A blue shift in the optical gap was found to occur with decreasing well width from which an effective mass for the electrons was estimated. Multilayer sp^2/sp^3 alternate structures have been directly observed from high-resolution TEM. A low electron (tunnel) effective mass in multilayer carbon systems was suggested. From optical absorption spectroscopy, an increase of the bandgap up to a few hundred meV with the reduction of the well width has been noticed as in previous reports. This is well supported by the temperature-dependent conductivity and the activation energy data. However, direct observation of negative differential conductance in multilayered carbon films remained difficult. Therefore, a double-barrier QW carbon structure was constructed by laser ablation (given in Chapter 2).

A double-barrier resonant tunnel diode (DB-RTD) based on amorphous carbon was attempted and a prominent signature of resonant tunneling with NDR has been shown together with quantized conductance and memory switching (Figure 5.8a–d).

A strong NDR signal with the highest peak-to-valley current ratio of about 2:1 and a maximum value of about $-0.1\,mA/cm^2$ at 77 K is recorded (Figure 5.8b). From the analysis of the tunneling current, the position of the resonant peaks was found to match well with the theoretically predicted values. The half-width at half-maximum of the NDR peak, a measure of the resonant state lifetime, is found to be about ~75 meV and the value of $m^* \sim 0.07 m_e$ estimated from this analysis yields a coherence length of ~7 nm. The value of m^* is found to agree with the previous report on a-C SL structures. The value of mobility (μ_e) in the extended states is calculated by equating the estimated inelastic diffusion length with $\sqrt{[(\mu_e kT/e)\tau\phi]}$ and assuming the inelastic scattering time $(\tau\phi)$ of 1 ps. The value of $\mu_e \sim 15\,cm^2/Vs$ of the present structure is much larger than in bulk DLC films (i.e., $10^{-4}\,cm^2/Vs$). From the NDR region of I-V curves, the cut-off frequency is estimated as 28 GHz. The dwell time of electrons in the well is found to be in the range of 10 ps. A switching behavior in the range between 50 and 110 GHz of these devices has been established.

Having described the possibility for future applications of carbon structure we admit that in practice there is no real application of these carbon SL structures. However, this subject can open a new field of study in carbon which can be compared

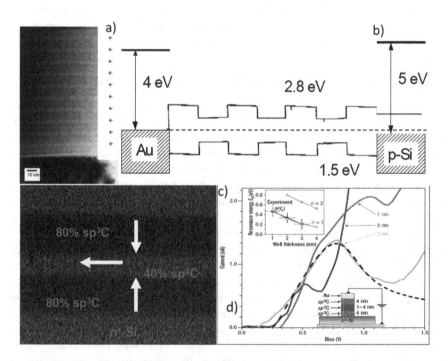

FIGURE 5.8 (a) Multilayers of amorphous carbon structure made of sp_3C- and sp_2C-bonded carbon layers. (b) The band structure of the periodic structure is modulated between 2.8 eV and 1.5 eV. [adapted from the work of Prof. SRP Silva, University of Surrey, UK]. (c) Artificial quantum well structures of carbon produced from sp_3C- and sp_2C-bonded carbon layers which exhibited (d) negative differential conductance (top inset shows theoretically calculated position of the NDR peaks; bottom inset shows the device structure).

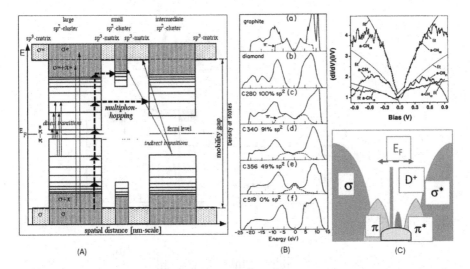

FIGURE 5.9 (a) A schematic diagram of band structures (the conduction and the valence bands) of disordered carbon. The densities of the σ-σ^* and π-π^* states are separated by the defect states (D+). The Fermi level shifts depending on the defect density between the conduction and valence bands. (b) Local density of states of graphite, diamond, and of the four random networks calculated by the recursion method [63]. (c) Normalized differentiated tunneling conductivity representing the surface local density of states as a function of bias voltage of carbon films having different nitrogen concentrations [64].

with the disordered SL structures of GaAs systems. In multilayered GaAs structures, the effect of the disorder was shown by adjusting the hopping parameters and adding random potential fluctuations. We believe that for a-C multi-layers such a theoretical treatment would be challenging since the electron transmission through the carbon structures, consisting of a wide distribution of sp^2 to sp^3 bonds, would depend on the connectivity of these bonds. We propose multi-layers of thin a-C films as disordered SL structures where the sp^2 clusters (wells) can be confined by the sp^3 carbon (barriers). In the proposed model, the disorder in the SL structure of carbon is quasi-one-dimensional instead of a real 3D SL structure. These structures are significantly different from molecular structures which use a periodic disorder of one type of bonds as well as from graphene where the effect of the disorder is very low (Figures 5.10 and 5.11). The effect of N incorporation on carbon microstructures is highlighted in the next subsection since the potential applications of N in carbon are suggested by many researchers.

5.1.7 METHODOLOGY: TIGHT-BINDING CALCULATION

In this work, a quasi-1D (nitrogenated) amorphous carbon superstructure is considered which can be presented by a network of narrow carbon nanoribbons with a mix of saturated and conjugated compound (sp^3-sp^2) areas, corresponding to σ and

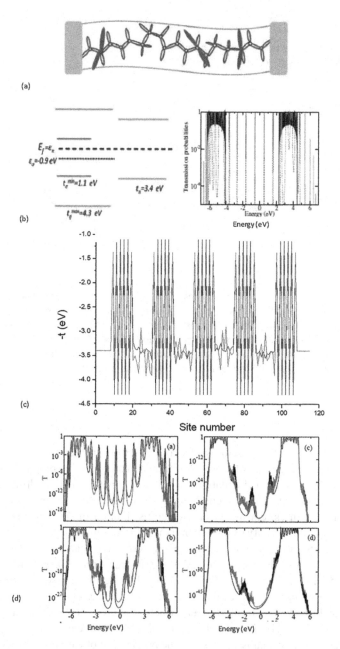

FIGURE 5.10 (a) The microstructure of a-C consists of σ and π bonded carbon atoms. (b) The band diagram shows the sp^3 hopping term alternation and the sp^2 hopping term level. (c) The longer dashed line corresponds to sp^2 on-site energy, which is taken as the zero level, the shorter dashed line corresponds to sp^3 on-site energy. (d) T(E) vs. incident electron energy (E) showed a number of quasi-bound states of mainly conjugated (dashed curve) compared to pure saturated (sp^3) structure (solid curve). The sp^2 bond concentration was varied from: (a) 72%, (b) 42%, (c) 24%, and (d) 7.8%. (see references for details) [65].

π electronic structures (Figure 5.10). Nitrogen atoms can substitute both sp^3 and sp^2-bonded carbon atoms and change their disorder level accordingly. To describe an electron confined in the q-1D a-C ribbons, a tight-binding Hamiltonian is used in the following form: $H = \sum_n \varepsilon_n c_n^\dagger c_n - \sum_n t_{n+1,n} \left(c_{n+1}^\dagger c_n + c_n^\dagger c_{n+1} \right)$

Here the on-site energy ε_n represents the atomic energy as well as an external potential and the operators of creation (c) and annihilation (c^+) of electrons. The hopping term between site n and $n+1$ is $t_{n+1,n}$, which takes into account nearest-neighbor hopping however can be different from site to site reflecting the structural change. The solution of the tight-binding lattice can be obtained with the known eigenvalue $E = \varepsilon_1 - 2t \cos k_1$, where the unit cell length is fixed to 1. t_1 and ε_1 are the hopping term and on-site energy corresponding to the left lead, respectively, and k_1 is an incoming wave vector. Details of the method and the proposed carbon structure have been given in our previous works. It appears important to know at least the most probable distribution of bonds in the a-C films and the ratio of the distorted trigonal to tetragonal bonds in addition to the coordinated to non-coordinated impurity atoms in carbon structures. One of the models could be measuring the local distortion of sp^2- to sp^3-bonded carbon structures based on the deformation potential of the respective bonds. An estimation of the number of sp^2-clusters in the system as a function of the number of atoms per cluster is given from previous detailed calculations, which is the main ingredient of our model. The percentage of sp^2-bonded carbon as a function of nitrogen incorporation has been estimated both experimentally and theoretically. So, the change in the percentage of sp^2 bond is considered a measure of nitrogen atomic concentration in the films (see Figure 5.10). From the wide range of previous works, the exact energetic position of nitrogen atoms was found to be difficult in a-C films. While disorder versus the band gap energy for low sp^3 percentage is estimated experimentally for a wide range of a-C films belonging to the high band gap (E_g) regions, the exact variation of disorder vs. E_g is unknown. It seems the disorder level can also increase further at high sp^3 regions by an increase of the nitrogen incorporation level, which is significantly different from undoped amorphous carbon films.

Proposed carbon 1D superstructures: Understanding the microstructures of amorphous carbon films consisting of sp^2 and sp^3 bonds appears complicated and therefore we would like to construct a SL of carbon to shed some light on the study (Figure 5.10a–d). For a long time, the variation of the line width of x-ray diffraction peaks with the size of graphitic clusters was considered to be a measure of the disorder in carbon specimens. More recently, extensive studies on the line width of the Raman G peak (ΔG) and the I_D/I_G ratio varying with the optical bandgap have been made. The degree of the disorder indirectly estimated from these two parameters was found to maintain a strong non-linear trend passing through a maximum at about 1.2 eV (Figure 5.10c). It was found that the size of the $sp^2 C$ clusters greatly influenced the bandgap. The optical absorption edge was described by a Gaussian plot based on the normal distribution of cluster size. The effect of q-1D filamentary channels on electronic transport has been observed in low-dimensional a-C films and related devices. Based on the SL model, the structure and their structural resistance were estimated through the calculation of transmission coefficients $T(E)$ and local density of states (LDOS) (Figure 5.10).

The newly proposed q-1D a-C superstructure has been constructed in the form of narrow nanoribbons which included some of the essential features specifically sp^2/sp^3 clusters and bond angle distortions (Figure 5.10a–c). This structure consists of several segments of the sp^2 structure with different widths depending on the phase percentage. The superstructure is a mixture of saturated and conjugated (sp^3–sp^2) areas, corresponding to σ and π bonds. Nitrogen atoms can substitute both sp^3 and sp^2 bonded carbon atoms and change their disorder level accordingly (Figure 5.10d). The first resonance peak position E_{res} (from the zero-energy level) is a function of the sp^2 phase percentage, which has an inverse square dependence on the percentage. To describe an electron confined in the q-1D a-C ribbons, a tight-binding Hamiltonian is used where alternating values of the resonance integral produce conduction and valence bands as well as a bandgap. The hopping term t corresponding to the sp^2 phase experiences distortion due to the structural disorder, which reflects the difference in the bond length via the deformation potential. To create the bandgap in these structures sp^3C has been added through alternating bonds and the distortions of sp^2C and sp^3C bonds are combined in several ways. The hopping between sp^3 and sp^2 hybridized C atoms is thought to be extremely significant due to the non-planar geometry of the sp^3 structure. There are several resonant peaks at high sp^2 concentrations corresponding to the relatively large sp^2 cluster size. Such dependence reflects a linear increase of the average sp^2 cluster size and a quadratic decrease of the energy associated with the resonance. As the concentration of sp^2 atoms decreases, there is only one resonant peak, which moves to the bandgap edge (Figure 5.10). The peak height decreases, while the width increases due to the structural disorder. The resonant energy decreases with the increase of sp^2 concentration (size of sp^2C cluster) (Figure 5.11a). When the concentration of the sp^2 structure is high, the peaks are close to the E_F, however, for the low sp^2 phase, the peaks move close to the band edges of the sp^3 structure (Figure 5.11b). For the mixed structure, there are several narrow peaks within the sp^3 bandgap corresponding to quasi-bound states of the sp^2 "wells".

The localization length is calculated as $L_{loc}(E_{res}) = 2L/\ln T(E_{res})$, where L corresponds to the length between the leads. The resistance of the structure was calculated by using the Landauer formula. L_{loc} shows an abrupt decrease till E_{res} reaches the value of about 2 eV. Then, it shows a slight increase with the decrease of bond distortion and as E_{res} approaches the sp^3 band edge (see Figure 5.11). The structural resistance R vs. sp^2 graph shows a sharp decrease followed by saturation in high sp^2 regions. The width of the first resonant peak showed a decrease with the increase of $sp^2C\%$ (Figure 5.11).

To analyze the experimentally observed resonant transmission in DLC SLs with NDR in the current-voltage (I-V) measurements, the one-dimensional model has been expanded to three-dimensional molecular structures of carbon mixing sp^2 and sp^3 carbon atoms. The interpretation of the data is strengthened by calculations of the I-V characteristics, which show that while DLC SL structures differ from their classical counterparts, the confinement of quantized states will be aided by the near-field structural order. Calculations of the I-V characteristics utilizing the Landauer-Büttiker formalism and a 1D model for structurally disordered SLs help to comprehend these temperature-dependent patterns in the obtained I-V characteristics. According to the computations, the size of the sp^2/sp^3-hybridized patches affects the potential barrier at the contact (see Figure 5.12).

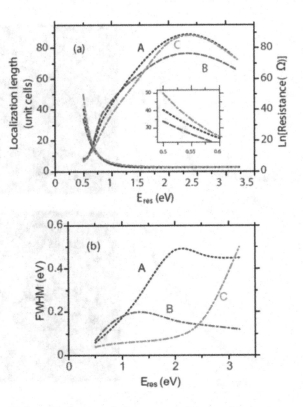

FIGURE 5.11 (a) Localization length (lower curves) and resistance (upper curves) calculated at E_{res} peaks, which are determined by the factor of disorder as well as the energy distance from the sp^3 structure band edge. The inset shows L_{loc} at low energies. (b) Variation of FWHM for the transmission peaks E_{res}. The three curves correspond to cases A, B, and C of hopping disorder parameter dependence on E [66].

5.2 MICROSCOPIC MODEL OF CORRELATED DISORDER IN NITROGENATED AMORPHOUS CARBON

Electronic (and related device) properties of carbon layers suffer severely from an inherent structural disorder that grows naturally as well as by incorporating impurities (e.g., nitrogen). Despite nitrogen incorporation in carbon producing numerous examples of dramatic improvement of electronic properties realization of electronic devices has appeared to be elusive. The origin of these problems lies in the poor understanding of the complexity of the electronic structure of a-C films consisting of sp^2 and sp^3 bonded carbon where the mechanism of controlling the energetic positions of nitrogen centers as well as the amount of structural disorder with the change of impurity concentration remains unknown.

Amorphous carbon showed some unique features of electronic properties including conductivity crossover which can be associated with disorder effects. Experimentally electronic spectra of carbon showed a shift of π-bands relative to the Fermi level (E_F) which may be correlated with the conductivity crossover of the films tuned by the concentration of impurities. From these results, two types of disorder working in a combination in a-C films can be revealed which may not be seen in

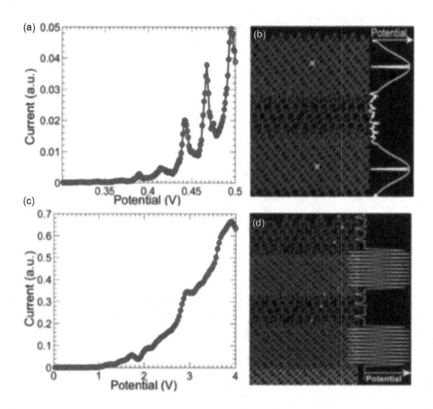

FIGURE 5.12 (a) Calculated current-voltage characteristics of a diamond-like carbon super-lattice with four barriers of 7 nm and 3 well of 7 nm with nitrogen trap states incorporated in the sp^3-C regions. (b) Schematic of the potential for quantum wells and barriers with nitrogen incorporated substitutionally in the sp^3-C regions (white dots as impurities). (c) Calculated current-voltage characteristics of a DLC superlattice with four barriers of 7 nm and 3 well of 7 nm assuming significant bond alternation in the sp^3-C regions. (d) Schematic showing bond alternation in the sp^3-C regions [66].

other semiconductors or metals. An inherent problem in carbon is that impurities (nitrogen) create a structural defect in carbon structure, unlike substitutional doping in semiconductors. It was thought that Anderson's random potential model should apply to this inhomogeneous system however an appropriate distribution of hopping parameters that depends on the local structure of carbon films is yet to be proposed. A unique problem in a-CN$_x$ system is that disorder in nitrogen sites does not change the disorder associated with carbon uniformly, it is rather strongly non-linear which is very difficult to control. From Raman spectroscopy, it was found for carbon nitride films that nitrogen addition caused an independent evaluation of sp^3 fraction and sp^2 clustering, which resulted in nonequilibrium of the sp^2 configuration for a given sp^3 content. So, the way of changing disorder from ta-C to ta-CN films can be different from ta-C or a-C films. Hence a microscopic model of disorder for a-CN$_x$ films is necessary. Studies on the effect of 'N' incorporation in q-1D SL structure of carbon can be seen from the tunnel features which can be generalized through the proposal of a microscopic model of disorder in carbon. Here we show how to control the disorder effectively even impurity increases disorder in a-C films. We note that earlier only sp^2 features were considered for interpreting electronic spectra of carbon however in combination with sp^3 bonds acting as thin tunnel barriers and describing the band gap energy of carbon films can provide a complete picture of the film's microstructure having considered both kinds of disorder. We show how a combination of bond alternation for sp^3-C added with hopping energy term associated with sp^2-C, which has a strong non-linear dependence with band gap energy can explain the conductivity and related electronic spectra of disordered carbon.

5.2.1   DISORDER MODEL OF GROWTH OF CARBON AND CORRESPONDING ELECTRONIC STRUCTURES

To generalize the problem as a first step we propose several (four) cases of the LDOS for C and N sites that cover all possibilities of CN structures (see Figure 5.12a–d). Nitrogen incorporation in carbon can increase sp^2-CN and sp^3-CN phases depending on the relaxation of structures. With the increase of $N\%$ in carbon, the tendency to decrease disorder level and ultimately form some ordered CN structures either sp^2-CN or sp^3-CN can be increased as shown in Figure 5.1a–d, which is explained by the structures shown in Figure 5.13a–d. In addition to the structural disorder in sp^2 bonds (δt_π), which determines the band gap (E_g) of the films, we consider disorder in sp^3 structures (Δt_σ) that controls the strength and stability of the sp^3 network (Figure 5.14a). Since a direct relationship between "N" disorder and N concentration is yet to be produced experimentally, we propose an empirical non-linear model of variation of δt_N with Δt_σ (or δt_π) based on the relationship between δt_N and $N\%$ as well as δt_σ vs. $N\%$ observed from experiments (Figure 5.14b). This model is made based on the fact that nitrogen incorporation in carbon can create both sp^3CN- and sp^2CN-bonds depending on the technique followed to deposit the films (Figure 5.14b). In this case, δt_N is varied with Δt_σ in a non-linear manner, however, correlated with δt_π. Hence, we introduce the concept of correlated disorder in carbon which is supported by the growth model of the film's microstructures and their stability. We discuss all possibilities of CN microstructural configurations in Figure 5.13a–d. Figure 5.13a shows formation of sp^2CN (assuming σ bond alternation as zero) bonds, which is similar to sp^2 bonded a-C films (b) nitrogen incorporation by high energetic beam sp^2 bonds can be transformed into sp^3 bonds and ultimately t-aCN$_x$ films can be formed. In Figure 5.12c for a very small amount of sp^2 bonds available in the films, nitrogen doping configurations can be realized since with the increase of N concentration, band gap increases. In that case, Δt_σ would increase with $N\%$; however, δt_π would be very small due to less availability of sp^2 bonds. In the regime of high N concentration, N–N centers can be formed yielding a barrier that increases the E_g of the materials shown in Figure 5.12d. All four cases can be described by the non-linear variations of Δt_σ and δt_π; however, they have different trends at different resonant energies. Hence, the proposed model includes all previous concepts of C-N microstructures as well as variation of disorder level which is summarized into four cases shown in Figure 5.15a–d.

(i) sp^2CN: Here (δt_π) decreases with $N\%$; however, Δt_σ increases since the bond alternation in remaining sp^3 carbon increases. In this case, the sp^3 bond alternation and sp^2 disorder show opposite trends when plotted as a function of E_g or $N\%$. This is justified as nitrogen concentration increases the bond distortion for sp^3CN structures.

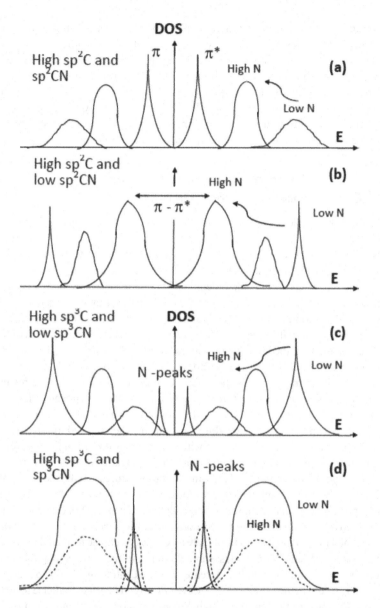

FIGURE 5.13 Proposed electronic spectra show the variation of bandgap with the disorder level. Different cases are considered. (a) For a high percentage of both sp²C and sp²CN the variation of π-π* peak from a sharp to broad feature with the increase of nitrogen percentage. (b) For a high percentage of sp²C and a low percentage of sp²CN the variation of π-π* peak from a broad to sharp feature as the nitrogen percentage increases.
(c) For a high percentage of sp³C and low percentage of sp³CN the variation of π-π* peak shows a very different trend than cases (i) and (ii) since the intensity of π-π* peak and N peak is low. The π-π* peaks show a change from sharp to intense and broad features as the nitrogen percentage decreases.
(d) For a high percentage of both sp³C and sp³CN the width of the π-π* peak is large and they are separated by a large bandgap. The intensity of the N peak changes with nitrogen percentage.

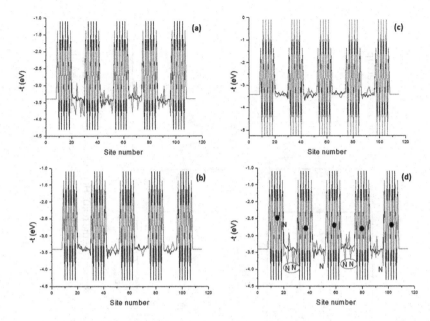

FIGURE 5.14 (a)–(c) Designing the disordered structures of carbon superlattices through a combination of bond alternation and hopping parameters. (d) The effect of nitrogen incorporation in these structures is shown.

The disorder starts to decrease beyond a certain concentration of N in these materials. In this case, Δt_σ and δt_π complement each other although their absolute values differ by an order of magnitude (Figures 5.13a and 5.15a). (ii) Above a certain energy both δt_π and Δt_σ decrease with N% because both sp^2 and sp^3 structures can lead to the stability of the films and maintain the hardness of the films (Figures 5.13b and 5.15b). (iii) Formation of sp^3N: Here sp^3 bonds become much more available and bond alternation will change (decrease) as N substitutes carbon in the sp^3 network. As a result, the bandgap increases with the increase of bond alternation. δt_π will decrease with the increase of the band gap of the materials (Figures 5.13c and 5.14c). (iv) At a very small sp^2C% δt_π can increase further with E_g in parallel to the increased bond alternation, that is, Δt_σ (Figures 5.13d and 5.15d). These four cases are summarized into the variation of δt_π and Δt_σ graphs which vary with E_{res} or E_g or N% (Figure 5.15a). Also, N-N clusters can form for high N% at lower values of the bandgap δt_N and Δt_σ (Figure 5.15b).

5.2.2 LDoS Description

Initially the first LDoS peak is relatively sharp due to the low disorder level which is shifted toward the bandage with the change of structure, particularly with the increase of sp^3C percentage (Figure 5.16a–d). The first resonant transmission peak corresponding to the origin of conductance in these structures can be shifted to the bandedge which explains the tunnel transport through nitrogen-incorporated, diamond-like carbon films. With the increase of sp^2% carbon, the resonant peaks

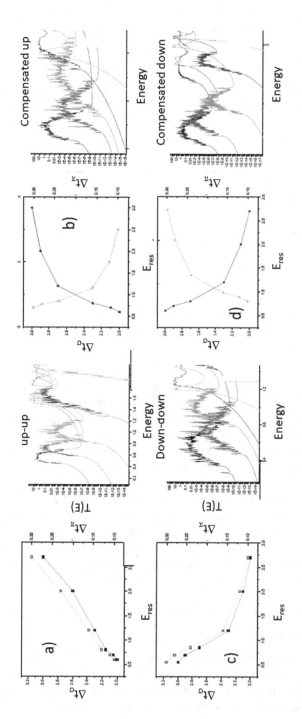

FIGURE 5.15 (a)–(d) Variation of the disorder in an uncorrelated and correlated manner is shown (see the text). Here we change the bond alternation and hopping parameters by either increasing (a) or decreasing (c) with the resonance energy. Also, they follow an opposite trend in (b) and (d). Right panel: The calculated density of states can be plotted as a function of energy.

become narrow and move toward E_F (Figure 5.15a–d). For the lowest sp²%, a weak peak appears very close to the band edge. It is interesting to note that the variation of FWHM of LDoS peaks is well controlled for the wide sp² range which can be used to construct carbon quantum wells (Figure 5.16a). Even for a high sp³%, FWHM of this peak could be sharp suggesting a short characteristic time of these structures.

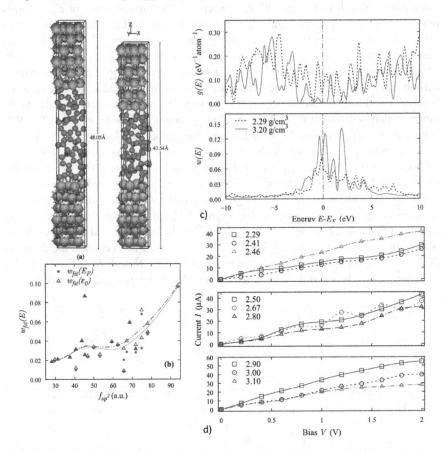

FIGURE 5.16 (a) Atomic structure of low-density (2.29 g/cm³) and (b) high-density (3.20 g/cm³) systems used in quantum transport calculations. Gray spheres represent Au and C atoms, respectively. (b) The top panel shows the total DOS, and the bottom panel shows the degree of localization w(E) of a-C samples with densities 2.29 (dotted line) and 3.20 g/cm³ (solid line). The Fermi level's position is marked with a vertical dashed line. (c): (a) Localization as a function of the density, and (b) localization as a function of the sp³ fraction fsp³. The triangles are the peak values of the fitted localization $w_{fit}\varepsilon_0$ and the dashed line is the fitted Bezier polynomial. The circles are the value of the fitted localization at the Fermi level $w_{fit}E_F$ and the solid line is the fitted Bezier polynomial. The localized states are located around E_F and three regimes of localization with q and fsp³ can be distinguished. (d) Current-voltage (I–V) characteristic curves of a-C thin films with different densities, calculated using DFT-NEGF self-consistent method and an SZP basis set. We identify three types of conduction: (a) semi-metallic for the low-density samples, (b) RT and VRH for middle-range densities, and (c) activated for the high-density samples [46].

The variation of the resistance of the model structure (R) in the low E_{res} regime is found to be similar for all cases. The lowest E_{res} values of R have to be minimum; however, with the increase of E_{res}, the localization length decreases which has to be stopped to achieve good conductance of the structures. By compensating the defects, we can make R constant over a wide range of energy in contrast to the uncompensated case where R increases with E_{res} except for the highest value of E_{res}. At this point, conductance near the band edge becomes dominant. If we disregard these two endpoints, we can see a trend that is nearly constant in energy. This observation is very different from pure sp^2-bonded carbon structures where R decreases smoothly with cluster size. This means not only sp^2-rich carbon but also sp^3-bonded disordered carbon can show resonant tunnel transport as the resonant peaks move toward the band edge and form a shallow doping level. R remains nearly constant over the whole range, which is different from disordered SL structures of GaAs systems. So DLC films can be used as a tunnel barrier having a long coherence length in general.

We consider two cases of 'N' incorporation which forms sp^{2}CN and sp^{3}CN networks (see Figure 5.12). For sp^{2}CN, $N\%$ is proportional to sp^{2}C% and for sp^{3}CN the relationship $N\%$ is proportional with sp^{3}C% $\propto$ (1 - sp^{2}C%) will be valid. These graphs establish the model that LDOS peaks get wider with the decrease of the intensity as it moves away from E_F as proposed in Figure 5.12a–d. It is possible when a phase transition from sp^{2}CN to sp^{3}CN takes place. Finally for the lowest $N\%$ intensity of N peak increases (see Figures 5.12d and 5.17a). Our model can explain the decrease of R with N% for sp^{2}CN and the increase of R for sp^{3}CN with $N\%$ (Figure 5.17b). The interplay of the sp^2 and sp^3 disorder levels yields different trends of the resistance

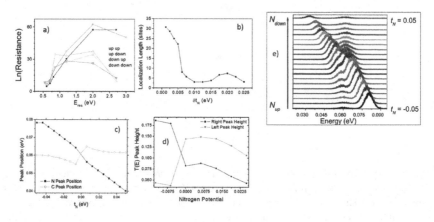

FIGURE 5.17 Electron transmission through a CN structure (a) consisting of some sp^{2}C wells and the hopping term profile with 3N centers. (b) Band diagram displays the level of the sp^2 and sp^3 hopping terms. The longer dashed line corresponds to sp^2 on-site energy, which is treated as the zero level, the shorter dashed line relates to sp^3 on-site energy. (c) Transmission coefficient as a function of incoming electron energy as the N potential is adjusted from N_{up} to N_{down}. The carbon and nitrogen transmission peaks' peak height and Full width at half maximum vary with t_N, or the N hopping parameter. Nitrogen functions as a potential barrier when t_N is negative. (d) A plot of the matching T(E) peak height vs nitrogen energy potential is shown. Right panel: (a)–(d) Transmission coefficient as a function of resonant energy shows how the midgap states are getting filled and peak broadening with the disorder. The electronic spectra of sp^{2}C-like to sp^{3}C-like filamentary structures are compared [67].

curves and also FWHM corresponding to the first resonant peak for C and N sites (Figure 5.17c). For a-CN_x films having a large E_g tunnel barrier can be produced.

Initially, we have discussed the formation of carbon structures via a nonequilibrium process where disorder Δt_σ or δt_π is not correlated. However, for a-C films grown at different conditions, it is possible to control the percentage of sp^3 and sp^2 phases and the level of disorder in respective phases. In the later part of the paper, we have compared the results with the correlated disorder where although $\Delta t_\sigma \neq \delta t_\pi$ but distortions for the different bonds very coherently. We choose the variation of $\Delta t_\sigma \approx \delta t_\pi$ in opposite directions (see Figures 5.12 and 5.13a) which can be an example of disorder compensation and explains the formation of sp^2CN and sp^3CN. Ultimately, we propose a generalized variation of $\Delta t_\sigma \approx \delta t_\pi$ (see Figure 5.12) where the low-energy part and the high-energy part correspond to sp^2CN rich and sp^3 rich phases, respectively. These two curves in Figure 5.14(a) (also Figure 5.14b) is decomposed into four separate cases, that is, either Δt_σ and δt_π varying in opposite directions or in parallel both in upward and downward directions (Figure 15a–d). For these configurations, the disorder is controlled by changing the structure which can be correlated with the band gap energy (see Figure 5.12). In that case, combined with the disorder effect in sp^2 clusters, the change of localization length can be made structurally invariant, at least well controlled. The LDoS is calculated for these configurations with varying $N\%$.

We consider almost zero sp^3 or for small sp^3 disorder to be maximum, however, it is compensated by nearly ordered large sp^2 clusters. For low $sp^2\%$ (small sp^2 clusters) the disorder decreases and at the smallest size of the clusters the disorder saturates and goes to zero for 0% sp^2-C. For the smallest sp^3 size the sp^3 disorder (originating from sigma bond alternation) goes to zero for 0% sp^3. From the decay rate of electronic wavefunctions, we suggest the relationship $\Gamma_b + \Gamma_w = $constant which gives or $\tau_{sp3\ barrier} + \tau_{sp2\ well} = $constant. In other words, $\Delta t_\sigma + \delta t_\pi = $constant, which produces $\Delta E_{\sigma g} + \delta E_{\pi g} = $const. This means for large sp^2 clusters δt_π is small but small sp^3 regions have large disorder. However, large sp^2-C regions try to compensate for the sp^3 disorder. For small sp^2 clusters, the sp^2 disorder is large and is placed between two sp^3 regions. A combination of distortion expressed as $\Delta t_\sigma + \delta t_\pi = $constant can make the films stable and explain the experimental observation of E_g vs. total disorder of the system and the change of transport properties in carbon films for different sp^2 and sp^3 percentages. For example, high-quality ta-C films contain over 10% sp^2 regions to make a stable structure. An increase of $sp^2\%$ relaxes the compressive stress in the films which is achieved by nitrogen incorporation. However, at high $sp^3\%$, the energy gap is largely due to the small contribution of disorder which does not contribute to the broadening of the band edge. The amount of sp^2 disorder could be high for intermediate $sp^2\%$, which creates localized states but does not create resonant tunneling. For high $sp^2\%$, the gap becomes small whereas sp^3 disorder could be large which changes the band gap. Anyway, to see the resonant tunneling the disorder level should be low however a tunnel (sp^3) barrier should exist. For two large clusters separated by a thin sp^3 layer (barrier), the interdot coupling is very strong. Also, the disorder for large clusters is small which suggests a large coherence time (or slow decay rate) within the clusters. If the barrier is thick although having a low disorder level the intercluster coupling is small, which gives a fast decay rate resulting in a broad LDOS. So coupling determines the decay rate. However, for a very small disorder of

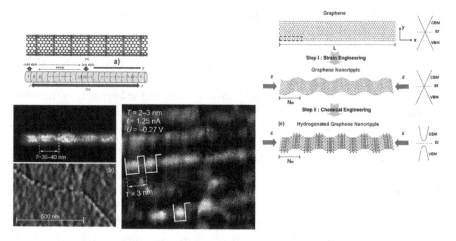

FIGURE 5.18 The SEM image (a) and AFM image (b) of CNT superlattice (period ~30–40 nm). The STM image the electron gas superlattice in CNT (period ~3 nm at $U = -0.27$ V) [68].

the sp³ barrier and wells, NDR can be seen. A particular range of sp²% can control the lifetime of the electrons for a particular E_g value. For making an effective SL structure the bandgap should be modulated properly, which is correlated with the disorder level of the sp² and sp³ phases of carbon. This will give the control of tunnel time through the barrier and dwelling time of electrons in the well. The LDOS for the barrier and the well should be distributed accordingly.

In another study, utilizing ab initio molecular dynamics in the isothermal-isobaric ensemble, microstructures were created as a function of the density of ultrathin films made of a-C. The density of states (total and projected), degree of localization of electronic states, and transmission function were computed using the density functional theory and nonequilibrium Green's functions technique. This method, which takes a different route from the tight-binding concept, can nonetheless account for the experimental findings in low-dimensional carbon systems.

The computed electron transmission demonstrated the localization of ultrathin films' electronic states at the contact electrode, which has an impact on the electron transport processes, using Kohn-Sham DFT and the NEGF formalism (see Figure 5.18). To create a quantitative model, more calculations must be made and advanced methodologies must be used.

5.3 SUMMARY

The aim of the work is to resolve the long-standing issue of conductivity variation in disordered carbon films incorporated with nitrogen, particularly the shift of E_F concerning the band edges. This is satisfied within the framework of a SL model however controlling the disorder level effectively by nitrogen incorporation. The localization length of electronic wavefunctions in these structures can be found to be constant over a wide range of energy by compensating for the disorder associated with sp² and sp³ bonds. Hence, the experimentally observed shift of the bands (compared to E_F) and the increase of resistance in sp³CN structures as well as the decrease of the resistance of the sp²CN structure with nitrogen percentage can be explained by the proposed theoretical model. To achieve a workable device of pure carbon system multilayer SL structures have been employed whose characteristics can be interpreted if the transport mechanism of these systems is known particularly the energetic position of N atoms and the amount of change of disorder in sp²/sp³ CN systems. The simulation of Quantum Tunneling will be discussed in Chapter 10.

Q. Calculate the band energy diagram of diamond and graphene

Main problems discussed in this chapter.

- What has been achieved?

In Chapters 1–4, we have not considered disorder (or randomness) in the structure which will be introduced in Chapter 5 both in diamond and graphite phases. The quantum wells are filled with disordered carbon which can be maximally entangled. In addition, space-time distortion in the maximally entangled structure can be compared with the theoretical model of a black hole. Tunneling between two disordered carbon quantum wells can be explained by a wormhole model in Chapter 10.

- What has not been achieved?

Chapter 5. Multilayer SL structures can be prepared artificially, which can show prominent resonant tunnel conductance. In many cases, we cannot control the resonance peaks (both position and width) through synthesis and engineering. Raman spectra of disordered carbon have not been understood fully since a proper theory is not available.

BIBLIOGRAPHY

1. J. Isberg, et al., High carrier mobility in single-crystal plasma-deposited diamond, *Science* 297, 1670–1672 (2002).
2. M. W. Geis, N. M. Efremow, and D. D. Rathman, Summary abstract: Device applications of diamonds, *Journal of Vacuum Science and Technology* A6, 1953–1954 (1988).
3. G. A. J. Amaratunga, A dawn for carbon electronics?, *Science* 297, 1657–1658 (2002).
4. K. S. Novoselov, et al., Electric field effect in atomically thin carbon films, *Science* 306, 666–669 (2004).
5. S. R. P. Silva, J. D. Carey, R. U. A., Khan, E. G. Gerstner, and J. V. Anguita, Review on amorphous carbon thin films. In *Handbook of Thin Film Materials*. H. S. Nalwa, Eds., 403–506, Academic Press, New York, 2002.
6. N. Mason, M. J. Biercuk, and C. M. Marcus, Local gate control of a carbon nanotube double quantum dot, *Science* 303, 655–658 (2004).
7. L. F. Lindoy, Optoelectronics: Marvels of molecular device, *Nature* 364, 17–18 (1993).
8. J. H. Burroughes, C. A. Jones, and R. H. Friend, New semiconductor device physics in polymer diodes and transistors, *Nature* 335, 137–141 (1988).
9. S. R. P. Silva, G. A. J. Amaratunga, C. N. Woodburn, M. E. Welland, and S. Haq, Quantum size effects in amorphous diamond-like carbon superlattices, *Japanese Journal of Applied Physics* 33, 6458–6465 (1994).

10. E. R. Brown, W. D. Goodhue, and T. C. L. G. Sollner, Fundamental oscillations up to 200GHz in resonant tunneling diodes and new estimates of their maximum oscillation frequency from stationary-state tunneling theory. *Journal of Applied Physics* 64, 1519–1529 (1988).

11. Y. -P. Zhao, B.-Q. Wei, P. M. Ajayan, G. Ramanath, T.-M. Lu, and G.-C. Wang, Frequency-dependent electrical transport in carbon nanotubes, *Physical Review B* 64, 201402–201405 (2001).

12. J. Chen, M. A. Reed, A. M., Rawlett, and J. M. Tour, Large on-off ratios and negative differential resistance in a molecular electronic device, *Science* 286, 1550–1552 (1999).

13. Y. L. He, G. Y. Hu, M. B. Yu, M. Liu, J. L. Wang, and G.Y. Xu, Conduction mechanism of hydrogenated nanocrystalline silicon films, *Physical Review B* 59, 15352–15357 (1999).

14. S. Miyazaki, Y. Ihara, and M. Hirose, Resonant tunneling through amorphous silicon-silicon nitride double-barrier structures, *Physical Review Letters* 59, 125–127 (1987).

15. J. M. Shannon, and K. J. B. M. Nieuwesteeg, Tunneling effective mass in hydrogenated amorphous silicon, *Applied Physics Letters* 62, 1815–1817 (1993).

16. S. Datta, *Electronic Transport in Mesoscopic System*, Cambridge University Press, Cambridge, 1995.

17. M. J. Kelly, *Low Dimensional Semiconductors*, Oxford University Press, Oxford, 1995.

18. J. Hajto, A. E. Owen, S. M. Gage, A. J. Snell, P. G. LeComber, and M. J. Rose, Quantized electron transport in amorphous-silicon memory structures, *Physical Review Letters* 66, 1918–1921 (1991).

19. E. G. Gerstner, and D. R. McKenzie, Nonvolatile memory effects in nitrogen doped tetrahedral amorphous carbon films, *Journal of Applied Physics* 84, 5647–5651 (1998).

20. S. Bhattacharyya and S. R. P. Silva, Demonstration of an amorphous carbon tunnel diode, *Applied Physics Letters* 90, 082105 (2007).

21. S. Bhattacharyya, O. Auciello, J. Birrell, J. A. Carlisle, L. A. Curtiss, A. N. Goyette, D. M. Gruen, A. R. Krauss, J. Schlueter, A. V. Sumant, and P. Zapol, Synthesis and characterization of highly-conducting nitrogen-doped ultrananocrystalline diamond films, *Applied Physics Letters* 79, 1441 (2001); S. Bhattacharyya, *Physical Review B* 70, 125412 (2004).

22. S. Bhattacharyya, Two-dimensional transport in disordered carbon and nanocrystalline diamond films, *Physical Review B* 77, 233407 (2008); S. Bhattacharyya, Observation of delocalized transport and low-dimensionality effects in disordered carbon thin films, *Applied Physics Letters* 91, 142116 (2007).

23. J. J. Mares, P. Hubik, J. Kristofik, D. Kindl, M. Fanta, M. Nesladek, O. Williams, and D. M. Gruen, Weak localization in ultrananocrystalline diamond, *Physics Letters* 88, 092107 (2006).

24. S. Bhattacharyya, F. Richter, U. Starke, H. Griesmann, and A. Heinrich, Thermoelectric power of nitrogen doped amorphous nitrogenated carbon, *Applied Physics Letters* 79, 4157 (2001).

25. S. Bhattacharyya and S. R. P. Silva, Transport in Low-Dimensional carbon, *Thin Solid Films* 482, 94 (2005).

26. C. Godet, Variable range hopping revisited: The case of an exponential distribution of localized states, *Journal of Non-Crystalline Solids* 299, 333 (2002); C. Godet, Electronic localization and Bandtail hopping charge transport, *Physica Status Solidi* 231, 499 (2002).

27. K. K. Choi, B. F. Levine, R. J. Malik, J. Walker, and C. G. Bethea, Periodic negative conductance by sequential resonant tunneling through an expanding high-field superlattice domain, *Physical Review B* 35, 4172 (1987).

28. A. Rakitin, M. Y. Valakh, N. I. Klyui, V. G. Visotski, and A. P. Litvinchuk, Possibility of a double-well potential formation in diamondlike amorphous carbon, *Physical Review B* 58, 3526 (1998); M. Pelton, S. K. O'Leary, F. Gaspari, and S. Zukotynski, The optical absorption edge of diamond-like carbon: A quantum well model, *Journal of Applied Physics* 83, 1029 (1998).

29. A. R. Merchant, D. R. McKenzie, and D. G. McCulloch, *Physical Review B* 65, 024208 (2001).

30. M. C. dos Santos and F. Alvarez, *Physical Review B* 58, 13918 (1998).
31. U. Stephan, Th. Frauenheim, P. Blaudeck, and G. Jungnickel, π bonding versus electronic-defect generation: An examination of band-gap properties in amorphous carbon, *Physical Review B* 49, 1489 (1994).
32. R. Gago, M. Vinnichenko, H. U. Jager, A. Yu. Belov, I. Jimenez, N. Huang, H. Sun, and M. F. Maitz, Evolution of sp2 networks with substrate temperature in amorphous carbon films: Experiment and theory, *Physical Review B* 72, 014120 (2005); A. A. Valladares and F. Alvarez-Ramirez, Bonding in amorphous carbon-nitrogen alloys: A first principles study, *Physical Review B* 73, 024206 (2006).
33. A. K. Savchenko, V. V. Kuznetsov, A. Woolfe, D. R. Mace, M. Pepper, D. A. Ritchie, and G. A. C. Jones, Resonant tunneling through two impurities in disordered barriers, *Physical Review B* 52, 17021 (1995).
34. M. Tamor and W. Vassel, Raman 'fingerprinting' of amorphous carbon films, *Journal of Applied Physics* 76, 3823 (1994).
35. A. A. Ferrari and J. Robertson, Interpretation of Raman spectra of disordered and amorphous carbon, *Physical Review B* 61, 14095 (1999).
36. G. Fanchini and A. Tagliaferro, Disorder and Urbach energy in hydrogenated amorphous carbon: A phenomenological model, *Applied Physics Letters* 85, 730 (2004).
37. F. Alibart, M. Lejeune, O. D. Drouhin, K. Zellama, and M. Benlahsen, Influence of disorder on localization and density of states in amorphous carbon nitride thin films systems rich in -bonded carbon atoms, *Journal of Applied Physics* 108, 053504 (2010).
38. O. A. Williams, M. Nesladek, M. Daenen, S. Michaelson, A. Hoffman, E. Osawa, K. Haenen, R. B. Jackman, Growth, electronic properties and applications of nanodiamond, *Diamond and Related Materials* 17, 1080 (2008).
39. S. R. P. Silva et al., Nitrogen modification of hydrogenated amorphous carbon films, *Journal of Applied Physics* 81, 2626 (1997).
40. J. N. Hart, F. Claeyssens, N. L. Allan, and P. A. May, Carbon nitride: Ab initio investigation of carbon-rich phases, *Physical Review B* 80, 174111 (2009).
41. J. Robertson, Electronic and atomic structure of diamond-like carbon, *Semiconductor Science and Technology,* 18, S12 (2003).
42. Th. Kohler, G. Jungnickel, and Th. Frauenheim, Molecular-dynamics study of nitrogen impurities in tetrahedral amorphous carbon, *Physical Review B* 60, 10864 (1999).
43. G. Fanchini, A. Tagliaferro, N. M. J. Conway, and C. Godet Role of lone-pair interactions and local disorder in determining the interdependency of optical constants of a–CN:H thin films *Physical Review B* 66, 195415 (2002).
44. J. Robertson and C. A. Davis, Nitrogen doping of tetrahedral amorphous carbon, *Diamond and Related Materials* 4, 441 (1995).
45. S. Stafstrom, Reactivity of curved and planar carbon–nitride structures, *Applied Physics Letters* 77, 3941 (2000).
46. S. Caicedo-Davila, O. Lopez-Acevedo, J. Velasco-Medina, and A. Avila, Density and localized states' impact on amorphous carbon electron transport mechanisms, *Journal of Applied Physics* 120, 214303 (2016).
47. B. Liam, R. V. Gorbachev, R. Jalil, B. D. Belle, F. Schedin, A. Mishchenko, T. Georgiou, M. I. Katsnelson, L. Eaves, S. Morozov, et al. Field-effect tunneling transistor based on vertical graphene heterostructures, *Science* 335, 947–950 (2012).
48. J. Yu, G. Liu, A. V. Sumant, G. Viyek, and A. Balandin Alexander, Graphene-on-diamond devices with increased current-carrying capacity: Carbon sp^2-on-sp^3 technology, *Nano Letters* 12, 1603–1608 (2012).
49. S. Takabayashi, S. Ogawa, Y. Takakuwa, H.-C. Kang, R. Takahashi, H. Fukidome, M. Suemitsu, T. Suemitsu, and T. Otsuji, Carbonaceous field effect transistors with graphene and diamond-like carbon, *Diamond and Related Materials* 22, 118–123 (2012).
50. P. A. Schultz, and C. E. T. Gonc'aleves da Silva, Disorder effects on resonant tunneling in double-barrier quantum wells, *Physical Review B* 38 10718–10723 (1988).

51. A. Onipko, Analytical model of molecular wire performance: A comparison of π and σ electron systems, *Physical Review B* 59, 9995–10006 (1999).

52. J. N. Hart, F. Claeyssens, N. L. Allan, and P. W. May, Carbon nitride: Ab initio investigation of carbon-rich phases, *Physical Review B* 80, 174, 111–113 (2009).

53. C. Mathioudakis, P. C. Kelires, Y. Panagiotatos, P. Patsalas, C. Charitidis, and S. Logothetidis, Nanomechanical properties of multilayered amorphous carbon structures, *Physical Review B* 65, 205203–14 (2002).

54. E. B. Halac, E. Burgos, and M. Reinoso, Amorphous carbon multilayered films studied by molecular dynamics simulations, *Physical Review B* 77, 224101–7 (2008).

55. J. P. Zhao, Z. Y. Chen, X. W. Wang, and T. S. Shi, Electron field emission from tetrahedral amorphous carbon films with multilayer structure, *Journal of Applied Physics* 87, 8098–8102 (2000).

56. T. Matsumoto, et al., Inversion channel mobility and interface state density of diamond MOSFET using N-type body with various phosphorus concentrations, *Applied Physics Letters* 114, 242101 (2019).

57. Y. Wu, Y.-M. Lin, A. A. Bol, K. A. Jenkins, F. Xia, D. B. Farmer, Y. Zhu, and P. Avouris, High-frequency, scaled graphene transistors on diamond-like carbon, *Nature* 472, 74–78 (2011).

58. H A Le, S Ta Ho, D C Nguyen, and V N Do, Optical properties of graphene superlattices, *Journal of Physics: Condensed Matter* 26, 405304 (2014); R. Banerjee et al., Strain modulated superlattices in graphene, *Nano Letters* 20, 3113–3121 (2020).

59. Z. Zhang, C. Chen, X. C. Zeng, and W. Guo, Tuning the magnetic and electronic properties of bilayer graphene nanoribbons on Si (001) by bias voltage, *Physical Review B* 81, 155428–155429 (2010).

60. F. Mohammed and S. Bhattacharyya, Resonant transport features of disordered graphene devices, *Europhysics Letters* 100, 26009 (2012).

61. A. Raul, P. Bruno, D. J. Miller, M. Bleuel, J. Lal, and D. M. Gruen, Diamond nanowires and the insulator-metal transition in ultrananocrystalline films, *Physical Review B* 75, 195431–195511 (2007); P. Nemeth, K. McColl, L.A.J. Garvie, et al. Complex nanostructures in diamond. *Nat. Mater.* 19, 1126–1131 (2020).

62. D. Churochkin, R. McIntosh, and S. Bhattacharyya, Tuning resonant transmission through geometrical configurations of impurity clusters, *Journal of Applied Physics* 113, 044305–044308 (2013); S. Bhattacharyya, and D. Churochkin, Understanding resonant tunnel transport in non-identical and non-aligned clusters as applied to the disordered carbon systems, *Journal of Applied Physics* 116, 154305–154310 (2014).

63. J. Robertson and E. P. O'Reilly, Electronic and atomic structure of amorphous carbon, *Physical Review B* 35, 2946 (1987).

64. S. Bhattacharyya, K. Walzer, M. Hietschold, and F. Richter, Nitrogen doping of tetrahedral amorphous carbon films: Scanning tunneling spectroscopy, *Journal of Applied Physics* 89, 1619 (2001).

65. M. V. Katkov and S. Bhattacharyya, Theoretical model of nano-electronic transport in structurally disordered carbon, *Europhysics Letters* 99, 37005 (2012).

66. R. McIntosh, S. J. Henley, S. R. P. Silva, and S. Bhattacharyya, Coherent quantum transport features in carbon superlattice structures, *Scientific Reports* 6, 35526 (2016).

67. M. V. Katkov, R. McIntosh, and S. Bhattacharyya, Tunnel transport model of nitrogen doped amorphous carbon superstructure, *Journal of Applied Physics* 113, 093701–093708 (2013).

68. B. Minsk, *28th International Symposium Nanostructures: Physics and Technology*, September 2020, Ioffe Institute, Russia.

6 Mesoscopic Phenomena
Electronic Transport in Low-Dimensional Carbon Films

So far, we discussed high-mobility (fast) highly ordered carbon systems in nanoscale devices where resistance decreases with temperature (Tables 6.1 and 6.2). A large variety of carbon structures have very low mobility like organic materials where conductivity decreases by lowering the temperature. How do we use their electronic properties in large-area electronics like solar cells, biosensors, etc., where hopping and diffusive transport dominate? It requires analysis of spectral density (DoS vs. E) for device applications. In addition, it is important to know the phase-coherent time and its temperature dependence for memory device applications. It can also be applied to a 3D system.

TABLE 6.1

Classification of Amorphous Carbon Based on Transport Properties

Type	Conductivity (Ohm·cm)$^{-1}$	Activation energy (eV)	Conduction Mechanism	Carriers (cm^{-3})/ Mobility (cm^2/(Vs))$^{-1}$
Diamond	10^{-10}	5.5, 1.7 (N)	Activated	Depends on doping
Undoped DLC/ta-C	10^{-6}	2–4	Activated	Unknown
Low doped DLC/ta-C	10^{-4}	0.015	Activated/hopping	10^{19}/10
Heavily doped a-C	10^{-1}	0.03	VRH	--/--/10
Alloyed a-C	10^{1}	0.05	VRH + Activated	$>10^{20}$
Pa-C/a-C:H	10^{-1}	0.05	Activated/ multiphonon	$\sim 10^{21}$
Ga-C	10^{3}	0.001	Delocalized + hopping	Unknown
n-a-C	10	0.1	Quantum dot	Unknown
Graphite	10^{4}	0	Extended state	--/--/few microvolts

DOI: 10.1201/9781003316411-6

TABLE 6.2

Doping of a-C films

Dopant/Film	Conductivity (Ohm·cm)$^{-1}$	Activation energy (eV)	Mechanism	Carriers/Mobility/ TEP/DoS
N (n) t a-C	10^{-2}	0.2	Activated	Low/10^{-4}
N (n) DLC	10^{-7}	--	Hopping	--
N (n) t a-C	10^{-1}	0.12	Activated (?)	--
N (n) a-CN	10^{+2}	0.008	VRH + Activated	Unknown
N(n) a-C:H	10^{3}	Unknown	Unknown	10^{20}/cm^{-3}/15 cm^2/(Vs)
P (n) a-C	10^{4}	0.1–0.3	Hopping	Low/-
B (n) t a-C	10^{-1}	0.2–0.3	VRH + Activated	--
B (n) DLC	10^{-6}	0.15	VRH	--
N(n) NCD	1.8×10^{2}	0.05	VRH + Activated	1.5×10^{20}/10 cm^2/(Vs) spin~10^{19}
N a-C:H	10^{3}	0.4	VRH > 300 K	/ ~μV/K (–ve to +ve) spin > 10^{20}

Based on a proper model one can estimate the minimum conductivity. The 3D electronic transport properties of ultrananocrystalline diamond (UNCD) films are much more complicated than 1D and 2D systems, although the interconnections of 1D or 2D GBs can be important. Although nitrogen incorporation enhances the conductivity of UNCD films by several orders of magnitude, it seems more than one kind of mechanism can dominate in the whole process at different temperatures. The Fermi level can move with the temperature and be aligned with the localized and delocalized states (Figure 6.1). Moreover, the transport properties of nitrogen-doped UNCD (N-UNCD) samples depend on the deposition techniques and the deposition parameters used by various groups. Nitrogen doping increased the electrical conductivity by four orders of magnitude when 20% N$_2$ was introduced in the (microwave) plasma CVD chamber. The GBs played the role of conduction channel; however, finding the atomic structure of the GB is very challenging. These electrical transport mechanisms are strongly dependent on the amount of nitrogen in the films, resulting in different densities of states giving different electronic structures. In general, temperature-dependent conduction can be explained using three different mechanisms, namely (i) activated conduction, (ii) hopping conduction, and (iii) weak localization (WL) corrections. The first two work in classical regime, whereas the third one in a quantum phenomenon which will be explained here. In this chapter, we mainly focus on the phase-coherent time (or length) in carbon devices, which will be useful for memory devices. We begin this chapter with an understanding of the (geometric) phase and the A-B effect, followed by coherent backscattering (WL). The coherence time associated with the scattering phenomena can be recorded from the low-temperature transport and MR measurement.

6.1   A-B EFFECT

An electrically charged particle experiences the A-B effect, a quantum mechanical phenomenon, even though it is contained in a space where both the magnetic field (B) and

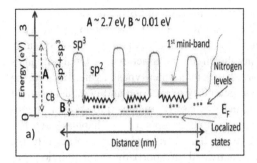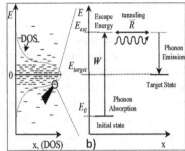

FIGURE 6.1 NV complexes have donor and acceptor sites (act) near the middle of the band gap. (a) The effective donor and acceptor band for an n-type UNCD film [25]. (b) An illustration of hopping mechanism, which can be from one energy level to another or same level through tunneling. Distribution of density of states shows the excitation of electrons from a lower energy lever to a higher energy level (W) and tunneling route (R) [26].

the electric field (E) are zero. Interference experiments are used to demonstrate the A-B effect because the fundamental process is the connection of the electromagnetic potential with the complex phase of a charged particle's wave function. The A-B solenoid effect the most well-known example occurs when the magnetic field inside a lengthy solenoid causes a phase shift in the wave function of a charged particle as it passes through.

Experimental observation of this phase shift has been made. In trying to generate quantum dots, nano rings were unintentionally created. They exhibit intriguing excitons- and A-B-effect-related optical characteristics. Geometric phases in mesoscopic rings are being analyzed and measured. With charge-density-wave (CDW) rings up to 85 μm in circumference above 77 K, several tests demonstrate A-B oscillations in CDW current vs. magnetic flux with the dominating period (h/2e). This behavior resembles that of quantum interference devices made of superconductors. The scalar electric potential affects the wave function's phase in the same ways as the magnetic vector potential affects it. A-B interference phenomenon from the phase shift is predicted by creating a scenario in which the electrostatic potential fluctuates for two trajectories of a particle across areas of zero electric field.

As the wave function can only be measured in absolute values, the A-B effect can be understood. There is no method to specify a wave function with a constant absolute phase, when this enables the detection of phase discrepancies using quantum interference experiments. In the absence of an electromagnetic field, one can get close by designating the wave functions relative to the eigenfunction "1" and declaring the eigenfunction of the momentum operator with zero momentum to be the function "1" (while disregarding normalization issues). There are many additional disciplines where effects with different mathematical interpretations might be found. A-B effects caused by a gauge field acting in a stochastic environment can be understood in classical statistical physics as quantization of a molecular motor motion.

6.2 GEOMETRIC PHASE

Understanding phase is crucial for understanding wave-related phenomena like interference and diffraction, which are crucial ideas in wave physics, quantum mechanics, and optics (related to wavefunctions). Phases are required in theoretical physics to comprehend the A-B effect. The adiabatic growth of a quantum system around a circuit results in the acquisition of the geometric phase, or Berry phase (path of rotation). Controlling the phase involves changing the circuit's characteristics. Quantum gates that rely on the geometric phase perform better than other types of gates because the Berry phase only depends on the global geometry of the loop and is robust to slight inaccuracies. We shall see how the geometric phase is connected to spin–orbit coupling (SOC), which can be simulated by a quantum computer (see Chapters 9 and 10).

Although we discuss the effect of strong localization (due to disorder effects), hopping transport, and band conduction (activated transport), the main attention is to collect geometric phase through the design of microstructure of carbon. We look for a 3D structure, like a cylinder that can be able to separate two phases by the alignment of two orbits in mutually perpendicular directions. The idea of phase separation is similar to the spin–orbit interactions or spin–charge separation, which can be implemented in carbon rings that are placed normally (or at high angles) to each other. With the change in temperature, the separation (or the coupling) can be altered. A phase transition can occur when these two phases in 3D are completely separated. Hence, we show how 3D WL can be realized in the 3D structure of UNCD films. It is important to understand the A-B effect in 1d CNTs and WL in 2D graphene.

According to classical physics, a material's resistance grows linearly as its length and the quantity of electron scatterers rise. In these circumstances, the material's phase relaxation length is always greater than the mean free path. In low-mobility materials and at low temperatures, the situation is different because the phase relaxation length can be shorter than the mean free path. A series of phase-coherent units with elastic scatterers can be seen as the material in this scenario. An electron's time-reversed path's interference between various scattering events causes greater backscattering, which causes the electron to become weakly localized. Yet, a tiny field with a typical value of 100 G can negate this impact. When this occurs, the conductance drops as the magnetic field increases, a phenomenon known as negative MR. Unlike other transport mechanisms, this effect is particularly sensitive to the length of the phase relaxation; therefore, it can be used to gauge that length.

6.3 WL PROCESS

The contemporary physics idea states that all tiny quantities have both corpuscular and wave characteristics (Figure 6.2a). The Hilbert space, whose vector describes a system state, must be introduced in order to execute this theory mathematically. Moreover, a superposition of base states can be used to represent the state in general. Several facets of physics have consequences for the concept of superposition. It describes the WL phenomenon in particular. There exist highly specific pathways for a tiny quantity that take the shape of a loop when it is scattered by a random assortment of scatterers. Then, using the superposition principle, one may create a superposition of states that corresponds to a specific loop's forward and backward traversal of time. As electrons can interact constructively under some circumstances if states are assigned to them. In that case an increase in sample resistance is anticipated. A magnetic field or spin's existence reduces or even inverts the impact of constructive interference, turning it into a weak anti-localization (WAL) event.

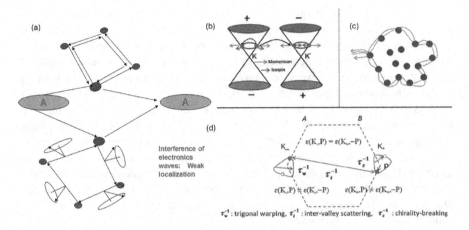

FIGURE 6.2 (a) An illustration of weak localization (top) without magnetic field and (bottom) with magnetic field. (b) Scattering events can occur either due to inter- or intravalley events, whichever is dominant in the device under test can be determined through magnetoresistance measurements. (c) The WL effect is a result of interference between particles with time-reversed paths, this is because, in the absence of strong spin–orbit coupling, in such an interference path both wavefunctions pick up the exact same phase and the inference is constructive. (d) Valley warping away from the Dirac cones leads to suppression of the WAL effect [27].

It should be noted that the appeared state that causes WL involves two coupled electron channels and may be modeled using the concept of Cooperon.

6.4 WEAK LOCALIZATION IN GRAPHENE FILMS

There is a great deal of interest in graphene research because of the mechanics of quantum interference effects studied at low magnetic fields. Once more, this results from the fermions' Dirac property. Strong limitations apply to the types of scattering events that can take place for electrons inside the same or separate band structure valleys due to the chirality and Berry phase of the carriers. The conservation of helicity eigenvalues prevents backscattering for intravalley scattering events, giving rise to the so-called Klein conundrum. On the other hand, scattering between various valleys (inter-valley scattering) might result in backscattering events because, in such a case, the helicity changes along with the sign of the momentum (Figure 6.2b).

This effectively means that the observation of either the WL or WAL effects was caused by extremely precise scattering occurrences in actual space. While short-range potential (point-like) scattering events caused by adatoms and vacancies lead to inter-valley scattering and are observed as the WL effect, long-range potential (point-like) scattering events (such as ripples in the graphene sheet, dislocations, and charged impurities) are responsible for intravalley scattering favoring the WAL effect. This makes it possible to determine how the disorder of the materials utilized affects the microscopic characteristics of graphene devices. As the WL effect has been observed to result from quantum interference between electrons traveling along time-reversed routes, it is possible to explore such events by breaking time-reversal symmetry using tiny magnetic fields.

Trigonal warping at energies close to but outside the Dirac point was also demonstrated to significantly suppress the WAL effect. As a result of the trigonal warping's effect on the band structure of the nearby K and K′ valleys, which would result in opposite signs in the various valleys, the WL effect is suppressed. Indeed, substantial observational work on graphene has revealed both WL and WAL. It is shown how single-layer graphene devices made utilizing the CVD fabrication process exhibit high MR. The intense negative MR makes it easy to see the mild localization effects. Since the CVD technique enables higher-scale development, CVD-generated graphene should be thoroughly explored in graphene. Yet generally speaking, the CVD process results in graphene that is less mobile and has larger fault density. Nonetheless, significant progress has been made in this area, and large-scale high-quality graphene has been produced. This synthesis process also provides the opportunity to substitute actively "dope" the graphene with substances like nitrogen and boron.

6.4.1 ZERO-FIELD CONDUCTIVITY

At low temperatures, all metallic systems with high resistivity should inevitably exhibit large quantum interference (localization) MR, eventually leading to the metal–insulator transition at $\sigma \approx e^2/h$. Until now, such behavior has been universal, but it was found missing in graphene. Even near the neutrality point, no significant low-field ($B < 1$ T) MR has been observed down to liquid helium temperatures and, although sub-100 nm Hall crosses did exhibit giant resistance fluctuations, those could be attributed to changes in the distribution of electron and hole puddles and size quantization. It remains to be seen whether localization effects at the Dirac point recover at lower T, as the phase-breaking length becomes increasingly longer, or the observed behavior indicates a "marginal Fermi liquid," in which the phase-breaking length goes to zero with decreasing E. Further experimental studies are much needed in this regime, but it is difficult to probe because of microscopic inhomogeneity. Away from the Dirac point (where graphene becomes a good metal), the situation has recently become reasonably clear. Universal conductance fluctuations (UCF) were reported to be qualitatively normal in this regime, whereas WL MR was found to be somewhat random, varying for different samples from being virtually absent to showing the standard behavior. On the other hand, early theories had also predicted every possible type of WL MR in graphene, from positive to negative to zero. Now it is understood that, for large n and in the absence of inter-valley scattering, there should be no MR, because the triangular warping of graphene's Fermi surface destroys time-reversal symmetry within each valley. With increasing inter-valley scattering, the normal (negative) WL should recover. Changes in inter-valley scattering rates by, e.g., varying microfabrication procedures can explain the observed sample-dependent behavior. A complementary explanation is that sufficient inter-valley scattering is already present in the studied samples, but the time-reversal symmetry is destroyed by elastic strain due to microscopic warping. The strain in graphene has turned out to be equivalent to a random magnetic field, which also destroys time-reversal symmetry and suppresses WL. Whatever the mechanism, theory expects normal UCF at high n, in agreement with the experiment.

6.5 ACTIVATED AND HOPPING CONDUCTION IN NANODIAMOND FILMS

Theoretical work by other researchers attempted to explain the above experimental data using a computational program. The model is based on determining the electronic structures of several nitrogen centers (defect centers) in nanodiamond films and then predicting the contribution of the center in conductivity. From the results, the conductivity in nanodiamond films was explained using three types of nitrogen centers formed in the GBs, namely complex structures of nitrogen with a dangling bond (N-DB), complex structures of nitrogen and π-bond (N–π) and lastly NV complexes. The last structures were predicted to form shallow donors, while N-DB and N–π complex are the compensation centers that do not contribute directly to the conductivity. This is because they lie deep in the band gap. It is the shallow donors (NV centers) that are activated into the conduction band and hence increase conductivity which illustrates this model (Figure 6.2). Tight binding studies also indicated the formation of new electronic states associated with carbon bonds and dangling bonds being introduced into the band gap and forming unoccupied states near the Fermi level, thus contributing to the conduction.

The schematic diagram (Figure 6.2) of the DoS shows that in films where we have NV complexes, we have acceptor and donor sites in the middle of the band gap. The size of this impurity band determines the transport properties in the films. In semiconducting films, the impurity band (act) is close to the conduction band and therefore activated conduction mechanism is predominating as shown in Figure 6.1 for 0.5% and 1% samples nitrogen. When the impurity band (act) is very close to the conduction band, the hopping mechanism dominates. Lastly, when there is a significant overlapping of the impurity band and the conduction band, a state known as delocalization, we have WL being the dominant mechanism (e.g., in semi-metallic/metallic films).

Heavily doped UNCD films with 10%–20% nitrogen (in the gas phase) were found to be metallic. Moderately doped films with 5%–8% nitrogen (in the gas phase) were semi-metallic and films with less than 5% nitrogen were semiconducting as seen in Figure 6.3. At high temperatures, thermally activated conduction is normally expressed in the form of the Arrhenius equation $\sigma(0, T) = \sigma_0 e^{-\Delta E/kT}$, where ΔE is the activation energy $\sigma(0, T)$ and σ_0 are the conductivities at temperature (T) and minimum temperature, respectively.

We thus do the Arrhenius plot, which is Figure 6.3. The linearity of this plot in the range from 28 to 89 K indicates that indeed we have activated conduction in this temperature region. The activation energy calculated from the plot is found to be approximately 0.128 meV. The low activation energy and the non-Arrhenius behavior at higher temperatures suggest the presence of an impurity conduction band slightly lower the conduction band, which implies that at higher temperatures the electrons are excited into the conduction band and thus we observe a semi-metallic or metallic behavior.

Since we are investigating a polycrystalline material, we might have amorphous components in the GBs, and hence possible contributions from hopping. One such model is the Variable Range Hopping (VRH) put forward by Mott and Davies and also Efros–Shklovskii (ES), which is observed in the presence of a coulomb gap. Under these models, the R–T variation is governed by the equation: $R(0) = R_0 e^{\left(-\frac{T_0}{T}\right)^{1/d+1}}$,

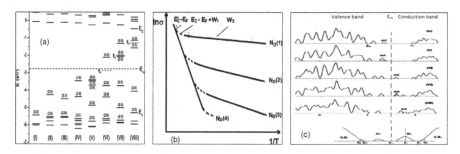

FIGURE 6.3 (a) Electronic structures of various N-centers formed in NCDs III and IV are N-DB complexes, V and VI are N–π, while VII and VIII are NV complexes. E_c is the bottom of the conduction band, E_m is the middle of the band, and E_v is the top of the valence band. (b) Illustration of the temperature and nitrogen concentration dependences of conductivity expected on model shown in this figure. W1 and W2 are activation energies. ND(1)–(4) are films with different nitrogen concentrations, with ND(1) having the highest concentration [28]. (c): (top) Schematic of the DoS of clusters in the conduction and valence band. NV complexes have donor and acceptor sites (act) near the middle of the band gap. (bottom) The effective donor and acceptor band for an n-type UNCD film [29].

where R_0 is the characteristic resistance, which may depend slowly on temperature. And d has a value depending on the nature of the hopping process, i.e., the dimension of the hopping. T_0 is the characteristics temperature coefficient which depends on the DoS ($N(E_F)$) at the Fermi level. $T_{0,\text{Mott}} = \dfrac{18}{k_B}\xi^3 N(E_F)$, $T_{0,\text{ES}} = 2.8e^2/k_B\xi\varepsilon$ in 3D, k_B is the Boltzmann constant, ξ is the localization length, ε is the dielectric constant, and e is the elementary charge of an electron. To check these models, we plot Ln R vs. T^{-x} where x can be 2, 3, or 4 for 1D/(ES), 2D, or 3D hopping, respectively. Figure 6.4 shows that this mechanism applicable for 1D and 2D systems does not work fully for N-UNCD films. There is, however, some 3D VRH contribution that needs to be confirmed with MR measurements.

We have seen that at higher temperatures we do not observe the hopping mechanism or activated conduction; we therefore check for band conduction mechanism or WL, given below.

6.5.1 WL IN NANODIAMOND FILMS

Heavily nitrogen-doped nanodiamond films are metallic as indicated earlier. In these materials, WL becomes the dominant process, and the conduction is diffusive. According to Drude's model, the resistance of a material increases linearly with the number of electron scatterers and consequently the length of the material. In this context, at high temperatures, the phase relaxation length of the material is always less than the mean free path. The picture is, however, different at low temperatures and in low-mobility materials; the phase relaxation length can be longer than the mean free path. And thus, the material can be viewed as a series of phase-coherent units with elastic scatterers as indicated in Figure 6.2. Interference between different scattering events of an electron's time-reversed path results in enhanced backscattering, and thus the electron becomes weakly localized. This localization can be 1D if it is within the same layer of the lattice; otherwise, it can be 2D or 3D. The effect can, however, be destroyed

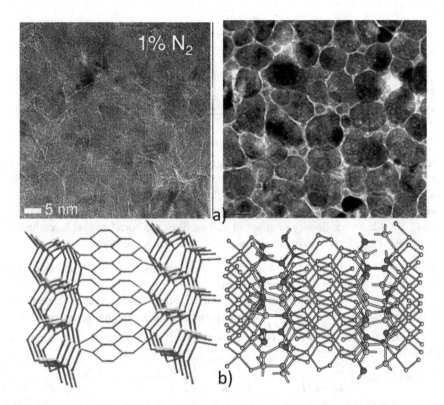

FIGURE 6.4 (a) High-resolution TEM micrograph of UNCD films and nitrogen-doped nanodiamond films [2]. (b) The proposed atomic model structure of grain boundaries of nitrogen-doped nanodiamond films is shown [25,29]. Schematic illustration of a superlattice-like structure that could exist in the GB of nanodiamond films. These structures can be compared to "polymorphs" of UNCD films particularly diaphite which are exclusively present in meteoritic nanodiamonds as a diamond-graphene hybrid.

by a small field – typical values of such a field can be 100 G. When this happens, it results in what is called negative MR, i.e., the conductance decreases with an increase in magnetic field. Because this effect is very sensitive to phase relaxation length, unlike other transport mechanisms it can be used to measure phase relaxation length.

The conductivity is governed by the following equation in both 2D and 3D isotropic systems: $\sigma(0,T) = \sigma_0 + \Delta\sigma(0,T)w_L + \Delta\sigma(0,T)_{e-e}$. The total conductivity is a sum of the classical Drude conductivity and WL and electron–electron interaction corrections. In 2D the WL is partly given by

$$\Delta\sigma(0,T) = e^2 \Big/ \pi h \ln\left(\frac{1}{L_{Th}}\right) = B\ln(T)(\text{ohm}\cdot\text{cm})^{-1}.$$ The electron–electron interac-

tions component is given by $\Delta\sigma(0,T) = \left(\dfrac{e^2}{4h\pi^2}\right)\left(2 - \dfrac{3}{2}F\right)\left(\ln\left(\dfrac{\tau k_B T}{\hbar}\right)\right)(\text{ohm}\cdot\text{cm})^{-1}$,

where F, τ, and k_B represent the electron screening factor in 2D, the relaxation time for the e–e interactions, and the Boltzmann constant, respectively. To test the applicability of this model we check for $\ln T$ dependence of the conductance. We do not

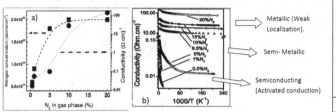
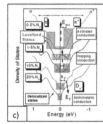

FIGURE 6.5 (a) Correlation between conductivity and nitrogen atomic concentration in the UNCD films as a function of nitrogen percentage in the gas phase. (b) Conductivity vs. inverse of temperature for different N-UNCD films grown with different nitrogen percentages. (c) An illustration of the variation of density of states with nitrogen percentage [2,30]. Electronic density of states of carbon films are correlated with conduction mechanisms e.g., activated, hopping, and semi- metallic conduction processes.

observe any logarithmic behavior in conductance, and this eliminates the 2D mechanism; thus we focus on the 3D mechanism.

In terms of this band conduction model, the conductance is given by $G(0,T) = \left[G_0 + a_1 T^{0.35} + a_2 T^{0.5} \right] \frac{s}{l}$, where the ratio s/l is the sample geometrical conversion to conductivity with s being the cross-sectional surface area and l is the length of the sample.

$$a_1 = e^2/2\pi\hbar L_\phi \left(\frac{C^2}{\text{kg}} \frac{s}{\text{m}^3} \right) \text{ and } a_2 = \left(\frac{e^2}{4\hbar\pi^2} \right) \left(\frac{1.3}{\sqrt{2}} \right) \left(\frac{4}{3} - \frac{3}{2} F \right) \sqrt{k_B/\hbar D} \left(\frac{C^2}{\text{kgm}^4} \frac{s}{k^{0.5}} \right)$$

From the fitting of the conductance graph, using this equation, we can see that the model approximates the material behavior well at the intermediate temperature range (40–180 K). The analysis in the high-temperature region becomes complex because so many factors come into play, such as phonon interactions. The fitting equation is however slightly modified with the temperature dependence of the WL component being $T^{0.33}$ and not $T^{0.35}$. The fitting parameters are $a_1 = 2.4 \times 10^{-5} \left(\frac{C^2}{\text{kg}} \frac{s}{\text{m}^3} \right)$, $a_2 = 9.3 \times 10^{-4} \left(\frac{C^2}{\text{kg m}^4} \frac{s}{k^{0.5}} \right)$, and $G_0 = 1 \times 10^{-4}$ (Ohm^{-1}). These parameters give us the diffusion coefficient $D = 1.202 \times 10^{-7} \text{m}^2\text{/s}$ and the temperature dependence of $L_\phi = 9.15 \times 10^{-8} T^{0.33}$ (m).

Generally, we have observed WL corrections to the classical Drude conductivity, and we have a little contribution from electron–electron interactions for the N-UNCD films compared to microwave films, as we shall see in the subsequent subsection. We will now turn to the MR analysis of these HFCVD-N$_2$ films.

From the graph in Figure 6.7, we observe a negative MR that is linear at high fields and with a B^2 variation at low fields. The negative MR is probably due to WL in our UNCD sample. Reports by other workers have indicated a $B^{1/2}$ or ln B dependence of MR in strongly localized systems, and this is not observed in our system; instead, we observe a linear dependence, which does not seem to saturate or turn upward into the positive regime. The magnitude of this MR decreases with an increase in temperature. In the graphs in Figures 6.7 and 6.8, we normalize this data and fit it

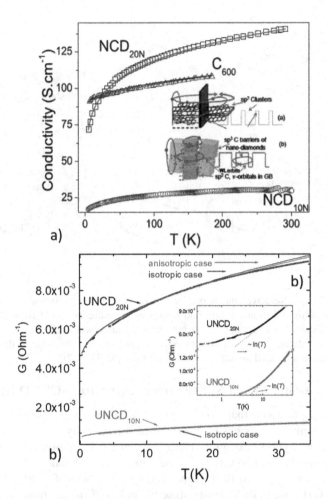

FIGURE 6.6 (a) A comparative study of temperature-dependent conductivity for NCD_{10N}, NCD_{20N}, and C_{600} samples fitted with the WL model. Inset shows the proposed superlattice model and WL orbits in NCD films [31]. (b) Conductivity vs. temperature for different N-UNCD films grown with different nitrogen percentages, which are fitted with anisotropic as well as isotropic WL model [32]. Inset: Conductivity is plotted with the weak localization [lnT] model.

with both isotropic and anisotropic models. The fitting equation for the anisotropic model is given as $\Delta\sigma(B)\big/\sigma(0,T) = \alpha\left[R(0,T)\left(\dfrac{e^2}{2\hbar\pi^2}\right)\left(\dfrac{eB}{\hbar}\right)1^{0.5}\right]f_3\left\{\dfrac{h/eB}{4D\tau_\phi}\right\}$.

The function f_3 is the Kawabata function proposed as

$$f_3(x) = \sum_{N=0}^{\infty}\left\{2\left[(N+1+x)^{0.5}-(N+x)^{0.5}\right]-\left(N+\frac{1}{2}+x\right)^{-0.5}\right\}.$$

N is the Landau quantum number to account for 3DWL in anisotropic cases. The coefficient α describes an isotropic transport in 3D, and the other symbols have their usual meanings. The isotropic model is described by the same equation without the α – the anisotropic coefficient.

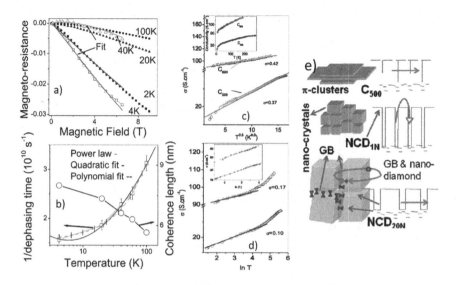

FIGURE 6.7 (a) Negative magnetoresistance of metallic and disordered carbon films. (b) Electron phase coherence length and scattering time are extracted from the MR data. Electrical conductance of conducting carbon films are fitted with (c) $T^{0.5}$ and (d) ln T dependence. (e) Superlattice structure of nitrogen incorporated nanocrystalline diamond films where grains are separated by grain boundary regions. [32,33,69,70].

6.5.2 THREE-DIMENSIONAL SL STRUCTURES APPLIED TO N-UNCD FILMS

It is interesting to notice that a superstructure of diamond grains together with the layers of graphitic planes has been described theoretically as ortho- or para-graphitic-diamond hybrid structures (Figures 6.4 and 6.5). In practice, UNCD films consist of materials of two different band gaps, e.g., sp^3-bonded carbon (nanodiamond crystals of size up to 10 nm) separated by GB regions of width ~1 nm of conducting sp^2C (Figure 6.4). A SL-like structure of UNCD films grown by microwave plasma deposition has been thought of earlier; however, a direct verification of the model from conductivity measurement was not available. Although transport in q-1D carbon systems has been analyzed thoroughly in previous subsections UNCD films analysis of 3D diffusive transport is needed to interpret the SL periodicity, which also includes the disorder parameter (controlled by nitrogen incorporation).

Transport in the SL model depends on the connection between SL layers with a period (c) and also on the dwelling time of an electron (τ_0) in a particular layer (Figure 6.7a). UNCD films after nitrogen doping may be considered as a semi-metallic layered structure with a translational quasi-periodicity, parallel to the surface of the layer (Figure 6.6). Based on the MR measurements, an anisotropic 3DWL effect was shown instead of a 2DWL approach. A possible reason for the anisotropy was suggested as the SL-like structure of UNCD films, which increased with the nitrogen doping level. The presence of the ordered layer-like GB sp^2 regions alternating with the sp^3C dielectric regions was achieved (Figure 6.5). From microscopic studies, many researchers found an increase in the width of the GB, as well as the size of the diamond crystals with increasing the N_2 gas percentage in the reaction chamber. In this process, the number of conducting monolayers per unit length in the samples

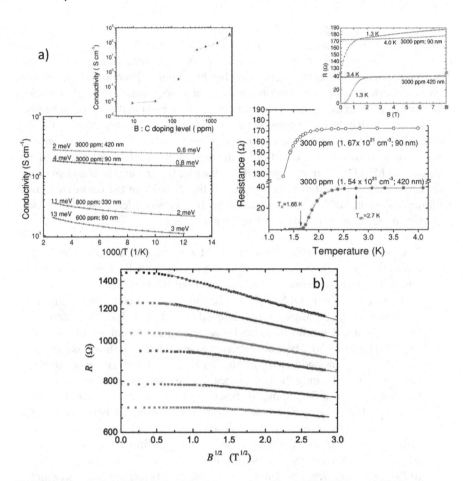

FIGURE 6.8 (left) (a) Electrical conductivity as a function of $1000/T$ for several B-doped nano-crystalline diamond films with thickness ranging from 80 to 420 nm. The samples show a variation of activation energy from 0.6 meV to 3 meV, respectively. The inset shows the plot of the room temperature conductivity for different B:C doping ratios. (right) Resistance vs. T shows a superconducting transition for high boron concentration at 1.66 K. Inset shows R vs. B variation, which shows a slow variation above a field of about 2T. The shadowed arrow shows the metallic B-concentration range for which the mobility could be determined [34]. (b) Magnetoresistance as a function of $B^{1/2}$ for several temperatures of a sample with 20% nitrogen concentration [34].

increases within a fixed distance as a graphite-diamond hybrid structure (Figure 6.4). The 2D carrier concentration in the first miniband estimated to be $>10^{15}$ cm^{-2} is relatively large compared to other conventional SL systems and disordered semiconductors. We have noticed that in several articles, based on the thorough investigation of both structural and transport properties of heavily nitrogen-doped UNCD films, the idea of nitrogen-induced (disordered) sheet-like character separated by almost non-conducting regions of the transport was marked. For a layered material having a long dephasing length (L_ϕ), the transport has been described by a diffusive Fermi surface (DFS) model, which can change into the propagative Fermi surface (PFS) model for strong interlayer coupling (t_z). As a first step, we check the validity of the SL model on the temperature (T) dependence of conductance over a wide range in NCD films at zero, as well as high magnetic fields (B). Two samples prepared with 10% and 20% N_2 in the plasma chamber labeled $UNCD_{10N}$ and $UNCD_{20N}$ have been studied thoroughly (Figures 6.6 and 6.7). From various reports, a long L_ϕ compared to the size of the diamond nanocrystals and their separation has been found. Therefore the incoherent DFS model can be applied to describe the present UNCD samples. If the number of layers increases with the increased incorporation of nitrogen in the reaction chamber, t_z increases, which induces coherent transport in $UNCD_{20N}$ films described by the PFS model. However, to reveal a concrete picture of the proposed SL compatible with the observed microstructure, the effect of localization on the angle-dependent magneto-resistance (AMR) for arbitrary degree of the disorder (Δ) and angles (θ and ϕ) should have been studied. For an arbitrary degree of the disorder in the disordered GaAs multi-layer system, MR has been explained by the Bryksin–Kleinert model by using two anisotropic parameters for a few fixed angles.

i. **Temperature-dependent conductivity at zero magnetic fields:** The DFS model in a quasi-2D system is used to find a correlation between 2D planes, which has been discussed in the context of coherent and incoherent trans-ports. In the coherent DFS regime, electrons change planes coherently, τ_ϕ (~100 ps) $\gg \tau_0$, so the planes are coupled (correlated). Two characteristic times, τ_ϕ and τ_0 determine the DFS transport in the coherent and incoherent regimes (Figure 6.7a). The main difference between the two limits is that the former includes the additional contribution from 3D trajectories, whereas the latter is defined by only 2D trajectories. In the strongly incoherent regime (i.e., for τ_ϕ (~1 ps) $\ll \tau_0$), the SL behaves as a set of independent planes. A general DFS model (based on the SL in Figure 6.7a) has been discussed and compared to the quality of the fit to conducting (σ) data (Figure 6.7b). This model included both coherent and incoherent limits and e–e interac-tion terms. It means that the total conductance is $G(0,T) = S[\sigma_0(0) + \delta\sigma_{||}(0,T) + \delta\sigma_{ee||}(0,T)]/l$, where S and l correspond to the cross-sectional area and length of the sample, respectively. The correction term $\delta\sigma_{||}(0)$ is expressed

as $\delta\sigma_{ll}(0,T) = -\left(\dfrac{e^2}{\pi hc}\right)\ln\left(\dfrac{\tau_\varphi}{\tau}\right) + \left(\dfrac{2e^2}{\pi hc}\right)\ln\left[\left\{\left(\dfrac{1}{2}\right) + \left(\dfrac{1}{2} + \dfrac{t_z^2\tau\tau_\varphi}{2\hbar^2}\right)^{0.5}\right\}\right]$.

The elastic relaxation time τ is a constant of temperature, which makes $[\sigma_0(0) - (e^2/\pi hc)\ln\{\beta/\tau\} + 2e^2/\pi hc\ln^2]$ also a constant. The function $G(0,T) = G_{0tot} + A(p\ln[T] + 2\ln[(1 + (1 + \gamma T^{-p})^{0.5})/2]) + S\delta\sigma_{ee||}(0,T)/l$, was found to fit the measured conductivity data (Figure 6.9a). The low value of "p" has been explained as an effect of non-Fermi liquid behavior in two-dimensional elec-tron gas (2DEG) systems. Also, a low value of "p" can be found in the 1D system and other disordered and granular systems. The temperature-depen-dent parameter γT^{-p}, which reflects the degree of decoherence in the DFS regime, becomes significantly less than unity at higher temperatures (above 100 K) shifting the $\sigma(0,T)$ for $UNCD_{10N}$ samples toward the weakly incoher-ent DFS limit. Overall, the DFS model did work quite well for $UNCD_{10N}$ samples, which made the SL picture more persuasive. The parameter γT^{-p} is almost three times smaller than for $UNCD_{10N}$ samples, reflecting the fact

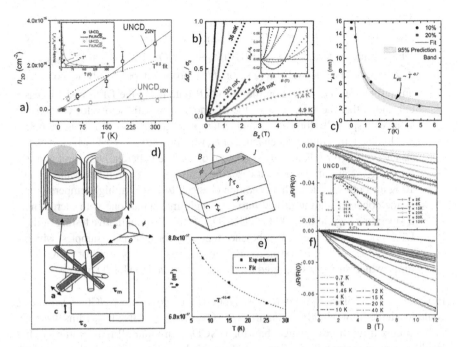

FIGURE 6.9 (a) Dephasing length of electrons in UNCD$_{20N}$ (black square) and UNCD10N (red circle) samples calculated from 3DWL PFS fit to data UNCD$_{20N}$ dashed line and UNCD$_{10N}$ dotted lines samples [32]. (b) Temperature-dependent magnetoresistance data (points) for the 10% N-UNCD sample, and corresponding BK fitting functions (solid lines) are shown from the lowest to the highest part of y-axis. For $T = 4.9$ (circle at the lowest part), 1.4 (triangle), 0.925 (triangle in reverse direction), 0.32 (diamond), and 36 mK (square). Inset: Enlarged portion of the main plot that more clearly demonstrates, by fitting to a second-order polynomial, the deviation from quadratic behavior past a very low threshold of magnetic field. (c) Temperature dependence of the phase-coherence length $L_{\phi,\parallel}$ of the 10% N-UNCD sample was fitted with a power law Tp with the exponent $p \approx -0.7$ [35]. (d) Diamond nanowires are covered by graphitic layers. Bottom: The geometry of the measurements and the SL (carbon layers) structures are drawn schematically. (e) Square of the dephasing length evaluated on the base of the experimental data presented. (f) Magnetoresistance of the NCD films fitted with superlattice model for UNCD$_{10N}$. Inset shows the DFS fit to MR data. 3D DFS fit to MR data of UNCD$_{20N}$ samples at different temperatures (bottom) [32,36].

that at low T (up to 10–20 K) UNCD$_{20N}$ samples are in the coherent regime rather than in the incoherent regime. The decrease of the SL period with the increase of the nitrogen gas concentration from 10% to 20% in the reaction chamber could be conceived as a rearrangement of the UNCD film's internal structures. Induced by nitrogen doping, this process leads to the emergence of numerous conducting layered-like inclusions, which is treated as a reduction of the period of an effective SL to a value of ~8 nm. This model can explain the main reason for the conductivity increase of UNCD$_{20N}$ samples by one order of magnitude over the UNCD$_{10N}$ samples without changing the nitrogen concentration within the films as reported earlier.

ii. **MR and 3D anisotropic DFS description:** The conductivity correction is given by $\sigma_{\parallel}^{3D}(B,T) - \sigma(0,T) = -(e^2/2\pi^2\hbar)(c)^{-1} f_3(x,x')$, in the strongly DFS regime, where $x' = \hbar/4eBD_{\parallel}\left[1/\tau_{\varphi} + 2t_z^2\tau/\hbar^2\right]$, $x = \left(\hbar/4eBD_{\parallel}\tau_{\varphi}\right)$, and $\tau_0 = \hbar^2/t_z^2\tau$. The Kawabata function f_3 is proposed previously to consider anisotropic 3DWL. Based on the fit to MR data using the DFS model it was concluded that the DFS limit of the SL model was satisfactory for UNCD$_{10N}$ samples, especially for $B < 8$ T and at temperatures up to 100 K (Figure 6.8a and b). The temperature dependence of L_{ϕ} was found to be $\sim T^{-0.35}$, which is consistent with $G(0,T)$ data analysis (Figure 6.8c). The coherent DFS limit works for $\tau_{\phi} \gg \tau_0$, and in the present case, the condition $\tau_{\phi} \geq \tau_0$ remains valid up to 10 K for both samples. This condition changed to $\tau_{\phi} < \tau_0$ at high temperatures as the coherent regime turned into a strongly incoherent one with increasing temperature, e.g., above 100 K in UNCD$_{10N}$ films. The qualitative testing of the DFS model ascertained the coherence tendency for UNCD$_{20N}$ samples, which produced $L_{\phi} \sim T^{-0.35}$; however, the PFS model worked very well for UNCD$_{20N}$ films, which strongly supported the SL model applied to UNCD films.

From the analysis of transport data, it was found that at low T for UNCD$_{20N}$ samples L_{ϕ} could reach the limit of 100 nm (Figure 6.7b). Overall L_{ϕ} for UNCD$_{20N}$ films was found to be almost three times greater than for UNCD$_{10N}$ films at all temperatures. These results were consistent with the analysis of $p = 0.7$ derived from $G(0,T)$ data in the previous subsection. The behavior of the $L_{\phi}(T)$ curves using DFS fit looked very different from the 2DWL model, proposed for 2D turbostatic graphitic carbon or pyrolyzed polymer chains, where τ_{ϕ} was strictly T^{-1}-dependent. Overall, the weak temperature dependence of the characteristic time has been found very similar to SL structures of GaAs systems at least qualitatively. To follow the changeover of the coherence of the electronic transport with temperature in the proposed SL structures, the corresponding parameter has been described $2\tau_{\phi}/\tau_0 = 2t_z^2\tau\tau_{\phi}/\hbar^2$. A large value of this parameter can confirm the coherent transport of electrons between layers like in an ideal 3D structure. Assuming τ_0 as a constant, the fit to the MR data for different temperatures was performed in the frame of the DFS approach that in turn defined the temperature dependence of the coherence parameter $2t_z^2\tau\tau_{\phi}/\hbar^2$ accurately. The desired evolution of the coherence parameter for UNCD$_{10N}$ and UNCD$_{20N}$ films is shown in Figure 6.7b, where the UNCD$_{10N}$ samples possess a coherence parameter three times larger than for UNCD$_{20N}$ films which are consistent with the analysis of $G(0,T)$ data.

Finally, the ratio $[t_z]_{UNCD10N}/[t_z]_{UNCD20N}$ can be expressed as $(v_F)_{UNCD10N}/(v_F)_{UNCD20N} \times (3.25/1.73) \approx 0.6$. As a consequence a lower value of the coupling parameter for UNCD$_{10N}$ films than for UNCD$_{20N}$ films was found, which agreed with the qualitative understanding of the SL conception because for the latter material lower value of the periodicity than for the former one was yielded. The parallel and perpendicular conductivity is represented by $\sigma_{\parallel}(0,T) = e^2 D_{\parallel} n(\varepsilon_F)$ and $\sigma_z(0,T) = e^2 D_z n(\varepsilon_F)$ evaluated by the corresponding diffusion coefficients $D_{\parallel} = v_F^2\tau/2$ $D_z = D_{\perp} = \left(t_z c/\sqrt{2}\hbar\right)^2\tau$. Considering the total quasi-2D DoS $n(\varepsilon_F) = m^* N/\pi\hbar^2$, $\alpha = \sqrt{\sigma_{\parallel}/\sigma_{\perp}} = \sqrt{D_{\parallel}/D_{\perp}} = v_F \hbar/t_z c$ was derived, where N, m^*, and ε_F represent the number of the wells, the electron effective mass, and the Fermi energy, respectively.

Analysis of experimental data finds $\alpha \sim 3$ for $UNCD_{20N}$ films at low temperatures, e.g., below 2 K, which can be reduced to 1.6 at higher temperatures. From the experimental value of $n(\varepsilon_F) \sim 10^{16} cm^{-2}$ (measured from Hall voltage) and the estimated value of $v_F \sim 10^5$ ms^{-1}, an expression for $c \sim 10^{-9}/t_z$ (eV) was derived. Assuming $t_z \sim 0.1$ eV, the value of $c \sim 10$ nm for $UNCD_{20N}$ films was derived, which matches the distance to GB layers separated by nanodiamond crystals.

6.5.3 PROPOSED TRANSPORT MODEL: DIFFUSIVE TRANSPORT IN LOW-DIMENSIONAL DISORDERED CARBON FILMS

A conductivity model showing the movement of the quasi-Fermi level and an enhancement of spectral conductivity in a partially delocalized π-state of low-dimensional disordered carbon is developed, which also explains the microscopic origin of the experimentally observed high n-type conductivity of nanocrystalline diamond films and diffusive mobility. The non-dispersive transport in carbon having a long coherence length was explained by resonant tunneling in these weakly localized and disordered systems within a modified 2D DoS by matching the E_F of the electrodes with the maxima of the disorder-induced spread-out resonant states. We redefine so-called WL as equivalent to the combination of (gapped) activated and (ungapped) hopping processes which also compete at different temperature ranges. Like spin–charge or spin–orbit separation, these two processes can be separated (which are coupled otherwise) yielding a resonant transition.

6.5.4 EXPERIMENT: MOBILITY AND [n] VS. TEMPERATURE, HOPPING, AND DELOCALIZED TRANSPORT

The room temperature values of μ and [n] of N-UNCD films are found to be more than 1 cm^2/Vs and 10^{19}/cm^3, respectively. The temperature dependence of μ and [n] of a-CN_{30} shows slightly activated transport at low temperatures (Figure 6.11b), which is remarkably similar to nitrogen-doped nanocrystalline diamond films. At high temperatures this activated nature of $\mu(T)$ and any dependence of [n] on μ can hardly be seen, which is just opposite to the hopping conduction ignoring the case for low disorder. Above all, the enhanced conductivity along with high values of μ and [n] of N-UNCD films is completely different from weakly conducting and undoped a-C films showing $\mu < 10^{-4}$ cm^2/Vs and [n] $< 10^{18}$/cm^3, where hopping conduction dominates. In addition, the hopping conductivity pre-factor (σ_0), which is inversely proportional to the correlation radius and the hopping (or localization) length R_h, decreases at high temperatures. In the presence of low disorder, as presently observed from $\mu(T)$, R_h becomes weakly temperature dependent and varies inversely with $N(E)$ showing a contradiction to hopping conduction. This also yields a high value of μ which is nearly temperature-independent. This weak temperature-dependent variation of μ indicates a very weak energy dependence of μ except very close to zero energy (see Figure 6.11c). Therefore, the temperature dependence of the conductivity of the samples is explained by the energy and temperature dependence of [n], which is very high as presently observed (Figure 6.11b and c). From the experimentally derived $N(E) \sim 10^{20}$/cm^3 and localization length > 1 nm the localization parameter

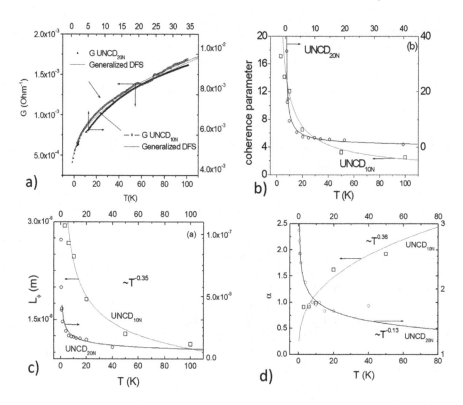

FIGURE 6.10 (a) Variation of conductance of $UNCD_{20N}$ and $UNCD_{10N}$ with temperature. (b) Coherence parameter as a function of T for $UNCD_{10N}$ and $UNCD_{20N}$ samples extracted from 3D DFS fit of MR data (see text). The best-fitted lines are shown. (c) Dephasing length of electrons in $UNCD_{10N}$ (square) and $UNCD_{20N}$ (circle) samples calculated from 3DWL DFS fit to data. (d) The anisotropy parameter plotted as a function of temperature [37].

$[H_0 = (Nv^3)^{-1/3}]$ has been estimated as 1–10 eV for which the hopping rate and mobility can be much higher than in diamond-like carbon films. The conduction is a tunneling-assisted diffusive process where both mobility (μ_D) and diffusion coefficient D decay exponentially with $[n]$ as μ (or D) ∞ $\exp(-C[n]^{-1/3}/\alpha)$, considering α^{-1} as the delocalization length. Although an exact exponential decay of μ vs. $[n]$ is not observed from Figure 6.10b, inset, a sharp decrease of the mobility at low temperature followed by a slow variation in the high-temperature region can suggest a constant nonzero value of $N(E_F)$, i.e., n_0 in the expression of the carrier density $n(E) = [n_0 + \exp(-E/E_0)]$, and a shift of E_F to the extended state which shows delocalized conduction together with the temperature-dependent conductivity (Figure 6.11a) (see Eq. 6.1). As E_F goes very close to zero energy, the Fermi distribution function $f(E)$ becomes ~1 with a weak temperature-dependent $N(E)$ and $\sigma(T)$ can be explained by considering $\mu(E) \sim E^{+0.1}$, which would show a slight decrease of μ at the vicinity of zero energy. This behavior finds similarities with high diffusive mobility in carbon nanotubes, graphite, and semiconductor SLs.

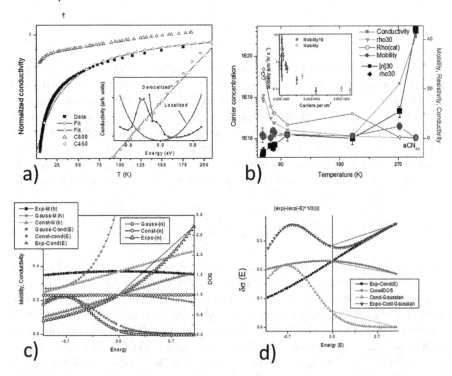

FIGURE 6.11 (a) Variation of experimental $\sigma(T)$ vs. T for a-CN_{30}, $UNCD_{10N}$, and $UNCD_{20N}$ films. Using Eq. (6.1), the simulated curves for $\mu(1 - 0.001$ eV$)$ are plotted, showing different trends of conductivity saturation as temperature tends to 0 K. (b) Variation of experimental $\mu(T)$, and $[n(T)]$ for a-CN_{30}, $UNCD_{10N}$, and $UNCD_{20N}$ films. Inset: Variation of μ with $[n]$, for a-CN_{30} and a-CN_{19} films. (c) Proposed electronic structures and spectral conductivity of a-C films for (i) semi-metallic conduction and (ii) activated and VRH conduction. (iii) A microstructure model. (d) Variation of calculated (i) spectral conductivity and (ii) mobility and carrier concentration with energy for a combined hopping and delocalized conduction. Strong (experimentally) and weak energy variations of $\mu(E)$ with energy in hopping and delocalized conduction system are shown. The corresponding $N(E)$ is also drawn [30].

From the experimentally observed variation of mobility (μ) and carrier concentration [n] with temperature, a model on the electron transport mechanism both for GB of UNCD and low-dimensional amorphous carbon can be proposed. Based on the resonant tunneling in the extended state conduction and modified by disorder an enhanced diffusive μ can be seen due to the shift of the E_F relative to the maxima of spectral conductivity, $\delta\sigma(E)$. Following the approach in disordered metallic alloys, spectral conductivity for carbon is established by adding the quantum mechanical effect resonant tunneling associated with localized states or due to some nanostructures or sp^2-bonded clusters (as discussed in this chapter).

6.5.4.1 Proposed Model(s)

6.5.4.1.1 Disordered Metal

A transport model applied to metallic amorphous carbon and doped UNCD films is presented, which was also used to explain the temperature-dependent conductivity of metallic alloys and polycrystalline thin films. $\sigma(T)$ depends on the electron concentration at the E_F, which is thermally accessible, $\sigma(E_F)$, and the change of spectral conductivity, $\Delta_{qm}\sigma(E)$, in the quantum mechanical energy scale or $\delta\sigma(E)$, which determines the sign of conductivity temperature coefficient. The DoS $N(E)$ and $\sigma(E)$ can be expressed as a combination of Lorentzians, where the height of each Lorentzians ($\beta\gamma_{1,2}$), its width ($\gamma_{1,2}$), and the position ($\delta_{1,2}$) relative to E_F can be found from the fitting parameters of

$$\sigma(T) \text{ such as } \sigma(E) = \frac{\beta}{\pi}\left\{\frac{\gamma_1}{\left(E-\delta_1\right)^2+\gamma_1^2}+\frac{\gamma_2}{\left(E-\delta_2\right)^2+\gamma_2^2}\right\}^{-1}. \tag{6.1}$$

Fitting $\sigma(T)$ in Eq. (6.2) with experimental data gives us parameters for $\sigma(E)$ and we can draw spectral conductivity for metallic disordered carbon in Figure 6.11d. For pure carbon only one Lorentzian and for a-CN$_x$ two Lorentzians (one for impurity) are applicable. The second Lorentzian is due to the hybridization effect of CN bonds, due to the formation of chemical shift expressed as $\mu(T)=E_F-ET^2$ that will determine the positive and negative temperature coefficient of resistivity (TCR) of $\sigma(T)$ vs. T curves. This model is very consistent with the formation of delocalized states of sharp peaks in N-UNCD films but a movement of E_F has not been described. Assuming diffusive transport the only conductivity mechanism, in these conducting carbon films conductivity can be expressed as $\sigma(E)=e^2 N(E) D(E)$ and $N(E)=N(E_F) \mu(E) \sigma(E)/\sigma_0$, where σ_0 is the lowest temperature conductivity, and $\mu(E)$ is the mobility as a function of energy in different regime. This picture can explain the WL effect of conductivity. However, a pseudogap at the E_F can be found, which is very applicable for carbon. As previously suggested, this pseudogap can be opened due to the interaction of the Fermi surface and the Brillouin zone, which is a long-range order effect or due to quantum confinement. But a complete picture of electronic structure, particularly, the presence of localized states, cannot be described by Eq. (6.1) and a modified $\delta\sigma(E)$ as given in Eq. (6.3).

Conductivity in sp^2-bonded clusters created around a nitrogen center can be described as a disordered metal alloy of C and N, where the disorder parameter is included. Based on the model of the Green–Kubo formula applied for metals, we try to understand the contribution from the delocalized sp^{2}C to the presently observed conductivity, which also includes the contribution from hopping transport due to all types of localized states, such as nitrogen or highly disordered sp^{2}C as

$$\sigma(T) = \int_{-\infty}^{\infty}\left[\delta\sigma^\pi(E)\frac{\partial f(E,\mu_F,T)}{\partial E}\right]dE = \int_{-\infty}^{\infty}\left[\delta\sigma_D^\pi(E)\frac{\partial f(E,\mu_F,T)}{\partial E}+\delta\sigma_h^\pi(E)\frac{\partial f(E)}{\partial E}\right]dE.$$

$$\tag{6.2}$$

The first and the second term represent delocalized and localized π states conductions with the spectral conductivity $\delta\sigma_D^\pi(E)[=\mu_D(E) N_D(E) f(E,\mu_F,T)$, similar to diffusive conduction, where $\mu_D(\sim E^{-0.1})]$ and $\delta\sigma_h^\pi(E)$ represent the extended state (or diffusive) and hopping conductivity, respectively. For a metallic state, σ_D varies with E and the chemical potential $\mu_F(T)$, a shift of E_F from zero energy, to maintain a finite conductivity at zero energy as well as zero temperature (see Figure 6.11d). For hopping conduction, $\delta\sigma_h{}^\pi(E)$ would increase exponentially (or behave like Gaussian distribution or constant with E, where the minimum conductivity matches with the zero energy yielding no conductivity at zero Kelvin. The other term in Eq. (6.1), $\partial f(\mu_F, E, T)/\partial E$ signifies the activation energy or the characteristic energy barrier, either temperature-dependent for the delocalized or independent for the localized π states, respectively. Combining the diffusive and hopping processes the elementary conductivity is described:

$$\delta\sigma^\pi(E) = \frac{e^2}{kT}\delta n_D(E)D(E) + e\mu_h(E)N_h(E)f(E)\delta(E)$$

$$= \left[e\mu_D(E)N_D(E) + \frac{4e^2 R_h^3 v_0}{3E_0 N_h(E)}\exp\left\{-2\nu(N_0)^{1/3}\exp\left(E/E_0\right)\right\}N_0\exp\left(E/E_0\right)\right]$$

$$f(E)\delta(E)$$

$$(6.3)$$

Here, $\delta\sigma(E)$ will rise up with E due to the increase of μ_D ($\sim E^{0.1}$) and $N(E_F)$ in the respective processes. As because $\mu_D \gg \mu_h$, revealed from the experimental results, σ_D becomes greater than σ_h, even though $N_h(E)$ could be comparable to $N_D(E)$. As a combination of $\delta\sigma_D{}^\pi$ and $\delta\sigma_h{}^\pi$, $\delta\sigma_{\text{Total}}{}^\pi(E)$ will be strongly energy-dependent but the minimum conductivity shifts from the zero energy (Figure 6.11c and d). The corresponding temperature-dependent conductivity is fitted in Figure 6.11a based on Eq. (6.3), expressed as

$$\sigma(T) = \int_0^1 \frac{\delta\sigma(E,\mu_F,T)}{\delta E}\delta E = \int_0^1 \frac{\left[eE^{0.1}(E)N_0 + \sigma_{0h}\left\{\exp\left(E/E_0\right) - H_0\exp\left(E/E_0\right)\right\}\right]}{1 + \exp\left(\dfrac{E - \mu_F}{kT}\right)}dE.$$

$$(6.4)$$

By adding the effect of localized states, we get

$$\sigma_{\text{total}} = \sigma_0 + \sigma_{\text{activated}} + \sigma_{\text{hopping}}$$

$$= \sigma_B(T) + \sigma_h(T) = e\mu_v N_v(T)\exp[-\mu_F(T)/kT] + \sigma_{0h}\exp\left[-\left(T_{0h}/T\right)^{1/4}\right]. \quad (6.5)$$

This case is similar to the distribution of DoS in DB-RTD structure having disorder. From the simulation, we can see that the slope of the $\sigma(T)$ curve and their convergence at low temperatures depends mostly on the shift of E_F from the zero energy (Figure 6.11a). The activation energy increases as E_F comes closer to the mid-gap or zero energy, i.e., when μ_F becomes smaller. However, for very small values of μ_F no significant changes of $\sigma(T)$ with μ_F can be found. Our analysis shows that $\sigma(T)$ is mainly governed by the movement of E_F even though the hopping formula for the conductivity can generally be considered. The value of the disorder parameter (H_0) does not contribute significantly to the activation energy but to the magnitude of the conductivity. If E_F is pinned at the mid-gap region hopping conduction dominates, however, an addition of the slowly varying diffusive term of conductivity can induce the conductivity saturation at low temperatures, which is described by Eqs. 6.1–6.3. This movement of E_F is similar to the shift of quasi-Fermi level, commonly observed in mesoscopic systems raises the conductivity.

From Figure 6.11a and d, we understand that nitrogen creates new states, which gives slight hopping conduction at high temperatures and adds to the total conductivity. The hump of conductivity is attributed to hopping at a low temperature that decreases and becomes zero at very low temperatures, and therefore, mobility at low temperature is governed by the delocalized tail of σ_D and not by σ_h. The decrease of DoS and $\delta\sigma_D(E)$ with energy is similar to graphite, where at E_F these two parameters become very small at the Fermi point (the junction of conduction band (CB) and valence band (VB)) yielding very small conductivity due to low carrier concentration at the E_F. But, in the case of nitrogen-doped a-C films, a finite DoS at zero energy and a shift of E_F toward high DoS guarantee the delocalized conduction in this system. As a result, through a negative TCR is observed the metallic nature at low temperatures, i.e., finite conductivity can be seen (Figure 6.11a). Nitrogen creates additional states close to E_F, which gives hopping $(\delta\sigma_h)$ conductivity but it also creates some delocalized states; as a result E_F moves to the extended states yielding finite $\delta\sigma_D$. We notice that the peak position (of the hump) and the intensity are close to the CB and very high for the low disorder $(E_0 < 0.1$ eV), which pushes E_F toward CB and yields high conductivity. For high disorder $(E_0 > 0.5$ eV), localized states are created at the mid-gap and with low current and for high disorder it does not change significantly. Here μ_F plays an important role for low E_0 $(0.05–0.1$ eV), but the contribution of μ_D, i.e., the shift of E_F from zero energy, is not significant for high E_0 $(> 0.1$ eV). Although nitrogen does not form a substitutional band with the sp³ network, it creates new states from localized π-clusters that contribute to the delocalized conduction except at very low temperatures.

6.5.4.1.2 Contribution of Resonant Tunneling to Spectral Conductivity

In addition to the exponential tail of the localized states which does not contribute to the spectral conductivity (a series of Lorentzians overlapping with each other) and 2D delocalized states contribution of 1D channels or quasi-1D channels should be evident. Superposition of all this DoS may result in conductivity maxima or minima depending on the sign of the power of energy-dependent on DoS. A similar approach can be found, where, for a 1D chain having several scattering centers, the current was expressed as $I(E,V,T) = e^2/\pi h \; \Sigma T_i(E,V) \otimes W(E,V) \otimes [-df(E,T)/dE]$ for the conductance through several channels, which is a convolution of the voltage window W near the Fermi surface and both energy- and temperature-dependent Fermi function. The Drude–Sommerfeld conductance formula in 1D (also in 2D and 3D) conductor can be considered as a quantum particle moving in a set of potential barriers. The mean free path appearing in the Drude formula can be considered as half of the average spacing between scatters in the diffusive limit of Landauer's conductance formula. In this case $e^2/\pi h \; R_{1D} = R_a/T_a + R_b/T_b + \ldots + 1$; a sum of the ratio of transmission and reflection coefficient from each scattering center or energy-dependent dimensionless resistance $(=R/T)$. The 1D conductance G varies with the chemical potential of the metal contacts and a deviation from an ideal step of $G(=e^2/\pi h)$ based on the phase coherence time can be seen, which is also followed in our model. In addition to the conductivity of the weakly disordered system, the strongly disordered system has been tested through resonant tunneling through several localized states in a row (or N the number of channels for the electrons resonantly tunneling through an $N + 1$ barrier structure) and found a similar expression to Breit–Wigner formula for double-barrier RTD as $\sigma = \{e^2/\pi h\}[\Gamma_n^L \Gamma_n^R/\{(\mu_F \, \varepsilon \, \Delta\varepsilon)^2 + \{(\Gamma_n^L + \Gamma_n^R)/2\}^2\}]$. The resonant tunneling will occur whenever $\mu_F = E \, \Delta E$, producing a maximum conductance [31]. Since the transmission through each barrier or well is different, the net conductance dependence depends on the coupling constants Γ_n^L, Γ_n^R of the site n; however, an experimental demonstration of coherence transport can support this model. However, from previous analysis, it was found that conductance for each subband in parallel wires is additive, particularly in low temperatures less than the energy spacing of subbands, and resonance effects for the balanced wells were preserved. The first test will be the temperature-dependent conductivity in tunnel structures. An accurate determination of the dimensionality from the temperature dependence of conductivity has been found difficult, where 1D transport in a 2DEG system can be found. A theory in 1D conductor based on resonant tunneling at low temperatures was proposed, showing its transition to Mott's VRH at high temperatures. In the intermediate temperature, both resonant and non-resonant conductance follow Mott's $<\ln G> \sim -(T_0 - T)^{1/2}$; however, at lower temperatures, they would differ but ultimately reach T independent in the valleys and have a universal metallic T dependence ($G \sim T^{-1}$) at the resonant peaks. It was suggested that resonant conductance will be seen as long as the width of the eigenstates is less than the energy level spacing of the sample. However, later it was suggested that inelastic scattering length will be important to observe the tunneling. Mott's VRH is always a competitor to RT, which occurs in a shorter time scale (10^{-13} seconds) but RT charge build-up and decay vary with time, through which resonance and non-resonance state appear. Therefore, the net conductance would be a sum of the RT and VRH process for a sample length comparable to the hopping length. However, at very low temperatures the variation of $\ln G$ vs. E_F as a function of T and how far E_F is away from the energy for good resonance E_r will determine the possibility of RT. If E_F is within kT of E_r or outside temperature-independent resonant conduction and an activated or hopping conduction will be manifested, respectively. Thus the position of the E_F as described in the proposed model is so important.

Carbon nanostructured films can be described as localized quantum dots or wells, which are separated by tunnel barriers. The delocalized part is modeled as a ring structure, which can be a graphitic ring or an extended structure that can be connected in a mutually perpendicular configuration or at an angle. The transport (as a function of magnetic fields) can be modeled as tunneling between the rings where phase coherence remains unchanged. A set of qubits can be employed to construct the rings, which are connected by tunnel barriers. Quantum simulations of these ring structures connected by tunnel barriers will be shown in Chapter 10. The qubits represent the carbon atoms or quantum dots, a two-level system can be operated by a microwave signal, which records the return probability of electrons and the phase from the coherent backscattering process. Hence, the change of the resistance with the scattering time or the detuning time of the qubits, equivalent to the magnetic fields, can be recorded. Although the qubits can be arranged in the form of a hexagon mimicking a graphene lattice, disordered structures can also be created effectively by linking the qubits with arbitrary coupling parameters. Resonant structures can be formed by adjusting the coupling between the qubits. All these features are discussed in Chapters 9 (partially) and 10. Although resonant tunneling and the WL effect can be simulated by a quantum computer, similar work for the VRH can be more challenging. We make a suggestion on the qubit operations and the analysis of the structure of the disordered carbon.

The VRH conduction can be described by the rotation of qubits from the ground to the excited states by absorbing phonons or microwave photons followed by a tunneling process, which can be modeled by the qubit rotations. Finally, the state is de-absorbed with the emission of phonons where the electrons reach a different state compared to the initial state. The emission and absorption can be compared with the "read" and "write" processes of a qubit. In the VRH process, the E_F does not shift and the MR varies very weakly with B. Since a qubit can be rotated at different angles and produce an "n" number of states, it can be compared with the VRH process, which can also have "n" states. A probability distribution of the state can describe the electronic properties of the carbon system with low mobility. It seems this multi-qubit system interaction between qubits produces "noise." Hence, the disorder-induced transport in amorphous carbon can be described, which is always present in addition to the activated or resonant tunnel transport. E_F can be shifted based on the strength of the electric fields, which is generated by spin–orbit interactions. By the accumulation of the geometric phase in a loop structure or by the holonomic operation of qubits, Rashba spin–orbit coupling can be created, which shifts the E_F based on the direction of the rotation or the sign of the geometric phase. If nitrogen addition created a loop structure or induces more ring-structured sp^2-bonded carbon, more geometric phase can be accumulated, which results in a shift of the E_F. From the shift of the minima of P_r as a function of the detuning frequency (time) or the B field the shift of E_F can be estimated. However, it is difficult to perform this simulation as a function of temperature directly. By changing the configuration of the qubits the decay of the time (or the detuning frequency) of P_r can be changed dramatically from the sharp decay to the slow decay process. This P_r (or the DoS) vs. frequency (or E) can be matched with the model of VRH transport, where this shows a flat response (around the E_F) to the delocalized transport with a resonant peak feature, which represents a short and a long scattering time, respectively. These DoS vs. E can explain the temperature-dependent conductivity (Figure 6.11a and d).

6.6   SUMMARY

From the fit of the $\sigma(T)$ data and $\sigma(E)$, the electronic structure close to E_F of the presently investigated materials can be understood, which suggests possible improvements of these materials. For very weakly conducting samples, $N(E_F)$ is almost zero with the conductivity as activated or hopping at the band edge because of the presence of a finite gap. As the defect, DoS rises and fills up the energy gap and conductivity of the samples is governed by VRH conduction at E_F. Evidence of the Coulomb gap can be found with the increase in carrier concentration. However, for a large overlapping of π–π^* states in the metallic region the conductivity $\delta\sigma$ of the samples depends on the shift of E_F from the mid-gap region. In the sp^2 carbon system, like in disordered metal alloys, the Hall coefficient changes with E_F, which cannot be attributed to true substitutional doping, however, produces high n-type conductivity. Therefore, a model used in disordered metals can be used.

In this chapter, several ways of electronic transport in disordered materials were explained. Depending on the ratio of the length of the sample and the localization length, we see exponential decay (sub-diffusive) or constant (ballistic) nature of G with length and the sign of temperature-dependent conductivity changes accordingly based on the coherence time. Our model is an alternative approach to WL, where delocalization of wavefunctions can be achieved but consistent with the results. However, we extract more information on mobility and [n] based our model in addition to the spectral conductivity. It also identifies the point where conductivity starts, thus giving a complete picture of the electronic structure of carbon beyond the simple explanation of the WL picture.

The GB structures of diamond films show a good example of hybrid structures. In cubic diamonds, a small space or vacancy can also be created by nitrogen incorporation (or boron), which can store immense information useful for quantum technology employing light–matter interaction. This concept of making carbon-based ultrafast memory devices can be achieved in diamond films combined with the graphitic carbon structure. We would like to explain the fundamentals of resonant tunneling and associated spectroscopy which measures the symmetry and disorder of a structure and probes the magnificence of complicated nanostructures within the diamond. The microwave interaction as a probe can exhibit the true usefulness of the microstructure in quantum processing. This is the basis of artificial atoms or qubits that can be rotated by microwave radiation or a laser beam.

Q. From the scattering matrix calculate the electronic conductivity in quantum wells.

- Main problem discussed in this chapter.

  In the low-dimensional and disordered carbon the low-temperature electronic transport is discussed based on various models and we would like to separate the quantum effects. In the presence of high magnetic fields the resistance of the samples show a nonlinear trend from which diffusion coefficient of electron can be determined.
- What has been achieved?

  Chapter 6 shows quantum transport in disordered carbon films and nanostructures. The WL effect can be created with two orbits (rotating in opposite directions), whereas the WAL can be established with three orbits (associated with one orbit and two spins). The interactions between zeros can explain spin–orbit couplings.
- What has not been achieved?

  Chapter 6. Nitrogen-incorporated nanocrystalline diamonds showed high metallic transport through the atomically thin grain boundaries. The conduction mechanism is not clearly understood, since the structure of the grain boundaries has not been established experimentally.

BIBLIOGRAPHY

1. H. C. F. Martens, O. Hilt, H. B. Brom, P. W. M. Blom, and J. N. Huiberts, Voltage-modulated millimeter-wave spectroscopy on a polymer diode: Mesoscopic charge transport in conjugated polymers, *Physical Review Letters* 87, 086601 (2001); H. C. F. Martens, I. N. Hulea, I. Romijn, H. B. Brom, W. F. Pasveer, and M. A. J. Michels, Understanding the doping dependence of the conductivity of conjugated polymers: Dominant role of the increasing density of states and growing delocalization, *Physical Review B* 67, 121203 (2003); F.-C. Hsu, V. N. Prigodin, and A. J. Epstein, Electric-field-controlled conductance of "metallic" polymers in a transistor structure, *Physical Review B* 74, 235219 (2006).

2. S. Bhattacharyya, O. Auciello, J. Birrell, J. A. Carlisle, L. A. Curtiss, A. N. Goyette, D. M. Gruen, A. R. Krauss, J. Schlueter, A. V. Sumant, and P. Zapol, Synthesis and characterization of highly conducting nitrogen-doped ultra-nanocrystalline diamond films, *Applied Physics Letters* 79, 1441 (2001); S. Bhattacharyya, Mechanism of conduction in nitrogen doped nanocrystalline diamond, *Physical Review B* 70, 125412 (2004).

3. O. A. Williams, S. Curat, J. E. Gerbi, D. M. Gruen, and R. B. Jackman, N-Type conductive ultrananocrystalline diamond films grown by hot filament CVD, *Applied Physics Letters* 85, 1680 (2004).

4. P. Achatz, O. A. Williams, P. Bruno, D. M. Gruen, J. A. Garrido, and M. Stutzmann, Effect of nitrogen on the electronic properties of ultrananocrystalline diamond thin films grown on quartz and diamond substrates, *Physical Review B* 74, 155429 (2006); P. Achatz, J.A. Garrido, and M. Stutzmann, O. A. Williams, D. M. Gruen, A. Kromka, and D. Steinmueller, Optical properties of nanocrystalline diamond thin films, *Applied Physics Letters* 88, 101908 (2006).

5. J. J. Mares, P. Hubik, J. Kristofik, D. Kindl, M. Fanta, M. Nesladek, O. Williams, and D. M. Gruen, Weak localization in ultrananocrystalline diamond, *Physics Letters* 88, 092107 (2006).

6. S. Bhattacharyya, O. Madel, S. Schulze, P. Häussler, M. Hietschold, and F. Richter, Investigation of local structure of nitrogenated carbon films by electron diffraction and imaging, *Physical Review B* 61, 3927 (2000).

7. S. Bhattacharyya and S. R. P. Silva, Transport in low-dimensional carbon, *Thin Solid Films*; S. Bhattacharyya, S. V. Subramanyam, D. L. Wise (Eds), *Electrical and Optical Polymer Systems: Fundamentals, Methods and Applications*, Marcel Dekker, New York, 1998, 201; S.B. Private communications.

8. C. Godet, G. Adamopoulos, S. Kumar, and T. Katsuno, Optical and electronic properties of plasma-deposited hydrogenated amorphous carbon nitride and carbon oxide films, *Thin Solid Films* 482, 24 (2005); C. Godet, Physics of bandtail hopping in disordered carbons, *Diamond and Related Materials* 12, 159 (2003); ibid, Variable range hopping revisited: the case of an exponential distribution of localized states, *Journal of Non-Crystalline Solids* 299, 333 (2002); ibid, Electronic localization and bandtail hopping charge transport, *Physical Status Solid* 231, 499 (2002); S. Kumar, C. Godet, A. Goudovskikh, J. P. Kleider, G. Adamopoulos, and V. Chu, High-field transport in amorphous carbon and carbon nitride films, *Journal of Non-Crystalline Solids* 338, 349 (2004).

9. E. A. Ekimov, V. A. Sidorov, E. D. Bauer, N. N. Mel'nik, N. J. Curro, J. D. Thompson, and S. M. Stishov, Superconductivity in diamond, *Nature* 428, 542 (2004); E. Bustarret, J. Kamarik, C. Marcenat, E. Gheeraert, C. Cytermann, J. Marcus, and T. Klein, Dependence of the superconducting transition temperature on the doping level in single-crystalline diamond films, *Physical Review Letters* 93, 237005 (2004).

10. G. Abrasonis, R. Gago, M. Vinnichenko, U. Kreissig, A. Kolitsch, and W. Moeller, Sixfold ring clustering in sp^2-dominated carbon and carbon nitride thin films: A Raman spectroscopy study, *Physical Review B* 73, 125427 (2006); E. Betranhandy and S. F. Matar, A model study for the breaking of cyanogen out of CNx within DFT,

Diamond and Related Materials 15, 1609 (2006); A. N. Enyashin and A. L. Ivanovskii, Structural models and electronic properties of cage-like C3N4 molecules, *Diamond and Related Materials* 14, 1 (2005).

11. C. V. Landauro, and H. Solbrig, Modeling the electronic transport properties of Al–Cu–Fe phases, *Physica B: Condensed Matter* 301, 267 (2001); ibid *Materials Science and Engineering* 294, 600 (2003); C. V. Landauro, E. Marcia, and H. Solbrig, Analytical expressions for the transport coefficients of icosahedral quasicrystals, *Physical Review B* 67, 184206 (2003); E. Macia, T. Takeuchi, and T. Otagiri, Modeling the spectral conductivity of Al-Mn-Si quasicrystalline approximants: A phenomenological approach, *Physical Review B* 72, 174208 (2005); C. V. Landauro, and T. Janssen, Study of the conductivity of thin quasicrystalline films and its relation with the electronic friction, *Physica B: Condensed Matter* 348, 459 (2004).

12. S. Adhikari, H. R. Aryal, D. C. Ghimire, A. M. M. Omer, S. Adhikary, H. Uchida, and M. Umeno, Optoelectronic properties of nitrogenated amorphous carbon films synthesized by microwave surface wave plasma chemical vapor deposition system, *Diamond and Related Materials* 15, 1894 (2006); Y. Hayashi, N. Kamada, T. Soga, and T. Jimbo, Characterization of amorphous carbon nitride by bottom-gated thin-film structure, *Diamond and Related Materials* 15, 1015 (2006).

13. S. Ababou-Girard, F. Solal, B. Fabre, F. Alibart, and C. Godet, Covalent grafting of organic molecular chains on amorphous carbon surfaces, *Journal of Non-Crystalline Solids* 352, 2011 (2006).

14. M. Rovere, S. Porro, S. Musso, A. Shames, O. Williams, P. Bruno, A. Tagliaferro, and D. M. Gruen, Low temperature electron spin resonance investigation of ultrananocrystalline diamond films as a function of nitrogen content, *Diamond and Related Materials* 15, 1913 (2006).

15. C. E. Nebel, R. A. Street, N. M. Johnson, and J. Kocka, High-electric-field transport in a-Si:HI Transient photoconductivity, *Physical Review B* 46, 6789 (1992); M. C. J. M. Vissenberg and M. Matters, Theory of the field-effect mobility in amorphous organic transistors, *Physical Review B* 57, 12964 (1998); R. Schmechel, Gaussian disorder model for high carrier densities: Theoretical aspects and application to experiments, *Physical Review B* 66, 235206 (2002); Y. J. Fei et al., Magnetoresistance of boron-doped chemical vapor deposition polycrystalline diamond films, *Diamond and Related Materials* 11, 49 (2002).

16. Y.-P. Zhao, B. Q. Wei, P. M. Ajayan, G. Ramanath, T.-M. Lu, G.-C. Wang, A. Rubio and S. Roche, Frequency-dependent electrical transport in carbon nanotubes, *Physical Review B* 64, 201402 (2001); K. Takai, M. Oga, H. Sato, T. Enoki, Y. Ohki, A. Taomoto, K. Suenaga, and S. Iijima, Structure and electronic properties of a non-graphitic disordered carbon system and its heat-treatment effects, *Physical Review B* 67, 214202 (2003).

17. K. K. Choi, B. F. Levine, R. J. Malik, J. Walker, and C. G. Bethea, Periodic negative conductance by sequential resonant tunneling through an expanding high-field superlattice domain, *Physical Review B* 35, 4172 (1987); Y. J. Mii, R. P. G. Karunasiri, and K. L. Wang, Electrical and optical properties of GaAs/AlGaAs multiple quantum wells grown on Si substrates, *Applied Physics Letters* 53, 2050 (1988); S. K. Kim, T. W. Kang, C. K. Chung, C. Y. Hong, and T. W. Kim, Electrical transport studies in disordered GaAs/AlAs superlattices, *Thin Solid Films* 257, 94 (1995).

18. S. Datta, *Electronic Transport in Mesoscopic System*, Cambridge University Press, Cambridge, 1995.

19. K. S. Novoselov et al., Electric field effect in atomically thin carbon films, *Science* 306, 666 (2004); K. S. Novoselov, A. K. Geim, S. V. Morozov, D. Jiang, M. I. Katsnelson, I. V. Grigorieva, S. V. Dubonos and A. A. Firsov, Two-dimensional gas of massless Dirac fermions in graphene, *Nature* 438, 197 (2005).

20. S. V. Kravchenko, G. V. Kravchenco, J. E. Furneaux, V. M. Pudalov, and M. D'Iorio, Possible metal-insulator transition at B=0 in two dimensions, *Physical Review* 50, 8039 (1994); M. Ya. Azbel, Quantum particle in a random potential: Exact solution and its implications, *Physical Review B* 45, 4208 (1992).

21. E. Abrahams, P. W. Anderson, D. C. Licciardello, and T. V. Ramakrishnan, Scaling theory of localization: Absence of quantum diffusion in two dimensions, *Physical Review Letters* 42, 673 (1979).

22. V. I. Kozub, and N. V. Agrinskaya, Metal-insulator transition in two dimensions: Role of the upper Hubbard band, *Physical Review B* 64, 245103 (2001).

23. S. Kugler and I. Laszlo, Connection between topology and π-electron structure in amorphous carbon, *Physical Review B* 39, 3882 (1989).

24. N. M. Goncharuk, The influence of an emitter accumulation layer on field emission from a multilayer cathode, *Materials Science and Engineering* A353, 36 (2003); Q.-A. Huang, J. K. O. Sin, and M. C. Poon, Field emission from silicon including continuum energy and surface quantization, *Applied Surface Science* 119, 229 (1997); Z. B. Li, X. W. Liu, N. S. Xu, S. Z. Deng, J. Chen, M. M. Wu, S. Ren, J. Chen, and F. L. Zhao, Resonant field emission through amorphous diamond thin films (a model study), *Ultramicroscopy* 95, 75 (2003).

25. M. J. M. de Jong, Transition from Sharvin to Drude resistance in high-mobility wires, *Physical Review* 49, 7778 (1994).

26. J. H. Schon, Ch. Kloc, and B. Batlogg, Low-temperature transport in high-mobility polycrystalline pentacene field-effect transistors, *Physical Review B* 63, 125304 (2001).

27. P. F. Bagwell and T. P. Orlando, Landauer's conductance formula and its generalization to finite voltages, *Physical Review B* 40, 1456 (1989).

28. J. Zang and J. hL. Birman, Theory of coherent transport through a strongly disordered system: Resonant tunneling in the one-dimensional tight-binding model, *Physical Review B* 47, 10654 (1993); T. Kawamura, H. A. Fertig, and J. P. Leburton, Quantum transport through one-dimensional double-quantum-well systems, *Physical Review B* 49, 5105 (1994).

29. T. J. Thornton, M. Pepper, H. Ahmed, D. Andrews, and G. J. Davies, One-dimensional conduction in the 2D electron gas of a GaAs-AlGaAs heterojunction, *Physical Review Letters* 56, 1198 (1986).

30. M. Ya. Azbel, A. Hartstein, and D. P. DiVincenzo, T dependence of the conductance in quasi one-dimensional systems, *Physical Review Letters* 52, 1641 (1984); B. Ricco and M. Ya. Azbel, Tunneling through a multiwell one-dimensional structure, *Physical Review B* 29, 4356 (1984).

31. A. D. Stone and P. A. Lee, Effect of inelastic processes on resonant tunneling in one dimension, *Physical Review Letters* 54, 1196 (1985).

32. K. Chun and N. O. Birge, Dissipative quantum tunneling of a single defect in a disordered metal, *Physical Review B* 54, 4629 (1996).

33. T. Kawamura, H. A. Fertig, and J. P. Leburton, Quantum transport through one-dimensional double-quantum-well systems, *Physical Review B* 49, 5105 (1994).

34. Y. L. He, G. Y. Hu, M. B. Yu, M. Liu, J. L. Wang, and G. Y. Xu, Conduction mechanism of hydrogenated nanocrystalline silicon films, *Physical Review B* 59, 15352 (1999); W. Pan, J. J. Lu, J. Chen, and W. Z. Shen, Resonant tunneling characteristics in crystalline silicon/nanocrystalline silicon heterostructure diodes, *Physical Review B* 74, 125308 (2006).

35. P. Sheng, E. K. Sichel, and J. I. Gittleman, Fluctuation-induced tunneling conduction in carbon-polyvinylchloride composites, *Physical Review Letters* 40, 1197 (1978); B. Fisher, K. B. Chashka, L. Patlagan, and G. M. Reisner, Inter-grain tunneling conductivity of Sr2FeMoO6: effects of doping and grain-boundary modification, *Current Applied Physics* 4, 518 (2004).

36. A. Onipko, Analytical model of molecular wire performance: A comparison of π and σ electron systems, *Physical Review B* 59, 9995 (1999).

37. I. I. Oleynik, M. A. Kozhushner, V. S. Posvyanakii, and L. Yu, Rectification mechanism in diblock oligomer molecular diodes, *Physical Review Letters* 96, 096803 (2006).

38. Roche, Electronic conduction in multi-walled carbon nanotubes: role of intershell coupling and incommensurability, *Physics Letters A* 285, 94 (2001).

39. P. A. Schulz and C.E.T.G. da Silva, Disorder effects on resonant tunneling in double-barrier quantum wells, *Physical Review B* 38, 10718 (1988).

40. J.-C. Charlier, X. Blasé, and S. Roche, Electronic and transport properties of nanotubes, *Reviews of Modern Physics* 79, 677 (2007).

41. G. Bergmann, Weak localization in thin films: a time-of-flight experiment with conduction electrons, *Physics Reports* 107, 1 (1984); D. Rainer and G. Bergmann, Multiband effects in weak localization, *Physical Review B* 32, 3522 (1985).

42. K. Kechedzhi, E. McCann, V. I. Falko, H. Suzuura, T. Ando, and B. I. Altshuler, WL in monolayer and bilayer graphene, *European Physical Journal* 148, 39 (2007).

43. Ya. M. Blanter, V. M. Vinokur, and L. I. Glazman, Weak localization in metallic granular media, *Physical Review B* 73, 165322 (2006).

44. B. L. Altshuler, A. G. Aronov, B. Z. Spivak, D. Y. Sharvin, and Y. V. Sharvin, Observation of the Aharonov-Bohm effects in hollow cylinders, *JETP Letters* 35, 588 (1982); *Pis'ma Zhurnal Éksperimental'noĭ i Teoreticheskoĭ Fiziki* 35, 476 (1982).

45. B. L. Altshuler, A. G. Aronov, B. Z. Spivak, The Aaronov-Bohm effect in disordered conductors, *JETP Letters* 33, 94 (1981); *Pis'ma Zhurnal Éksperimental'noĭ i Teoreticheskoĭ Fiziki* 33, 101 (1981).

46. P. A. Lee and A. D. Stone, Universal conductance fluctuations in metals, *Physical Review Letters* 55, 1622 (1985).

47. V. V. Bryksin and P. Kleinert, Anderson localization in anisotropic systems at an arbitrary orientation of the magnetic field, *Zeitschrift für Physik B* 101, 91 (1996).

48. S. S. Pershoguba and V. M. Yakovenko, Shockley model description of surface states in topological insulators, *Physical Review B* 86, 075304 (2012).

49. C. Broholm, R. J. Cava, S. A. Kivelson, D. G Nocera, M. R. Norman, and T. Senthil, Quantum spin liquids, *Science* 367, 263 (2020).

50. P. Moses and R. H. McKenzie, Comparison of coherent and weakly incoherent transport models for the interlayer magnetoresistance of layered Fermi liquids, *Physical Review B* 60, 7998 (1999).

51. E. Svetitsky, H. Suchowski, R. Resh, Y. Shalibo, J. M. Martinis, and N. Katz, Hidden two-qubit dynamics of a four-level Josephson circuit, *Nature Communications* 5, 5617 (2014).

52. B. I. Shklovskii and A. Efros, *Electronic Properties of Doped Semiconductors*, Springer-Verlag, Berlin, 1984, pp. 228–244.

53. N. F. Mott, *Conduction in Non-Crystalline Materials*, Oxford University Press, Oxford, 1987, pp. 27–29.

54. M. S. Dresselhaus, Down the straight and narrow, *Nature* 358, 195 (1992).

55. P. A. Lee and T. V. Ramakrishnan, Disordered electronic system, *Reviews of Modern Physics* 57, 287 (1985).

56. G. Bergmann, Weak localization in thin films: a time-of-flight experiment with conduction electrons, *Physics Reports* 107, 1 (1984).

57. T. Klein, P. Achatz, J. Kacmarcik, C. Marcenat, F. Gustafsson, J. Marcus, E. Bustarret, J. Pernot, F. Omnes, B. E. Sernelius, and C. Persson, Metal-insulator transition and superconductivity in boron-doped diamond. *Physical Review B* 75(16), 165313 (2007).

58. A. Kawano, H. Ishiwata, S. Iriyama, R. Okada, T. Yamaguchi, Y. Takano, and H. Kawarada, Superconductor-to-insulator transition in boron-doped diamond films grown using chemical vapor deposition, *Physical Review B* 82(8), 085318 (2010).

59. K. W. Lee, and W. E. Pickett, Superconductivity in boron-doped diamond, *Physical Review Letters* 93(23), 237003 (2004).
60. H. J. Xiang, Z. Li, J. Yang, J. G. Hou, and Q. Zhu, Electron-phonon coupling in a boron-doped diamond superconductor, *Physical Review B* 70(21), 212504 (2004).
61. L. Boeri, J. Kortus, and O. K. Andersen, Three-dimensional MgB2-type superconductivity in hole-doped diamond, *Physical Review Letters* 93(23), 237002 (2004).
62. G. Chimowa, D. Churochkin, and S. Bhattacharyya, Conductivity crossover in nanocrystalline diamond films: Realization of a disordered superlattice-like structure, *Europhysics Letters* 99, 27004 (2012).
63. R. Schmechel, Gaussian disorder model for high carrier densities: theoretical aspects and applications to experiments, *Physical Review B* 66, 235206 (2002).
64. S. V. Morozov, K. S. Novoselov, M. I. Katsnelson, F. Schedin, L. A. Ponomarenko, D. Jiang, and A. K. Geim, Strong suppression of weak localization in graphene, *Europhysics Letters* 97(1), 016801 (2006).
65. Y. Dai, D. Dai, C. Yan, B. Huang, and S. Han, N-type electric conductivity of nitrogen-doped ultrananocrystalline diamond films, *Physical Review B* 71, 075421 (2005).
66. P. Zapol et al., Tight- binding molecular-dynamics simulation of impurities in UNCD diamond grain boundaries, *Physical Review B* 65, 045403 (2001).
67. S. Bhattacharyya, Novel electronic structure and transport properties of confined disordered carbon layers, *Physica Status Solidi B* 246, 1056 (2009).
68. S. Bhattacharyya, Observation of delocalized transport and low-dimensionality effects in disordered carbon thin films, *Applied Physics Letters* 91, 142116 (2007).
69. K. V. Shah, D. Churochkin, Z. Chiguvare and S. Bhattacharyya, Anisotropic 3D weakly localized electronic transport in nitrogen-doped ultrananocrystalline diamond films, *Physical Review B* 82, 184206 (2010).
70. S. Bhattacharyya, Two-dimensional transport in disordered carbon and nano-crystalline diamond films, *Physical Review B* 77, 233407 (2008).
71. L. W. van Beveren, D. L. Creedon, N. Eikenberg, K. Ganesan, B. C. Johnson, G. Chimowa, D. Churochkin, S. Bhattacharyya, and S. Prawer, Anisotropic three-dimensional weak localization in ultrananocrystalline diamond films with nitrogen inclusions, *Physical Review B* 101, 115306 (2020).
72. S. Bhattacharyya and D. Churochkin, Polarization dependent asymmetric magneto-resistance features in nitrogen-incorporated nanocrystalline diamond films, *Applied Physics Letters* 105, 073111 (2014).
73. D. Churochkin and S. Bhattacharyya, Tuneable anisotropic transport in nitrogen-doped nanocrystalline diamond films: Evidence of a graphite-diamond hybrid superlattice, *Europhysics Letters* 100, 67004 (2012).

7 Superconductivity in Boron-Doped Diamond and Related Systems

7.1 SUPERCONDUCTIVITY

When certain materials are cooled to a specific (transition) temperature, electrical resistance vanishes, and magnetic flux fields are ejected from the material. This phenomenon is called superconductivity. Metals, ceramics, organic materials, or strongly doped semiconductors that conduct electricity without resistance allow electrons to be transferred without losing energy to heat. An electron possesses a charge as well as a spin. An electron's spin can connect to the spin of another electron, and the same is true for the atomic lattice in which the electrons are placed. Because atoms may move, they can produce superconductivity through lattice fluctuations (Figure 7.1).

REMARKABLE PHYSICS PHENOMENA IN CARBON

- Superconductivity in fullerenes (AC_{60})
- Spin–charge decoupling in polyacetylene
- Quantum Hall effect in graphene
- SC in 4 Å nanotubes
- Single-electron tunneling and Coulomb blockade in quantum dots on nanotubes
- Wigner crystallization in nanotubes and GNR
- Klein tunneling in GNR and nanotubes
- Superconductivity in twisted graphene
- Spin-triplet superconductivity in diamond.

Transport properties correlated to boron concentration include (i) the temperature-dependent resistance showing the insulator–metal transition with increasing doping concentration, (ii) the conductivity as a function of boron concentration, and (iii) the critical temperature as a function of boron concentration.

DOI: 10.1201/9781003316411-7

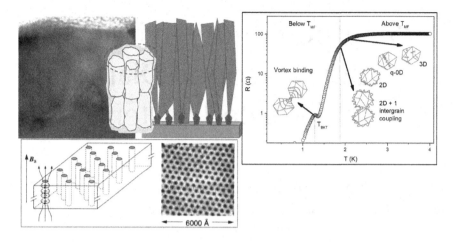

FIGURE 7.1 (top left) Diamond films show columnar growth where boron atoms can be incorporated. (right) Unconventional superconductivity shows the reentrant bosonic anomaly with a peak showing vortex–antivortex interactions (bottom). The temperature-dependent resistance in the various temperature regions where dimensionality crossover occurs with the q-0D phase has a Cooper pair size comparable to the grain diameter, the intermediate 2D state represented as a surface state around the circumference of the diamond grain, and lastly the 2D + 1 phase where grain coupling sets in and Cooper pairs can travel between grains [70].

7.1.1 BCS SUPERCONDUCTIVITY

The BCS theory named after John Bardeen, Leon Cooper, and John Robert Schrieffer is the first microscopic theory of superconductors. After the discovery of superconductors by Heike Kamerlingh Onne in 1911, no microscopic theory was accepted for half a century till 1957 when the BCS theory was proposed.

The cause of superconductivity according to the BCS theory is the electrons forming a binding state called Cooper pairs which conduct a supercurrent. The electrons in these pairs are correlated due to the Pauli exclusion principle. This correlation gives rise to an important property; the energy required to break one Cooper pair is the same as the energy required to change the state of all other pairs. This results in an energy gap for single-particle excitation unlike in the electron case where a minuscule energy could be used to excite an arbitrary electron. The energy is highest at absolute zero and decreases with increasing temperature until it vanishes at the transition temperature. Due to the energy gap, the specific conductivity of a superconductor is exponentially suppressed at small temperatures as the thermal excitations vanish. In fact, the BCS theory gives the ratio between the energy gap at absolute zero and the critical temperature (T_c) as $\dfrac{E(T=0)}{K_B T_c} = 1.764$, which is independent of the material. Here, K_B is the Boltzmann constant. The theory also gives quantitative relations between the energy gap that increases with attractive interaction and to the normal-phase single-particle density of state at the Fermi level. It also describes the change in the density of states on entering the superconducting state where there are no electronic states at the Fermi level. This energy gap can be observed through microwave reflections from a superconductor and tunneling experiments. The original BCS theory results described an *s*-wave (net spin zero) superconducting state, which is the rule in low-temperature superconductors but not in many unconventional superconductors, such as *d*-wave (a spin-triplet state with net spin = 1) high-temperature superconductors. The BCS theory extensions exist to address these other instances, but they are insufficient to fully characterize the known properties of high-temperature superconductivity. The BCS theory not only holds for superconducting materials but also describes any phenomenon with a weak interaction between electrons. It holds for weak coupling conditions like low-temperature superconductors or the pairing interaction between nucleons in an atomic nucleus.

7.2 NON-BCS SUPERCONDUCTIVITY: POSSIBLE PAIRING MECHANISMS

Just as the BCS theory works with phonon-mediated pairing, a spin-mediated pairing mechanism has been proposed. Reports have mentioned possible quantum interference effects between weakly localized holes that initiate superconducting phase transition. In diamond-like structures, the overlap between two holes of adjacent boron sites required a spin flip due to the Pauli's exclusion principle. The energy cost for forming such a pair through a spin-flip event is $\frac{\Delta}{2} = \frac{\mu_0 \mu_B^2}{b^3}$, where μ_0 is the permeability of the vacuum, μ_B is Bohr's magneton, and b is the effective (C–C≈C–B) bond length. Using the value obtained for the energy gap for hole pairing, $\Delta = 3.6 \times 10^{-4}$ with the Cooper relation $2\Delta = \Lambda KT_c$, the constant Λ comes out to be 5.1, which is comparable to the one predicted for weak metals which is 3.52.

Another alternative is the Cooperon propagator method, which tries to explain superconductivity. The advantage of this method is that it allows quantum interference effects and superconductivity to be explained through a single method. The method introduces a novel pairing mechanism relevant to Mott's metal related to WL interference effects but lacks a proper theoretical description (see Chapter 8).

7.2.1 Vortex

A vortex is a decoupling of a plane from space into two or more planes through the creation of imaginary space in-between. Vortices are created from multiple reflections or retro-reflection, allowing for time-reversal symmetry breaking. Vortices cause spin–charge separation by creating space at their core. It is proposed as a suitable model to explain how two vortices interact with each other in a medium such as a spin liquid. This model shows that a bosonic liquid placed in a potential initially maintains time-reversed symmetry. Due to this nature, the system becomes "locked" like an insulating state. To unlock this system one has to apply a π-pulse to one of the components to cause the time-reversal symmetry to break. This looks like two magnets or two dipolar molecules in opposite orientations but connected with flux lines. Two *hc/e* vortices aligned on the *x*-axis and two other *hc/e* vortices aligned on the *y*-axis form a quadrupole. This would look like a quadrupolar structure of a *d*-wave of the order parameter. The spin–charge separation model or the holon–spinon model could not explain triplet transmission in the superconducting spin valves (at least this field theory model was not used and it did not connect any other known interactions such as Rashba spin–orbit coupling [RSOC] or the Kondo resonance). Vortices have the potential to describe certain interesting phenomena such as WL, WAL, A-B oscillations, SOI, and the Kondo effect. WL is a phenomenon found when an electron travels through a weakly disordered medium. This results in a backscattering of the electron as well as the forward scattering. Both the paths can interfere constructively or destructively. The Kondo effect is characterized by the gradual flipping of a spin in a spin-triplet and also happens as a result of the Berezinskii–Kosterlitz–Thouless (BKT) transformation. RSOC occurs when the spins are initially confined in the vortices coupled with different levels such as S or F sublevels and this allows the transfer of energy. Based on the experimental observation of reentrant superconductivity MR peak, ZBCP, and spin-valve effect in a disordered superconductor or a spin-valve structure, we suggest a phenomenological model for a general understanding. Features of recent spin-triplet structures can be explained by a generalized dual vortex theory without the effect of a well-defined magnetic layer. This general approach can explain the formation of triplet superconductivity from the superposition of dual vortex fields and two spin currents. We have extended the original model prediction of vortex core by further introducing the geometry of a system and explaining the experimental data, particularly the zero-bias conductance peak, angle and magnetic field-dependent oscillatory MR, and an insulator peak in temperature-dependent resistance.

7.3 SUPERCONDUCTING CARBON: DIAMOND

7.3.1 TRANSPORT FEATURES OF BORON-DOPED DIAMOND

In terms of transport studies, most of the initial work was conducted on investigations relating to the mechanism of Cooper pair formation. One of the initial aspects that enjoyed a great deal of research attention was the MIT in single crystal samples. The MIT is a well-documented phenomenon and allows for an ideal system to test theoretical models relating to doping-induced superconductivity in semiconductors; essentially it allows for the doping concentration to be used as an experimental parameter to drive the system between phases. It was observed that at critical boron concentrations of $n_c \sim 4.5 \times 10^{-20} \mathrm{cm}^{-3}$, the MIT occurs (Figure 7.2). The superconducting phase has been observed only in samples on the metallic side of the transition, i.e., $n_B > n_c$. On the insulating side of the transition, the system exhibits transport properties closely following VRH; however, as the boron concentration is increased, the VRH temperature decreases until the metallic phase is reached where the normal state resistance is best fitted to transport models considering quantum interference effects. In this regime, the conductivity was found to scale as $\sigma_0 \propto \left(\dfrac{n_B}{n_c} - 1 \right)^{\upsilon}$, where the critical exponent was determined to be $\upsilon = 1$; this behavior is similar to other disordered metals and semiconductors, a clear indication that WL effects play a crucial role in the transport of the normal state. Studying the experimentally accessible parameters of MIT (such as T_c behavior) allowed for the development of several theoretical investigations. Most of the founding theoretical work focused on studies related to electron–phonon coupling, specifically involving the coupling of phonons to holes in the top of the σ bonding bands.

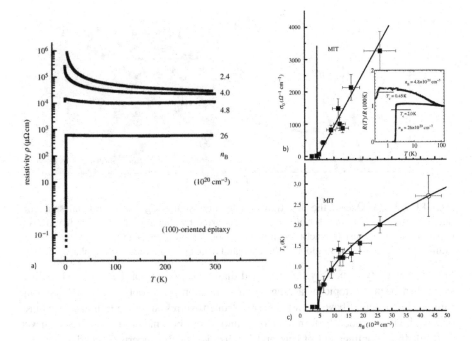

FIGURE 7.2   Various experimental findings regarding the transport properties correlated to boron concentration: (a) the temperature-dependent resistance showing the insulator–metal transition with increasing doping concentration, (b) the conductivity as a function of boron concentration, and (c) the critical temperature as a function of boron concentration [71].

Accordingly, the critical temperature was experimentally shown to roughly follow a ($n_B/n_c - 1)^{1/2}$ behavior and was explained as a result of a slow decrease in the coupling constant with increasing dopant concentration, which is counterintuitive. As the electron–phonon coupling is expected to increase approaching the MIT, this contradiction raises questions on the validity of such phonon-mediated transitions.

- In the study by Ekimov (2004), boron-doped diamond under pressure becomes a type II superconductor with $T_c = 4$ **K** and upper critical field 3.5 T [76].
- T_c varies between **1 and 10 K** based on dopant level [77].
- Signatures up to **25 K** for more crystalline diamonds [78].
- The acceptor state was considered as a Kramers' doublet.
- Magnetic field causes Zeeman splitting of the doublet due to SOC of Hole state.
- Lifting degeneracy: Spontaneously broken time-reversal symmetry.
- Degeneracy of energy level of Hole bound to boron acceptor state.
- Static Jahn–Teller effect lifts the degeneracy at low temperatures.

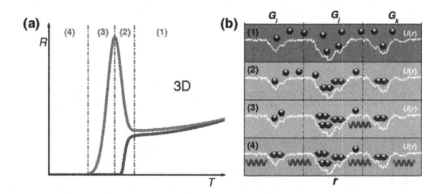

FIGURE 7.3 (a) One of the first transport features signifying unconventional superconductivity was the reentrant bosonic anomaly reported by Zhang et al. (b) This feature was
explained with in a resistive network model composed of bosonic islands within the diamond
grains that eventually reach global coherence and hence the superconducting phase [72].

Superconductivity in heavily boron-doped diamonds was established in 2004 and has
since then been the topic of a range of investigations. The poly and nanocrystalline
samples in particular have been the focus of intensive research due to unusual features
observed. This includes the formation of an anomalous bosonic insulating phase believed
to result from confinement effects and localization of the superfluid condensate. The
granularity has also been linked to a range of dimensional crossovers within the fluctuation regime and even the possibility of a BKT transition (Figure 7.3). One aspect of this
system that has received an enormous amount of attention is the pairing, which has been
investigated in terms of phonon, spin, and even resonating valence band mediation.

7.3.2 MICROSTRUCTURE ANALYSIS

Earlier, high-resolution transmission electron microscopy (HR-TEM) studies have
revealed that, in granular samples, particularly in samples with micrometer grain
size, the boron aggregates and forms triangular pockets between diamond crystal
grains (Figure 7.4).

Other studies claimed the intergranular boron network was responsible for the
superconductivity instead of the diamond; however, there was no proof since the
measurement represented some very localized structure. Nevertheless, the formation
of C–B bilayer sheets along the {111} planes was suggested. This included high-order,
SL reflections in X-ray diffraction and Laue patterns which unambiguously showed
an incommensurately modulated structure due to the displacement of boron atoms.
This model is based on the substitution of two carbon atoms in the (1/2, 1/2, 0) and
(1/4, 3/4, 3/4) positions of the diamond unit cell by two boron atoms (Figure 7.5).
Since the B–C bonds (1.6 Å) are longer than the C–C bonds (1.54 Å), the boron atoms
are shifted toward each other along the [111] direction and their displacement provides the elastic strain relaxation. The boron pair changes the cubic ABCABC stacking sequence on the hexagonal CACA stacking sequence.

Following the observation of Polyakov et al., we assumed that the superconductivity of the boron-doped diamond bears characteristics of surface superconductivity (old

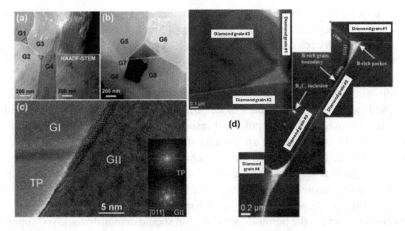

FIGURE 7.4 (a)-(c) HR-TEM images of the granular structure and the triangular boron pockets can clearly be seen at the diamond corners. (d) Four diamond grains are studied. The grain boundary regions are found to be boron rich. Some intergranular boron inclusions are identified near the diamond edge.

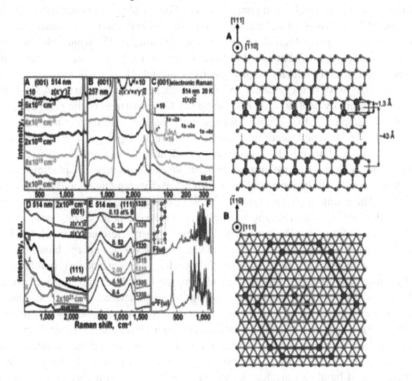

FIGURE 7.5 (left) Electronic Raman spectroscopy has been used to investigate the phonon modes across the insulator–metal transition with increasing boron concentration. On approaching the metallic phase (and hence superconducting phase) bands corresponding to an impurity level due to B-C have been achieved. Transitions between spin–orbit split valence bands and this impurity band have been mapped. (right) A 2D phase corresponding to B-C layers composed of boron dimers in the diamond lattice has also been suggested from XRD data [28].

Ginzburg idea). The specificity of diamond in that context is that the superconductivity phenomenon occurs locally on the surface area of boron-doped crystallites. This type of superconductivity is assumed to be BCS type. Along with this mechanism, there is an extra mechanism for superconductivity of non-BCS type. Indeed, the local appearance of BCS-type superconductivity suggests that there are always interface regions of a BCS/insulator (deformed by boron doping, part of a diamond crystallite) type. On the other hand, according to Fu–Kane's model, the deformation of a diamond lattice in the (111) direction may lead to a topological insulator state of a diamond lattice. It is the (111) direction reconstruction of a diamond lattice by boron doping that was claimed by Polyakov's group. Based on those observations, we assume that the type of mentioned interface might be BCS/topological insulator state. If it is a reasonable hypothesis, then, as it was shown theoretically, such types of interfaces can contribute to the overall superconductivity but with a p-type order parameter due to a specific interface scattering. In other words, the boron-doped diamond should demonstrate p-type superconductor phenomena in addition to the common BCS-type superconductor observations.

The ideas about topological transformations are the following. It is assumed that the topology of the initial pure diamond Fermi surface is strongly affected by the boron doping process and that finally leads us to a modified Fermi surface, corresponding to the 3D topological insulator state. The problem is to establish a particular type of boron-induced deformation and, subsequently, make a connection with the corresponding topological phases. To be specific, in Fu–Kane–Mele's paper, there was a sequence: "Though the 4-band diamond lattice model is simple, it is probably not directly relevant to any specific material. However, it may give insight into the behavior of real crystals. Consider a sequence of crystal structures obtained from a diamond by continuously displacing the fcc sublattices in the (111) direction: diamond – graphite ABC – cubic: Starting with the diamond, the 111 nearest-neighbor bonds are stretched, leading to the 0;(111) Weak Topological Insulator (WTI) phase. As the sublattice is displaced further both sublattices eventually reside in the same plane with a structure similar to ABC stacked graphite. Displacing further, the lattice eventually becomes cubic. At this point, the gap closes, and the system is metallic. The s-state model remains in the WTI phase up to the cubic point."

Based on electron microscopy another study of the diamond surface states known as "Pandey's chain" can be found. In Pandey's symmetric chain model, the bond length of the (111) plane was like graphite yielding the surface energy band, whereas two surface bands were derived from the bonding and antibonding combinations of dangling orbitals along the chain producing the bulk band gap (Figure 7.6). The calculated Fermi surface was found to be flat and degenerate and nearly degenerate along with J–K directions with the presence of electron–hole packets and unstable against Jahn–Teller distortions. Overall, this chain finds a remarkable similarity with polyacetylene with alternating single and double bonds like a dimer which yields a gap at the Fermi surface due to this asymmetry. Such a directionally flat (completely dispersionless) band can produce a straight-line-shaped Fermi surface as observed in type-III Dirac cones and tested from molecular-orbital representation. Starting from a square-lattice model in 2D for spinful fermions this model has also been extended to the diamond lattice model where a type-III Weyl semimetal has been constructed. Although a modified 2D SSH model was discussed in this paper we

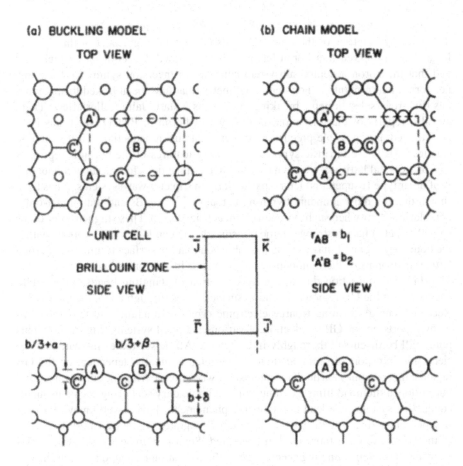

FIGURE 7.6   The surface geometry of the diamond (111) – (2×1). (a) The buckling model: The alternate rows of surface atoms (largest circles) are displaced in and out of the surface altering the vertical separations (relative to atoms C and C′ in the second layer) of the two surface atoms (A and B) in the unit cell, from their ideal value of $b/3$, where b is the ideal bond length [61]. (b) The chain model: The topology of the zig-zag chain structure of the top two layers is similar to that of the ideal (110) surface. The chains run along ABA′. Only the first layer atoms have broken bonds. All bond lengths have their ideal values except for bonds along the chains in the top layer. For the symmetric chain model, the bond lengths are equal and there is a reflection plane through A-C′. For the dimerized model, the two bond lengths differ and the reflection symmetry is lost.

think that a possible extension in 3D SSH on a diamond lattice structure particularly in the <111> direction can be useful in explaining our observed results. This zigzag chain can support the generalized SSH model, which is developed on the alternating single-double (zigzag) bonding arrangements as originally proposed in the Shockley model; however, it is treated with a topological phase. Such as unique configuration with arbitrary adiabatic deformation can produce a zero-energy edge model, which can explain the ZBCP features (explained in Chapter 8). It is well known that CVD diamond films frequently show twinning, where the crystal symmetries of the twins strongly affect the properties of the GBs. The effect of high boron concentrations on the lattice symmetries of the diamond is also an interesting aspect of this system. It has been observed that boron has a preferred growth direction in the diamond and that the boron acceptor subsystem can lead to inversion symmetry breaking, i.e., formation of points of non-centrosymmetry, that have been linked to spontaneous time-reversal symmetry breaking, as well as a static Jahn–Teller effect. There are also reports of the formation of a bilayer of boron in BNCD which has been investigated as a possible precursor to interfacial superconductivity in this system (Figure 7.5). These intrinsic symmetry-breaking features of the boron acceptor, as well as a crystal lattice, are known to have huge implications for the superconducting Δ of other type II superconductors, particularly in triplet p-wave systems. It has been shown that the most abundant twinning is that of $\sum = 3$ GBs and that higher-order boundaries occur when such symmetries meet (Figure 7.7). These higher-order twins ($\sum = 9$, 27, etc.) have a stronger lattice mismatch and can lead to dislocations within the boundary region. It has also been established that the surface termination (strong lattice mismatch) of the diamond grain can result in the formation of an extended π^* orbital configuration due to the hybridization of dangling bonds. The heightened stress at the GB region can lead to surface states through a modification of the electronic energy forming frontier electronic orbitals. In addition to this, there are many reports on the GB conduction of various diamond systems (Figure 7.6). This point will be discussed thoroughly in Chapter 8. All these studies indicate that the electronic transport of granular diamond systems is greatly dependent on the GB regions; this becomes particularly interesting when considering the superconducting boron-doped diamond films as granularity and boundary scattering events in superconducting systems can lead to interesting phenomena. This is well known for type II superconducting systems such as the high T_c cuprates where transport properties of the GBs have been thoroughly investigated. Such systems can exhibit zero-bias resonances in their tunneling spectra due to the formation of bound states at the GB junction. These bound states are the result of the anisotropic nature of the pair potential (p-wave) and the sign change that occurs as the pairs scatter from the GB. The unconventional superconductivity in BNCD films can be explained by strong SOC generated from the electric fields at the GB regions.

7.4 SPIN–ORBIT COUPLING

This part has been discussed in Chapter 1. The SOC causes a shift in the atomic energy levels of the electrons moving in the finite electric field of the nucleus due to the electromagnetic interaction between the spin of the electron and the electric field. In the rest frame of the electron, there exists a magnetic field created by the interaction of the angular momentum of the electron and the electric field of the nucleus. Based on the notion of the effective magnetic field, it will be straightforward to conceive that SOC can be a natural, non-magnetic means of generating spin-polarized electron current. The effect of SOC in nanotubes has been discussed, which we summarize as follows.

The production of spin-polarized currents in pristine carbon nanotubes with RSOC has been shown to be very sensitive to the symmetry of the tubes and the geometry of the setup. The generalized idea of this effect is space–time distortions

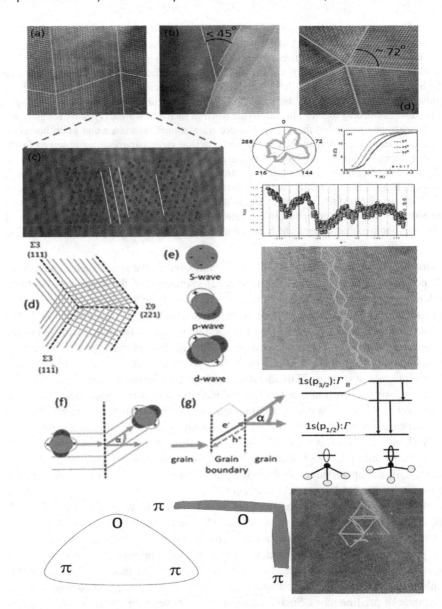

FIGURE 7.7 Microstructure analysis. (a) UHR-TEM imaging of the nanocrystalline dia-mond films. The individual grains are highly ordered with clearly defined lattice planes. Grain boundaries are regular planes that span the length of the individual grains and define planes of crystal twinning. (b) Intersection of the grain boundaries occurs at acute angles as small as 45° (when two twinned planes meet) but also shows higher-order intersections of fivefold symmetry oriented at approximately 72 apart. (d) HAADF imaging of the grain boundary region shows the interface is composed of stacking fault layers where translational symmetry is clearly broken. (e) Angle-dependent resistance at a fixed field shows pronounced anisotropic features, with 72 as well as smaller periods, indicating grain boundary conduction. The critical

(Continued)

FIGURE 7.7 (*Continued*) temperature of the system is shifted to lower temperatures upon rotation in a small field. (c) HAADF imaging allows for a more detailed structural analysis; as shown here: the dots indicate atomic positions. The extended lattice is then dominated by stacking faults terminated by crystal twins. The high proportion of $\sum 3$ boundaries allows for higher-order twinning such as the $\sum 9$ and, this is schematically shown in (d). (e) Structural anisotropy is strongly related to the parity of Cooper pairs, particularly features such as mid-gap bound states can only manifest in anisotropic pairings such as d-wave and p-wave states. (f and g) A schematic indicating the effect of lattice translational invariance and grain boundary mismatch can have on an anisotropic superconducting order parameter, as observed in the high T_c cuprates, flipping of the pair potential can occur, and this gives rise to bound states at the junction area. A triangular shape can introduce a π-phase of the Cooper pairs. However, this can happen from the (111) plane (see discussion in the text). Microstructure analysis of CVD-grown diamond films oriented close to the [110] zone axis, which shows the modulation in width. (d) The boron acceptor can be present in the GB, which removes the degeneracy of the $1s(p_{3/2})$ spin–obit split level, and hence creates a strong splitting of the energy levels [11,74].

associated with the curvature of the space compared to flat space such as graphene. A pentagon–heptagon breaks the hexagonal symmetry of flat graphene sheets by adding a curvature vertical, which is a lateral space–time distortion. Further, distortions happen by vertical space–time distortion in a vortex structure which adds helicity to the space. In both cases, an excited (p) state is created from the distortion of the ground (s-) state. The height of the distortion or the helicity corresponds to the level of excitation of the p-state. More breaking of symmetry creates more p-like states, hence producing stronger SOIs in a disturbed system (Figure 7.8a and b).

Space–time distortion is a very general idea that works in a black hole or a quantum vortex. This is a maximally entangled system or extremely disordered system. In that case, amorphous carbon should show a high value of SOC and a signature of a WAL effect which has not been observed experimentally due to the dominance of extremely random and high scattering rates. The topological defects can be observed in the GBs of superconducting diamond films which have a thin atomic layer, hence suppressing the random scattering effect.

It was claimed that the role of defects on the spin quantum conductance of metallic carbon nanotubes was due to an external electric field that results from a polarized state. The geometry of the flat space is changed due to absorbed hydrogen atoms or pentagon–heptagon pairs, which also break the time-reversal symmetry. As a result, the Rashba spin-polarized current increases in the nanotubes. Moreover, this enhancement takes place for energies closer to the Fermi energy as compared to the response of pristine tubes. Such increments can be even larger when several equally spaced defects are introduced into the system as the density of p-states increases. It was claimed that spin-valve devices at the nanoscale may be achieved via defect engineering in carbon nanotubes which can be explained in detail as follows.

It is commonly understood that being the second lightest material, free carbon cannot have a significant amount of SOC since the splitting $3P_0 - 3P_1$ is only ~ 2 meV. Also, it is almost completely suppressed near the Dirac points in flat graphene, which relies on the symmetry of graphene. However, this symmetry is broken by curvature in carbon nanotubes, leading to a coupling up to a few meV between the spin and orbital moment of electrons. This coupling, first detected by Kuemmeth et al. [66], is the key to electrically controlling spins in nanotubes. The effect of this coupling is to mix single-particle states with opposite spin from different orbitals, such as $jp_z\uparrow i$ and $jp_x\downarrow i$. Whether this leads to SOC in the band structure depends on how it affects hybridization between orbitals in different atoms, which in turn depends on the crystal structure.

The contrasting situations in flat and curved graphene are illustrated in Figure 7.8a, which shows the atomic orbitals for two adjacent atoms A and B. Any effect on the band structure arises through the combination of intra-atomic SOC and interatomic hopping. In flat graphene (Figure 7.8b), symmetry forbids direct hopping from a p_x state on one atom to a p_z state on another because p_x and p_z orbitals have opposite parity under z-inversion. Therefore, atomic SOC between, e.g., $jpA_z \uparrow i$ and $jpA_x \downarrow I$ state does not introduce any non-spin-conserving hybridization between $jpA_z \uparrow i$ and $jpB_z \downarrow i$, and thus SOC in the π band is second order and in practice negligible. This situation is changed in the presence of curvature, which breaks the z-inversion symmetry on which the above suppression relies (Figure 7.8b). In the curved coordinate basis the radial, circumferential, and axial directions are denoted as f_r, c, and t_g, respectively. The π band is composed predominantly of hybridized p_r orbitals. Since the p_{Ac} and p_{Br} orbitals are not orthogonal, hopping between them is allowed, leading to an indirect hybridization between $j_pA_r\uparrow i$ and $j_pB_r\downarrow i$ and consequently a SOC in the π band. The effective hopping matrix element between p_{Ar} and p_{Br} now contains

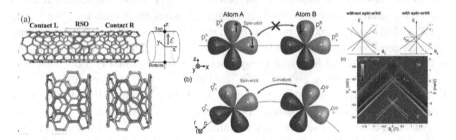

FIGURE 7.8 (a) Schematic view of the device geometry. Left (L) and right (R) contacts are pristine CNTs without RSO interaction. The central part of the device is a conducting CNT of length L, with RSO interaction induced by the presence of a uniform electric field E in the z-direction (arrow). The directions x, y, and z are shown in the right part of the figure; the electric field points into the $+z$-direction. The top and bottom locations of the CNT are defined with respect to the electric field, as indicated. Examples of the defects considered as a Stone–Wales topological defect [69]. (b) In flat graphene, atomic spin–orbit coupling and interatomic hopping between the p-wavefunctions consisting of p_z and p_x orbitals for two adjacent atoms A and B are shown, with the sign of the wave function in each lobe marked. Intra-atomic spin–orbit coupling mixes opposite-spin states involving different p orbitals in the same atom. However, this does not mix spin states in the band structure, because hopping between different p orbitals is forbidden by symmetry. (b) In a nanotube, curvature breaks the up-down symmetry, meaning that direct hopping between different orbitals on adjacent atoms becomes possible. The combination of atomic spin–orbit coupling, and interatomic hopping therefore mixes opposite-spin states on adjacent atoms, leading to a spin–orbit splitting of the π band [68]. [right] (c) Observation of spin–orbit coupling in a nanotube. (top) Expected spectra without and with spin–orbit coupling. (below) Conductance as a function of magnetic field and gate voltage across the 0-1e transition at $V_{SD} = 2$ mV, allowing high-bias spectroscopy of the lowest four one-electron states [66].

both a direct and a spin-flip term. The interference between these terms causes a spin precession about y-axis, and a corresponding splitting of the two spin states within a given valley as though by a magnetic field B_{SO} directed along the nanotube. The spin–orbit splitting is defined as the Zeeman splitting due to this field Δ_{SO}.

Figure 7.8c shows the consequences of SOC for the band edges. Without SOC, the zero-field levels are fourfold degenerate but are split in a magnetic field through a combination of Zeeman and orbital coupling. SOC splits each fourfold degenerate level into a pair of twofold degenerate levels (Figure 7.8c); each element of the pair comprises a Kramers' doublet, as required by time-reversal symmetry. The sign of Δ_{SO} determines whether the parallel or antiparallel alignment of spin and valley magnetic moments is favored. For $\Delta_{SO} > 0$, the magnetic moments of spin and valley of the lowest (highest) edge of the conduction (valence) band add, whereas they subtract for $\Delta_{SO} < 0$. For $\Delta_{SO} > 0$, as drawn here, SOC favors the alignment of the spin and valley magnetic moments. The lower doublet, therefore, comprises the states $f_{K0}\uparrow$; $K\downarrow g$, for which both magnetic moments have the same sign, while the upper doublet comprises the states $f_{K0}\downarrow$; $K\uparrow g$. Figure 7.8c shows excited-state spectroscopy of the first electron shell of an ultraclean nanotube as a function of the magnetic field [66]. The positions in gate voltage of the first four conductance peaks provide a map of the lowest four energy levels [which also explains how to convert from gate voltage (left axis) to energy (right axis)]. The dependence of the energy levels on the magnetic field arises from the combination of valley and spin magnetic moments in each level but taking account of the spin magnetic moment. From the line slopes, valley and spin quantum numbers can be assigned to each level (as in Figure 7.8c). The key signature of SOC is the separation of the four spin-valley levels at zero fields into two doublets. The magnitude of the splitting gives the SOC strength $\Delta_{SO} = 0.37$ meV, corresponding to $B_{SO} = 3.1$ T. From the spin and valley assignments deduced from the line slopes, in this case, SOC favors states with parallel spin and valley magnetic moments.

Although RSOC was successfully explained for transport in 2D semiconductors, the exact role of SOC in interface superconductivity has not been fully understood. Strong SOC can overcome the disorder effect related to coherent backscattering, so-called WL which becomes a WAL effect with a minimum of the WL peak as observed in the field-dependent MR. To explain the interference effect in superconductivity due to SOC, several theoretical models were suggested particularly to explain the experimentally observed π-phase shift of the current at the spin-active (SA) interface of superconducting grains (and quantum dots) or qubits. In the RSOC model, the strength of the SOC depends on the Coulomb potential which binds two opposite spins through a repulsive interaction.

The symmetry of pair wavefunction is given by momentum $\otimes$ spin $\otimes$ frequency. A key parameter to describe the symmetry breaking of the grains producing the GB region is defined as $\sum = [C_1 \cdot (C_2 \times C_3)]/[a \cdot (b \times c)]$, which finds similarity with symmetry breaking of the angular momentum vector in introducing the helicity, popularly known as the Rashba SOI described by Edelstein as interface SO: $H_{SO} = \alpha (p \times c) \cdot \sigma \delta (c \cdot r)$, where c is one of the two nonequivalent normal unit vectors and the δ function describes the interface potential with a position vector (r).

The original description of this effect involved the Hamiltonian term: $H_{SO} = \alpha (\sigma \times p) \cdot n$. Here p is the 2D quasi-momentum, σ are the Pauli matrices, and

n is the unit vector normal to the surface. The SOI ensures that electrons in the plane ($\perp n$) will have spin aligned perpendicular to the momentum p. Although this effect was first described for purely 2D systems, subsequent variations have been formulated for interfacial systems of superconducting materials: $H_{SO} = \alpha(p \times c) \cdot \sigma \delta(c \cdot r)$.

Here c is the two nonequivalent normal unit vectors and $\delta(x)$ is the interfacial potential between such surfaces. Transport in systems such as granular superconductors rely on tunneling through the GB interface, thus electrons moving between grains results in a double electric layer on the same scale of the screening length. Charge carriers conducting through this layer would thus be subjected to an inequivalent electric field defined by the junction potential $(\delta(x))$, and thus a SOC of the Rashba type. The inequivalence of the two normals ensures that inversion symmetry along the boundary is broken and thus the spin component of Cooper pairs traversing the interface will no longer be well defined, allowing for mixed states to occur, essentially converting singlet Cooper pairs to the short-ranged triplet.

Cooper pairs conducting through this atomically thin layer would be subjected to a nonequivalent electric field (defined by the interface potential as a step function), and thus a SOC of the RSOC followed by spin-triplet superconductivity arises. In Edelstein's model, the interface was described as a double electric layer, which is similar to a vortex structure. However, details of the microstructure of the interface and their interconnectivity remain unclear until today. The periodic potential variation in the GB from the mismatch of the lattice planes gives rise to the modulation of Δ. The maxima of misalignment take place at the angle of 45°, which gives the peak in the angle-dependent MR. Such phenomena cannot occur in s-wave superconductors due to the isotropic pair potential and thus their observation has been used to classify superconducting systems (Figure 7.7).

Therefore, the proposed interface RSOC needs to apply to the diamond lattice to achieve the topological phase which needs crossing or braiding of vortex lines or superposition of multiple bound states. Here we find the well-known Shockley model to describe diamond on a topological insulator. In Chapter 8, we shall see how SOC influences the properties of a superconductor and turns it into a spin-triplet state.

7.4.1 JOSEPHSON TUNNELING

When two superconductors are separated by a thin insulating barrier, supercurrent flows through the junction due to the tunneling of Cooper pairs. This tunneling effect was predicted by B. Josephson in 1962 and is called Josephson tunneling. When a voltage is applied across this junction, an oscillating current is found flowing through the barrier, this is the AC Josephson effect. If the two superconducting electrodes have a phase difference, then a DC current is found flowing and this is called the DC Josephson effect. The junction has a localized discontinuity at the barrier and therefore a weak link in the order parameter of the superconducting electrodes. In discontinuity, the dissipationless Cooper pair transport is controlled by a macroscopic quantum-phase difference across the junction. The Josephson junction (JJ) is a nonlinear element because it combines low dissipation with extremely high nonlinearity – the energy of a single photon can vary the junction inductance by an order of unity. Magnetic fields near superconductors can influence the Josephson effect. Therefore, it can be used to detect extremely small magnetic fields.

The Kondo effect in superconducting diamond films (Magnetoresistance) has been explained in Chapter 4 (also see Chapter 8).

7.5 ANDREEV REFLECTION (AR) AND ANDREEV BOUND STATES (ABS)

In diamond, GB accommodating boron atoms or holes can produce charging effects in the layers, which also creates Andreev bound states (ABS) (Figures 7.8c and 7.9). Besides, we have claimed a bound state formation in heavily boron-doped diamond (BNCD) films and associated with the charge-Kondo effect in the 2D layered structures within the diamond. Moreover, we have shown spin-triplet superconductivity in the diamond interface due to RSOC created by breaking the translational symmetry. However, in a multifunctional and inhomogeneous system such as BNCD films where evidence of non-s-wave Δ has been claimed, a more general idea of the exact nature of Δ can be developed through the vortex structures and superposition of bound states at the interface where additional symmetry-breaking operation can take place. RSOC at the GB produces spin channels; hence, an Andreev reflection (AR) at the SA interface can undergo a π-phase change or a sign inversion Δ for d-wave superconductors. A repulsive force from the interface can explain this SAAR, which can effectively form a vortex structure associated with a chiral Δ or a p-wave Δ. The origin of SAAR is not very clear in p-wave and triplet superconductors. Unlike d-wave, a p-wave superconductor is much less studied although intrinsic or spontaneous π-junction such system has been suggested. Nevertheless, a combined effect of RSOC and SAAR or chiral vortex effect can show direct evidence of odd frequency Δ in the hybrid structures made of bulk diamond and 2D layered of GB. An oscillatory (exchange-type, e.g., Kondo) interaction associated with the magnetic component (q vector) can introduce a strong orbital effect and hence produce the well-defined vortex structures in the GBs aligned in the form of a lattice or a SL structure. The model can be extended to a generalized SL structure like in the Fu–Kane–Mele and the Shockley model consists of multilayers and an array of vortex lines in the presence of bound states (such as ABS) coupled to each other (Chapter 8). The topological phase needs a modulation of the order parameter in a hybrid structure arising from the atomic potential that can be found in the p-wave GB structures coupled to an s-wave superconducting diamond crystal. The superposition of ABS having multifaceted properties can work as a superposition of reflected and transmitted Cooper pairs, which can be connected to create a nonlocal state. Hence, the previously predicted an odd frequency superconductivity can be realized in a diamond hybrid SL structure.

RSOC can arise from the excited state, as well as the degeneracy of the energy level of the Hole bound to the boron acceptor state. The acceptor state can be described by a Kramers' doublet. The Zeeman splitting of the doublet due to the SOC of Hole state lifting degeneracy related to spontaneously break time-reversal symmetry as seen in Figure 7.9. Raman spectra work as a tool to find the SOC in the boron-doped diamond films. See also Figure 7.10 for RSOC.

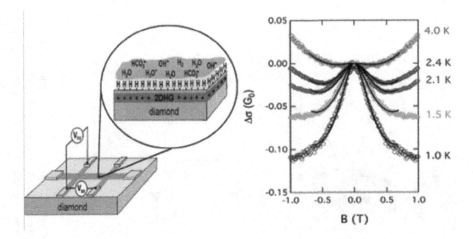

FIGURE 7.9 (left) Hydrogenated diamond possesses a unique surface conductivity because of transfer doping by surface acceptors, which is shown in a diamond device. The (100) surface of diamond, when functionalized with hydrogen, supports a p-type spin-3/2 two-dimensional surface conductivity with a spin–orbit interaction of 9.74 ± 0.1 meV. (right) Weak antilocalization effects in magneto-conductivity measurements at low temperature is observed. Fits to 2D localization theory yield a spin relaxation length of 30 ± 1 nm and a spin relaxation time of ~0.67 ± 0.02 ps [67].

7.5.1 ELECTRICAL CONDUCTIVITY

In our previous report, we have shown a BKT transition in 2D layered structures of BNCD films, which needs further explanation. In the charge-BKT transition, vortex (kink) and antivortex (anti-kink) pairs bind to form an insulating (localized) phase, namely an ABS where this pair corresponds to the bound electron–hole system. The ABS is known to exist in the *ab*-plane of layered superconductors and at superconductor–normal metal interfaces. It is thus generally a surface state having the low-dimensional character necessary for the BKT transition to occur. We explain the occurrence of the BKT transition and hence 2D transport of the films resulting from ABS present at GB regions which have layered structures. It is analogous to reports on BKT JJ arrays which are commonly used to model granular superconductors. A BKT transition has been reported in organic molecules and carbon nanotubes (see Chapter 8), which can be compared with the transport properties of BNCD films.

7.5.2 EVIDENCE FOR RSOC

RSOC effect in these samples is confirmed by the bias current-dependent R-T behavior which shows a pronounced peak before reaching the superconducting onset as the sample temperature is lowered. The intensity of the so-called boson-insulator or a charge-Kondo peak decreases exponentially with the bias current or the density of charge (electrons) supplied to the system. This insulator peak results from a bound

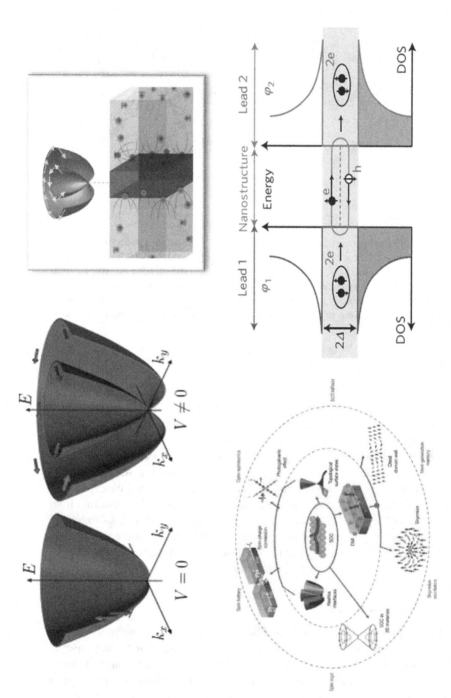

FIGURE 7.10 Structural inversion asymmetry (i.e., breaking of translational symmetry like at crystal surface, junction interfaces, and heterostructures) leads to Rashba SOC. (top, left) Rashba SOC causes a splitting of the spin bands, i.e., electronic field can cause magnetism in 2DEG. (bottom, left) Rashba SOC can lead to a number of very interesting effects with potential for quantum information technology. Josephson tunneling of Cooper pairs through a barrier or nanostructure and Andreev reflection is shown in right panels. Also, very interesting effect when combined with superconductivity, can lead to exotic pairing such as *p*-wave superconductivity and odd frequency superconductivity.

state formed at the GBs of the diamond and produces RSOC in the superconducting diamond films. Since the lattice symmetry in this diamond system is broken in the GB region, a level splitting occurs, yielding RSOC. However, the increase of current populates the spilled levels and widens the band which starts overlapping the level, superconductivity is regained, and the peak is depleted. By decreasing the supply of the carriers to the system, the effect of level splitting and the RSOC becomes prominent. With the increase in the rate of scattering from the SOIs, the resistance of the samples also increases.

To obtain an insight into the phase transition, we first define three critical points which we use to construct a phase diagram in Figure 7.11, namely $T_{c\text{-global}}$ on the left-hand side of the peak, $T_{\text{BI-peak}}$, $T_{c\text{-local}}$ on the right-hand side of the bosonic insulator peak, and $T_{\text{BI-peak}}$ the bosonic insulator peak location. Here, $T_{c\text{-global}}$ is the global critical temperature measured from the point the resistance diverges from zero to resume finite values, whereas $T_{c\text{-local}}$ is measured at a point on the right-hand side of the peak where a crossover from a nearly flat insulating phase. The $T_{c\text{-local}}$ can be considered as the onset temperature for superconductivity for the $R(T)$ in a high-bias regime where ordinary metal–superconductor transition is observed.

MR of the sample is measured at different temperatures between 300 and 1266 mK (Figure 7.12) with the field applied normally to the film. We observe hysteresis at all temperatures. Qualitatively, this means the resistance increases sharply with the increasing field up to a maximum (sharp peak) and then decreases again, as the field is further increased till it reaches the background curve. The hysteresis behavior has been observed in the R vs. B at different temperatures below or around the T_c (Figure 7.12), which is a new observation in these structures where magnetic impurities have not been introduced, therefore explaining the effect of RSOC in this system.

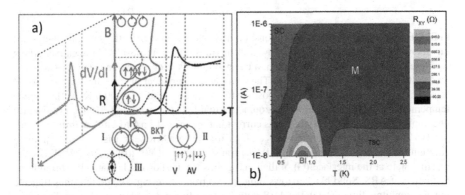

FIGURE 7.11 (a and b) A map shows various phases as current and temperature are varied. It is seen that as the bias current is reduced, the bosonic insulator peak increases sharply but does not shift in position. Phase diagram constructed based on the temperature- and current-dependent resistance data showing various phases: SC: superconductor phase, FL: the Fermi liquid, and QCP: The quantum critical point. It shows phases M: metal, BI: the bosonic insulator, and SC: superconductor. A plot $T_{c\text{global}}$, $T_{c\text{mid}}$, $T_{c\text{local}}$, and T_{BIpeak} as a function of the magnetic field show various phases. (b) Evolution of metal–insulator–superconductor transition with a current variation [75].

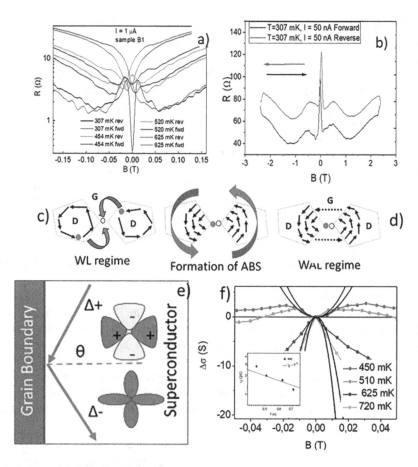

FIGURE 7.12 Low-field MR transitions: (a) magnetocondutance isotherms at temperatures far below the critical point (3 K) show a clear transition from positive to negative. Along with the positive to negative transition, a clear hysteresis is observed between forward- and reverse-field sweeping directions. (b) The transverse component of resistance of sample B4 around $B = 0$ changes the slope from positive to a negative trend (at ~600 mK) as the sample temperature rises. The region $B \to 0$ is zoomed to highlight the switching effect. (c) Resistance of sample B1 measured at different bias current in the range of 1 nA to 1 mA within the range of ± 0.5 T in the Hall configurations (R_{XY}). It clearly shows a MR peak arising at $B = 0$ point in addition to two other satellite peaks. (d) The transition from a WL (triplet) to a WAL (singlet pairing) state at the interface (G) of diamond grains (D) is shown schematically through the formation of ABS. Solid black arrows indicate electron scattering, and dotted lines indicate the electron tunneling through the ABS state. The left figure describes two WL paths of spin strictures, and the orbit-like ABS is formed by connecting the WL paths around a fixed point which is not accessed by the spins. Effectively this configuration can be formed by twisting a ring into another ring. The middle figure represents a RSO configuration where the spin and orbits are mutually perpendicular to each other, which can also be described with two rings making interconnecting loops. This configuration can be described by a vortex structure with ABS at the core as shown in the figure at the right. (e) The scattering of d-wave vector from the GB shows a sign reversal. (f) These isotherms can be neatly fitted to the HLN formula for a weak localization to weak antilocalization transition. The inset shows the temperature dependence of the dephasing time, which follows a T^{-1} dependence [75].

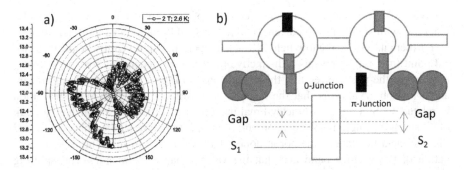

FIGURE 7.13 (a) R vs. T measured while rotating at different angles in a field of 0.1 T. Angle-dependent MR shows oscillations with a periodicity of $\pi/4$. (b) The proposed model of the microstructure describes formation of a superconducting quantum interference pattern in nanodiamond films, where 0-junction and π-junction are formed. We have two order parameters S1 and S2 which describe a non-s-wave symmetry [75].

The anomalous peak MRs of two kinds were observed, for example, one with a smooth transition and the other with asymmetric resonant peaks in the transition region (asymmetric concerning the scan direction) (Figure 7.13). This can be attributed to setting the combination of bias current and temperature to some point of instability regarding switching between superconducting and insulating states. These features can be attributed to the properties of the π–Josephson junction. The central peak at $B=0$ and two satellite peaks (measured at low current) appearing from an insulating phase (I) are described by an S_2-I-S_1 model of granular superconductors (S_1/S_2).

The peak feature near the BKT transition temperature was also observed to be greatly enhanced when conducting the measurement at lower bias; this feature is most notable in the transverse resistance as shown in Figure 7.11. To further investigate the nature of this insulator, peak MR is measured in the same temperature range. As shown in Figure 7.12, a transition from a positive to negative MR is observed, as temperature is increased; this clearly shows a crossover from WL (corresponding to a triplet state) to WAL (singlet state) above 500 mK.

The occurrence of the BKT transition is investigated through the current–voltage characteristic in Chapter 8 and explained to vortex pinning in BNCD films with smaller grain sizes. The vortex core energy determined through analysis of the resistance temperature curves was found to be anti-correlated to the BKT transition temperatures. It is also observed that the higher BKT temperature is related to an increased vortex–antivortex binding energy derived from the activated transport regions. Further, the magnetic field-induced superconductor–insulator transition shows the possibility of the charge glass state (see Chapter 8).

7.5.3 MR TRANSITION BELOW AND ABOVE THE CRITICAL POINT

7.5.3.1 MR at Low Fields: Interference Effects

The hysteresis behavior observed in the R vs. B plots at different temperatures below or around the T_c (Figure 7.12) has not before been reported in this system. It can be attributed to the strong inhomogeneity of the material bringing the hysteresis behavior of two-phase media, which consists of strongly dissipative intergranular junctions with small critical currents and fields in addition to intra-granular effects.

Further analysis of the low-field MR at a temperature significantly below the critical point (mK range) reveals a distinctive temperature-dependent crossover from a negative to the positive regime, this is indicated in Figure 7.12. What is significant about this observation is that the negative to positive i.e., WL to WAL transition is related to SOC and symmetry-breaking effects. Although carbon has a weak SOC strength, angle-resolved photoemission spectroscopy studies have indicated that the metallic state of the boron-doped diamond is due to substantial hole occupation of the valence band and that the valence bands show spin–orbit splitting in different directions of the Brillouin zones. Similar low-field MR features have also been observed in hole-doped diamond surfaces gated with ionic liquids. It was reported that the WAL transition occurred because of either an enhanced SOC from the inversion symmetry (Dresselhaus type) or asymmetry from the confinement potential (Rashba type). As WAL is a well-established 2D effect, it is reasonable to relate the effect to the GB junction. The likely origin is the hole carriers associated with the boron acceptor states, particularly those located at the surface of the GB. Hole states located at the diamond grain surface or junction area would thus exhibit an enhanced SOC due to the reduced dimensionality; this would lead to a spin-locking effect (symmetry breaking) and thus allow for the π-junction discussed earlier. This is furthermore supported by the reports on spontaneous time-reversal symmetry breaking of the boron accepter state in diamonds due to a static Jahn–Teller effect. To analyze such low-field features, a fitting is done in line with the Hikami–Larkin–Nagaoka (HLN) formalism where the magnetoconductance is described by $\Delta\sigma_{\mathrm{HNL}}(B) = \frac{\alpha e^2}{\pi h}\left[\left(\Psi\left(\frac{\hbar}{4eBL^2}\right) + \frac{1}{2}\right) - \ln\left(\frac{\hbar}{4eBL^2}\right)\right]$, where is Ψ the digamma function, α is a fitting parameter related to the positive or negative resistance (Berry phase) and L is the effective phase coherence length. In the low-field regime, however, the HLN formula reduces to a simple parabolic function given by

$$\Delta\sigma(B) = \frac{e^2}{24\pi h}\left(\frac{4eDB}{\hbar}\right)^2\left(1 - \frac{1}{(1 + 2\tau_\varphi/\tau_i)^2} - \frac{2}{(1 + 2\tau_\varphi/\tau_i + 2\tau_\varphi/\tau_*)^2}\right),$$ where D is

the diffusion coefficient and τ_φ are the dephasing time. τ_i, τ_* are inter and intravalley scattering times, respectively. This formula has before been used in low-dimensional carbon (graphene) and can be used to determine the temperature-dependent dephasing rate as shown in Figure 7.7 and follows a trend like what has before been reported for 2D hole gas on the surface diamond. This interpretation of the MR also thus extends the argument of the surface-state scattering events as a possible source of topological nontrivial property of the system.

The nature of AR for an M/S junction is like a resonant state, which makes a sign change of the pair potential on the Fermi surface, i.e., makes a π-phase shift and therefore was described as a mid-gap Andreev resonant state (MARS). This ABS or MARS can be created or activated below the superconducting onset, which exhibits a transition from the WAL effect depending on the temperature dependence of the spin–orbit scattering (explained in our previous paper). Below 500 mK the MR becomes positive at low B range, which shows a clear signature of RSOC. As the bias current becomes very low (50 nA), the SOI effect becomes very prominent as seen from the sharp peak at $B \rightarrow 0$. It is well known that diamond in the bulk superconductor behaves as a singlet a triplet state that arises at the interface, and a mixture can still show ZBCP (for ungapped d- or p-wave) and satellite peaks (gapped s-wave). Start with the application of the B field, which can break the symmetry and allow an interplay of singlet and odd-parity by breaking the singlet or triplet and even parity by fusing or joining two odds to make an even parity state.

To explain such a transition, we again rely on the granular structure of BNCD films (Figure 7.1) and the formation of the ABS is believed to exist in the boundary regions which act as transport links between grains. This is shown schematically in Figure 7.7, where, in the WL regime, electrons are coherently backscattered within the grain, the formation of the ABS allows for tunneling between the grains, and finally in the WAL regime electrons can freely tunnel between grains. At low fields, the singlet state is favored as evident by the positive MR. As the ABS linking the grains is degenerate concerning spin, electrons with either spin up or down can tunnel between grains via the ABS. Applying the field, however, removes the degeneracy of the intermediate ABS causing a preferred spin orientation for electron tunneling, essentially forming a spin-polarized tunneling current as observed from the hysteresis. At higher fields the resistance is observed to first decrease, then increase slightly, and then eventually saturate as the field strength is increased further; hence, the system is believed to be in the triplet state which is robust to further increasing the field.

7.5.3.2   Angle-Dependent Transition

Figure 7.13a shows R vs. T while rotated at different angles in the presence of a 0.1 T field. The critical temperature of the sample is observed to change as a function of the angle of the applied field. Such behavior is frequently seen in superconducting spin valves, where a long-ranged spin-triplet state is induced due to spin mixing at magnetic interfaces. The MR is highly anisotropic with a periodicity of 90°. Similar results have before been observed in layered superconductors such as La(O,F)BiSeS and are directly related to an anisotropic order parameter. To further probe the anisotropic transport features, the low-field MR is measured at 0°, 45°, and 90° concerning B. At $\theta = 0°$ (i.e., perpendicular to the sample) the MR shows the WL features previously observed; however, upon a rotation of 45° (corresponding to a minimum in the angle-dependent magnetoresistance, AMR), the WL peak is inverted to a WAL peak and then back to a WL feature with further rotation to a 90° orientation (corresponding to the peak in the AMR).

As we know the different MR regimes (WL or WAL) correspond to the scattering processes, an insulating state (WL) occurs due to coherent backscattering, whereas the WAL state corresponds to enhanced conduction. The angle-dependent change in the MR can thus be qualitatively explained if we consider superconducting pairs with anisotropic pairing potential scattering off spin-polarized boundaries. The rotation of the sample in the magnetic field changes the orientation of the spin-polarized boundary concerning the anisotropic Cooper pair; this orientation will result in either backscattering (WL) or transmission (WAL) of the Cooper pair. In Figure 7.7, the proposed p-wave scattering shows sign change of Δ accounts for the transition which needs simulation and theoretical modeling (discussed later).

MR measurements clearly show the superconducting gap ±1 T which closes as the temperature increases (Figure 7.12a). However, we concentrate at low-field regions at different temperatures and record some oscillatory and negative MR features. If we accept the layered structure of the superconductivity in the present heavily boron-doped diamond films (see Figure 7.7), then several observed features can at least be qualitatively explained by a 0–π JJ hypothesis. The model of Spivac and Kivelson developed a few decades before can be helpful in qualitatively explaining our results. This model includes a resonant level and is explained to be due to S-I-S structures. The key concept is based on explaining observations such as negative MR in terms of negative Josephson coupling. This model has been used for granular high T_c materials in the vicinity of superconductor–insulator transition. Negative MR features can be seen only at low-field regions below T_c by applying a low bias current. Although negative MR features have been reported by several researchers, a detailed explanation has not yet been found. The negative resistance can also be related to the JJs where the negative superfluid density arises from the random distribution of coupling between grains in disordered media. Here we propose a microstructure model, which consists of closed conduction loops through which flux is penetrating. This effectively leads to the appearance of the random distribution of the positive and negative supercurrents, giving rise to a negative MR contribution, whereas the superconducting quantum interference device (SQUID) structures form an elementary unit allowing for the oscillatory behavior of the MR. To investigate the properties of the oscillatory behavior of the MR deeper, we rely on the model developed in Ref. [50] that explains the oscillations arising from the coexistence of both π and 0 JJ s effectively described by the following equation:

$$V = \left(\frac{R_1}{2}\right)\sqrt{I^2 - \left(2I_{c1}\cos\frac{\pi SH}{\varnothing_0}\right)^2} + \left(\frac{R_2}{2}\right)\sqrt{I^2 - \left(2I_{c2}\sin\frac{\pi(2S)H}{\varnothing_0}\right)^2}, \quad (2), \text{ where } I,$$

I_{c1}, and I_{c2} {= 0.2 I} correspond to the measuring current and critical currents of the JJ in 0 and π SQUIDs, respectively. R_1 and R_2 represent the single junction resistances and S represents the effective area of 0-0 SQUID. The oscillations are subjected to a fast Fourier transform (FFT) to determine the magnetic period and consequently the dominant effective area. This is shown in Figure 7.13b. It is observed that the dominant amplitude of oscillations is greatly dependent on the temperature, at temperatures around 2.6 K (i.e., within the superconducting regime). We find that the lower frequency oscillations (smaller orbits) are less pronounced. This suggests that the oscillatory behavior of MR is due to closed paths formed by larger orbits comprised of multiple linked grains. As shown in Figure 7.13a, the oscillation amplitude is significant up to frequencies of 20 T^{-1} corresponding to an effective area with a radius up to 117 nm, indicating SQUID loops of approximately three to four grains in radius. The best fit to data is obtained using the effective area obtained from the dominant peaks in the FFT. This leads to a situation with a dominant 0-junction character with minimal π-junction behavior. The qualitative features of the oscillations, however, change as the temperature is decreased. In this case, the dominant FFT peaks are concentrated at the lower oscillations; these correspond to smaller orbits. The effective area of such orbits is found to correspond to the average grain size of the sample as determined from the TEM microstructure study. This indicates a temperature-dependent crossover from multi-grain tunneling to single-grain transport regime which is also marked by an increase in the π-junction character. We can thus relate the anomalous MR features to the microstructure of the films where JJ arrays consist of neighboring grains having different values of order parameters.

Recently, it has been shown that a π-SQUID can be achieved based on only geometrical and symmetry arguments. Indeed, if the order parameter of the superconductor is anisotropic (as for many other type II superconductors), then, at some angles, ABS can contribute to the transport through one junction of the SQUID leading to the π-phase shift as the reflected particles suffer from the sign change of the pairing potential. At the same time, the other junction of the SQUID can still be in a zero-phase-shift state due to different GB properties (thicker) between its superconducting constituents.

- **Main problem discussed in this chapter**
- Anisotropic features of superconductivity in BNCD films explained.
- **What has been achieved?** Correlation between the microstructure and the anisotropic features of unconventional superconducting properties of diamond has been established.
- **What has not been achieved?** The observed superconducting properties in boron-doped diamond is mixed with unconventional features due to defects or GBs. The superconducting pairing mechanism in carbon is not known.

7.6 SUMMARY

In this chapter, we have introduced some basic concepts of unconventional superconductivity through the symmetry breaking of the crystal structure of diamonds at the GBs. We have discussed SOIs and the Kondo resonance as the main reason for the observation of phase transitions in temperature-dependent resistance and MR including the angle-dependent MR. However, the phase transition-related features will be explained in great detail in Chapter 8 through (i) quantum spin liquid and (ii) vortex structures, vortex–antivortex interactions, and the BKT transition. This discussion will be extended through the FFLO and SSH model followed by quantum simulations. This also supports the modulated Δ as suggested in the FFLO model which is also formed at the temperature where an array of vortices and a π-junction is formed.

BIBLIOGRAPHY

1. S. Leger et al., Observation of quantum many-body effects due to zero point fluctuations in superconducting circuits, *Nature Communications* 10, 5259 (2019).
2. Z. H. Peng, S. E. de Graaf, J. S. Tsai, and O. V. Astafiev, Tuneable on-demand single-photon source in the microwave range, *Nature Communications* 7, 12588 (2016).
3. T. Yamashita, K. Tanikawa, S. Takahashi, and S. Maekawa, Superconducting π qubit with a ferromagnetic Josephson junction, *Physical Review Letters* 95, 097001 (2005).
4. P. Nemeth, K. McColl, L. A. J. Garvie, C. G. Salzmann, M. Murri, and P. F. McMillan, Complex nanostructures in diamond, *Nature Materials* 19, 1126 (2020).
5. K. Tanigaki, H. Ogi, H. Sumiya, K. Kusakabe, N. Nakamura, M. Hirao, and H. Ledbetter, Observation of higher stiffness in nanopolycrystal diamond than monocrystal diamond, *Nature Communications* 4, 2343 (2013).

6. V. M. Edelstein, Influence of an interface double electric layer on the superconducting proximity effect in ferromagnetic metals, *JETP Letters* 77, 182 (2003); L. P. Gor'kov, and E. I. Rashba, Superconducting 2D system with lifted spin degeneracy: Mixed singlet-triplet state, *Physical Review Letters* 87, 037004 (2001).

7. L. P. Gor'kov, and E. I. Rashba, Superconducting 2D system with lifted spin degeneracy: Mixed singlet-triplet state, *Physical Review Letters* 87, 037004 (2001).

8. T. S. Tinyukova and Yu. P. Chuburin, The role of Majorana-like bound states in the Andreev reflection and the Josephson effect in the case of a topological insulator, *Theoretical and Mathematical Physics* 202, 72–88 (2020).

9. S. S. Pershoguba and V. M. Yakovenko, Shockley model description of surface states in topological insulators, *Physical Review B* 86, 075304 (2012).

10. D. Mtsuko, C. Coleman, and S. Bhattacharyya, Finite bias evolution of bosonic insulating phase and zero bias conductance in boron-doped diamond: A charge-Kondo effect, *EPL* 124, 57004 (2019).

11. S. Bhattacharyya, D. Mtsuko, C. Allen, and C. Coleman, Effects of Rashba-spin-orbit coupling on superconducting boron-doped nanocrystalline diamond films: Evidence of interfacial triplet superconductivity, *New Journal of Physics* 22, 093039 (2020).

12. T. Mizushima and K. Machida, Multifaceted properties of Andreev bound states: Interplay of symmetry and topology, *Philosophical Transactions of the Royal Society A* 376, 20150355 (2018).

15. H-B. Leng, C. Li, and X. Lin, Intrinsic Andreev π-reflection and Josephson π-junction for centrosymmetric spin-triplet superconductors, *arXiv:2009.12891v*.

16. A. Haim, Spontaneous Josephson π junction with topological superconductors, *Physical Review B* 100, 064505 (2019).

17. V. L. Berezinskii, New model of the anisotropic phase of superfluid He³, *JETP Letters* 20, 287 (1974); A. Balatsky and E. Abrahams, New class of singlet superconductors which break the time reversal and parity, *Physical Review B* 45, 13125 (1992).

18. Y. Tanaka, M. Sato, and N. Nagaosa, Symmetry and topology in superconductors–odd-frequency pairing and edge states, *Journal of the Physical Society of Japan* 81, 011013 (2012).

19. Y. Tanaka and A. A. Golubov, Theory of the proximity effect in junctions with unconventional superconductors, *Physical Review Letters* 98, 037003 (2007).

20. M. Amundsen and J. Linder, Quasiclassical theory for interfaces with spin-orbit coupling, *Physical Review B* 100, 064502 (2019).

21. M. D. Croitoru and A. I. Buzdin, In search of unambiguous evidence of the Fulde–Ferrell–Larkin–Ovchinnikov state in quasi-low dimensional superconductors, *Condensed Matter* 2, 30 (2017).

22. Z. Zheng, M. Gong, Y. Zhang, X Zou, C. Zhang and G. Guo, FFLO superfluids in 2D spin-orbit coupled Fermi gases, *Scientific Reports* 4, 6535 (2014).

23. W. Zhang and W. Yi, Topological Fulde–Ferrell–Larkin–Ovchinnikov states in spin-orbit-coupled Fermi gases, *Nature Communications* 4, 2711 (2013).

24. K. W. Song and A.E. Koshelev, Quantum FFLO state in clean layered superconductors, *Physical Review X* 9, 021025 (2019).

25. S. Sugiura, T. Isono, T. Terashima, S. Yasuzuka, J. A. Schlueter, and S. Uji, *npj Quantum Materials* 4, 7 (2019).

26. T. Yoshida, M. Sigrist, and Y. Yanase, Complex-stripe phases induced by staggered Rashba spin-orbit coupling, *Journal of the Physical Society of Japan* 82, 074714 (2013); R. Masutomi, T. Okamoto, and Y. Yanase, Unconventional superconducting phases in multilayer films with layer-dependent Rashba spin-orbit interactions, *Physical Review B* 101, 184502 (2020).

27. Y. Matsuda and H. Shimahara, Fulde–Ferrell–Larkin–Ovchinnikov state in heavy fermion superconductors, *Journal of the Physical Society of Japan* **76**, 051005 (2007); T. Yokoyama, M. Ichioka, and Y. Tanaka, Theory of pairing symmetry in Fulde–Ferrell–Larkin–Ovchinnikov vortex state and vortex lattice, *Journal of the Physical Society of Japan* 79, 034702 (2010).

28. S. N. Polyakov, V. N. Denisov, B. N. Mavrin, A. N. Kirichenko, M. S. Kuznetsov, S. Y. Martyushov, S. A. Terentiev, and V. D. Blank, Formation of boron-carbon nanosheets and bilayers in boron-doped diamond: Origin of metallicity and superconductivity, *Nanoscale Research Letters* 11, 11 (2016).

29. C. C. Agosta, Inhomogeneous superconductivity in organic and related superconductors, *Crystals* 8, 285 (2018).

30. K. J. Sankaran, et al., Significance of grain and grain boundary characteristics of ultrananocrystalline diamond films and tribological properties, *Surface and Coatings Technology* 232, 75 (2013).

31. J. Narayan and A. Bhaumik, Novel phase of carbon, ferromagnetism, and conversion into diamond, *Journal of Applied Physics* 118, 215303 (2015).

34. J. Narayan, Dislocations, twins, and grain boundaries in CVD diamond thin films: Atomic structure and properties, *Journal of Materials Research* 5, 2411–2422 (1990).

35. M. P. Alegre, D. Araujo, A. Fiori, J. C. Pinero, M. P. Villar, P. Achatz, G. Chicot, E. Bustarret, and F. Jomard, Critical boron-doping levels for generation of dislocations in synthetic diamond, *Applied Physics Letters* 105, 173103 (2014).

36. A. K. Ramdas, and S. Rodriguez, Spin-orbit coupling, mass anisotropy, time-reversal symmetry, and spontaneous symmetry breaking in the spectroscopy of shallow centers in elemental semiconductors, *Solid State Communications* 117, 213–222 (2001).

37. H. Kim, Vogelgesang, A. K. Ramdas, and S. Rodriguez, Electronic Raman and infrared spectra of isotopically controlled blue diamonds. *Physical Review Letters* 79, 1706–1709 (1997).

38. H. Kim, A. K. Ramdas, S. Rodriguez, M. Grimsditch, and T. R. Anthony, Spontaneous symmetry breaking of acceptors in blue diamonds. *Physical Review Letters* 83, 4140–4143 (1999).

39. C. Kallin, and J. Berlinsky, Chiral superconductors, *Reports on Progress in Physics* 79, 054502 (2016).

40. J. R. Morris, C. L. Fu, and K. M. Ho, Tight-binding study of tilt grain boundaries in diamond, *Physical Review B* 54, 132 (1996).

41. H. Ichinose, and M. Nakanose, Atomic and electronic structure of diamond grain boundaries analysed by HRTEM and EELS, *Thin Solid Films* 319, 87 (1998).

42. R. Yu, H. Wu, J. D. Wang, and J. Zhu, Strain concentration at the boundaries in 5-fold twins of diamond and silicon, *ACS Applied Materials & Interfaces* 9, 4253 (2017).

43. S. Bhattacharyya, O. Auciello, J. Birrell, J. A. Carlisle, L. A. Curtiss, A. N. Goyette, D. M. Gruen, A. R. Krauss, J. Schlueter, A. Sumant, and Zapol, Synthesis and characterization of highly conducting nitrogen-doped ultra-nanocrystalline diamond films, *Applied Physics Letters* 79, 1441 (2001).

44. J. Birrell, J. E. Gerbi, O. Auciello, J. M. Gibson, D. M. Gruen, and J. A. Carlisle, Bonding structure in nitrogen doped ultrananocrystalline diamond, *Journal of Applied Physics* 93, 5606 (2003).

45. L. Alff, A. Beck, R. Gross, A. Marx, S. Kleefisch, Th. Bauch, H. Sato, M. Naito, and G. Koren, Observation of bound surface states in grain-boundary junctions of high-temperature superconductors, *Physical Review B* 58, 11197 (1998).

46. S. Kashiwaya, Y. Tanaka, M. Koyanagi, H. Takashima, and K. Kamimura, Origin of zero-bias conductance peaks in high-TC superconductors, *Physical Review B* 51, 1350 (1995).

47. J. W. Ekin, Y. Xu, S. Mao, T. Venkatesan, D. W. Face, M. Eddy, and S. A. Wolf, Correlation between d-wave pairing behaviour and magnetic-field-dependent zero-bias conductance peak, *Physical Review B* 56, 13746 (1997).

48. C. R. Hu, Midgap surface states as a novel signature for d2xa -x2b wave superconductivity, *Physical Review Letters* 72, 1526 (1994).

49. J. E. Mooij, B. J. van Wees, L. J. Geerligs, M. Peters, R. Fazio, and G. Schon, Unbinding of charge-anticharge pairs in two-dimensional arrays of small tunnel junctions, *Physical Review Letters* 65, 5 (1990); R. Fazio, and G. Schon, Charge and vortex dynamics in arrays of tunnel junctions, *Physical Review B* 43, 7 (1991).

50. H.-Z Lu and S.-Q Shen, Weak antilocalization and localization in disordered and interacting Weyl semimetals, *Physical Review B* 92, 035203 (2015); S. Hikami, A. Larkin, and Y. Nagaoka, Spin-orbit interaction and magnetoresistance in the two dimensional random system, *Progress of Theoretical Physics* 63, 707 (1980); E. McCann and V. I. Fal'ko, Landau-level degeneracy and quantum hall effect in a graphite bilayer, *Physical Review Letters* 96, 086805 (2006).

51. B. Z. Spivac and S. A. Kivelson, Negative local superfluid densities: The difference between dirty superconductors and dirty Bose liquids, *Physical Review B* 43, 3740 (1991); S. A. Kivelson and B. Z. Spivak, Aharonov-Bohm oscillations with period $hc/4e$ and negative magnetoresistance in dirty superconductors, *Physical Review B* 45, 10490 (1992).

52. P. V. Leksin, et al., Evidence for triplet superconductivity in a superconductor-ferromagnet spin valve, *Physical Review Letters* 109, 057005 (2012).

53. L.Y. Zhu, Yaohua Liu, F. S. Bergeret, J. E. Pearson, S. G. E. te Velthuis, S. D. Bader, and J. S. Jiang, Unanticipated proximity behavior in ferromagnet-superconductor heterostructures with controlled magnetic noncollinearity, *Physical Review Letters* 110, 177001 (2013).

54. C. R. Hu, Midgap surface states as a novel signature for d2xa-x2b-wave superconductivity, *Physical Review Letters* 72, 1526 (1994).

55. T. E. Baker, A. Richie-Halford, and A. Bill, Long range triplet Josephson current and 0-π transitions in tunable domain walls, *New Journal of Physics* 16, 093048 (2014).

56. P. Marra and M. Nitta, Topologically nontrivial Andreev bound states, *Physical Review B* 100, 220502(R) (2019).

57. L. Zhao et al., Interference of chiral Andreev edge states, *Nature Physics* 16, 862 (2020).

58. C-X. Liu, J. D. Sau, and S. D. Sarma, Distinguishing topological Majorana bound states from trivial Andreev bound states: Proposed tests through differential tunneling conductance spectroscopy, *Physical Review B* 97, 214502 (2018).

59. F. S. Bergeret, A. F. Volkov, and K. B. Efetov, Long-range proximity effects in superconductor-ferromagnet structures, *Physical Review Letters* 86, 4096 (2001).

60. S. Tamura, S. Kobayashi, L. Bo, and Y. Tanaka, Theory of surface Andreev bound states and tunneling spectroscopy in three-dimensional chiral superconductors, *Physical Review B* 95, 104511 (2017).

61. K. C. Pandey, New dimerized-chain model for the reconstruction of the diamond (111)-(2x1) surface, *Physical Review B* 25, 4338 (1982).

62. B. Pamuk and M. Calandra, Competition between exchange-driven dimerization and magnetism in diamond (111), *Physical Review B* 99, 155303 (2019).

63. B. Pamuk, J. Baima, F. Mauri, and M. Calandra, Magnetic gap opening in rhombohedral-stacked multilayer graphene from first principles, *Physical Review B* 95, 075422 (2017).

64. Y. G. Lu, S. Turner, E. A. Ekimov, J. Verbeeck, and G. Van Tendeloo, Boron-rich inclusions and boron distribution in HPHT polycrystalline superconducting diamond, *Carbon* 86, 156 (2015).

65. M. S. Anwar, F. Czeschka, M. Hesselberth, M. Porcu, and J. Aarts, Long-range super-currents through half-metallic ferromagnetic CrO2, *Physical Review B* 82, 100501 (2010).

66. F. Kuemmeth, S. Ilani, D. Ralph, et al., Coupling of spin and orbital motion of electrons in carbon nanotubes, *Nature* 452, 448 (2008).

67. M. T. Edmonds et al., Spin-orbit interaction in a two-dimensional hole gas at the surface of hydrogenated diamond, *Nano Letters* 15, 16 (2015).

68. E. A. Laird, F. Kuemmeth, G. A. Steele, K. Grove-Rasmussen, J. Nygård, K. Flensberg, and L. P. Kouwenhoven, Quantum transport in carbon nanotubes, *Reviews of Modern Physics* 87, 703 (2015).

69. H. Santos, L. Chico, J. E. Alvarellos, and A. Latgé, Defect-enhanced Rashba spin-polarized currents in carbon nanotubes, *Physical Review B* 96, 165401 (2017).

70. C. Coleman and S. Bhattacharyya, Signatures of two dimensional in superconducting nanocrystalline boron-doped diamond films, *EPL* 122, 57004S (2018).

71. E. Ekimov, V. Sidorov, E. Bauer, N. N. Mel'nik, N. J. Curro, J. D. Thompson, and S. M. Stishov, Superconductivity in diamond, *Nature* 428, 542 (2004).

72. G. Zhang, M. Zeleznik, J. Vanacken, P. May, and V. Moshchalkov, Metal-bosonic insulator-superconductor transition in boron-doped granular diamond, *Physical Review Letters* 110, 077001 (2013).

73. R. W. Dubrovinskaia, J. Wosnitza, T. Papageorgiou, H. F. Braun, N. Miyajima, and L. Dubrovinsky, An insight into what superconducts in polycrystalline boron-doped diamonds based on investigations of microstructure, *Proceedings of the National Academy of Sciences* 105, 11619 (2008);

74. S. Bhattacharyya, Microstructure and anisotropic order parameter of boron-doped nanocrystalline diamond films, *Crystals* 12, 1031 (2022).

75. S. Bhattacharyya, Unconventional superconductivity of the grain boundaries in boron-doped nanocrystalline diamond, (unpublished).

76. E. Ekimov, V. Sidorov, E. Bauer, N.N. Mel'nik, N. J. Curro, J. D. Thompson, and S. M. Stishov, Superconductivity in diamond, *Nature* 428, 542 (2004).

77. E. Bustarret, Superconducting diamond: An introduction, *Physica Status Solidi (a)* 205, 997–1008 (2008).

78. H. Okazaki et al., Signature of high Tc above 25 K in high quality superconducting diamond, *Applied Physics Letters* 106, 052601 (2015).

8 Carbon Hybrid System Odd-Frequency Order Parameter and Vortex Phase

Topological superconductors have been claimed in artificially grown compound/hybrid materials; however, in a system made of a single element, this can also be achieved in the well-aligned grain boundaries of a superconducting diamond. Here we show the signature of odd-frequency superconducting order parameter (Δ) in heavily boron-doped diamond films by breaking the structural symmetry to yield layered microstructure and enabling a RSOC. The superlattice-like structure in diamond describes the modulation of Δ (energy gap) which explains strong peak features observed in temperature-dependent resistance, oscillatory magnetoresistance, and differential conductance spectra, particularly a ZBCP with satellite peaks. A possible mechanism for creating FFLO-type state and chiral vortex lines from the superposition of multiple (Andreev) bound states due to boron acceptors predominantly at the well-aligned grain boundaries is discussed. Overall, the interface states of the diamond films can be explained by the well-known Shockley model describing the layers connected by tunnel barriers, hence forming a topological insulator on the diamond lattice. Superconductivity of the heavily boron-doped diamond can be compared to inhomogeneous superconductivity in disordered organic materials similar to bi- or trilayer graphene maintaining ABA stacking where π orbitals work as electric field normal to the planes and bind the layers just like vortices in the layered superconductors as a bilayer charge column as described in the generalized Shockley model.

8.1 DUALITY IN SUPERCONDUCTIVITY (MIXED WITH IMPURITIES)

Among all forms of dualism concepts used in physics electric-magnetic duality is the most famous one where electric (B) and magnetic fields are related to the symmetry through the Faraday–Maxwell equations $\nabla * B = \dfrac{dE}{dt}, \nabla \times E = -\dfrac{dB}{dt}$. The symmetry $E \to B$ and $B \to -E$ is known as duality, which although still holds in the presence of charge and currents in quantum field theory to find such symmetry between electric and magnetic charges was difficult. Later Montonen and Olive found a symmetry that would exchange electric and magnetic charges must exchange the quantum of electric charge with a multiple of the quantum of magnetic charge. This symmetry has

DOI: 10.1201/9781003316411-8

appeared similar to Faraday–Maxwell's equations where the symmetry must exchange elementary quanta with collective excitations since, for weak coupling, electric charges arise as elementary quanta and magnetic charges arise as collective excitations.

This model was applied to the condensed matter as vortex-boson duality (Figure 8.1). However, a precise formulation for this duality is lacking. Vortex–antivortex interactions were shown through the Kosterlitz–Thouless (KT) transition in 1+1 dimensions; however, this is not firmly established in 3+1 or higher dimensions (2+1d is still being developed). It is generally suggested that a vortex interacts with an antivortex via a spin-wave fluctuation in a superfluid. The vortices act as sources and sinks of supercurrents, and therefore supercurrent is no longer in the vortex condensate. Models claiming charge–vortex duality based on statistics (or transmutation of electrons) involving binding of electrons to magnetic fluxes or vortices was suggested in the frame of *anyon* superconductivity. A formal duality transformation between particles and vortices forms a two-dimensional (2D) Bose system, which was believed to explain the origin of high-temperature superconductivity. However, vortex features were identified in conventional superconductors before the development of the quantum spin liquid model.

The physics of superconductivity in confined 2D space is expected to result in a range of exotic phases and quantum dual phases. To study duality phenomena both the disorder such as doping and magnetic (electric) field-induced, superconductor-to-insulator transitions have been used.

One of the leading theories for explaining the SIT in disordered materials is that of the charge–vortex duality. In this model, when the system is in the superconducting phase, Cooper pairs are condensed into a superfluid leading to a decrease in resistance, whereas bosonic excitations in the form of vortices are delocalized and lead to a finite resistance. On the insulating side of this transition, the Cooper pairs become localized, leading to a rise in resistance, and the vortex states condense into a macroscopic state of zero conductivity. The charge–vortex duality has been applied to a range of different systems including Josephson junction arrays. In this system, the insulating state is a result of a Coulomb blockade or repulsion between superconducting regions analogous to granular material. One model predicts a universal scaling

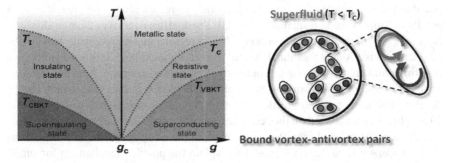

FIGURE 8.1 Vortex–antivortex can be described as clockwise and anticlockwise rotations. The strength of vortex–antivortex coupling is plotted as a function of temperature. Starting from T_1 in the insulating state, it goes to T_c in the resistive state through a quantum critical point. The T_{BKT} separates the superconducting state from the non-superconducting state. The BKT transition is a class of transitions that can have dual phases including a dual charge–BKT analog involving the binding of charge and anticharge. (Diagram from Ref. [23].)

behavior of the resistance as a function of the applied magnetic field. Also predicted by the model are universal values of longitudinal and transverse resistivity predicted to have a value of $h/2e$ at the SIT; these universal values are, however, difficult to verify except in well-defined patterned structures. One of the consequences of this duality is the dual BKT transition called the charge–BKT transition. This occurs in the insulating phase where the vortices which are bound during the superconducting phase unbind and form the electron-glass quantum phase. The charge–BKT effect has found widespread success in the analysis of Josephson Junction arrays, where the insulating state was determined to result from Coulomb forces acquiring a 2D character (i.e., act logarithmically) and resulting in the so-called super-insulating phase.

8.1.1 Ginsburg–Landau Theory and Vortex Structure

For fermion pairing and boson condensation BKT transition occurs. In other words, the gauge field becomes massive due to the Higgs mechanism. The most intriguing issue here is the quantization of the magnetic flux. Because the boson has charge e, while the fermion pairing is $2e$, the question arises whether the hc/e vortex is type A or B. (i) Type A: the fermion pairing order parameter psi vanishes at the core with its phase winding around it. The boson condensation does not vanish and the vortex core state is the Fermi liquid. The flux quantization is $hc/2e$. (ii) Type B: The Bose condensation is destroyed at the core and fermion pairing remains finite. Then the vortex core state is the spin-gap state. The flux quantization is hc/e.

8.1.2 Quantum Spin Liquid

It is like a resonance valence bond state that likes to change the direction of spin-like domain walls of a magnetic material that has some duality or symmetry. In the beginning, all spins were aligned in one direction in any magnetic material and formed a ground state as proposed by Heisenberg. However, if the alternating spins are aligned in two opposite directions, then antiferromagnetic (AFM) states can arise as proposed by Neel. With the dualistic nature of the spin liquid, the symmetry is broken spontaneously. Although most of them may have one direction or no direction due to the absence of symmetry, dualistic nature can always be present at least locally to support the non-dualistic liquid. If the dualism overtakes the non-dualistic liquid, then a nonequilibrium is broken. How to regain equilibrium by destroying the dualistic nature? A new rotational system or vortex should help in this regard, which is nothing and creates a space or void with a resonating structure. It means that two opposite rotating particles or currents can be superposed in one void, which can be developed with the resonating valence bond (RVB) as proposed by Phil Anderson in 1973. This idea was based on the resonantly single and double carbon–carbon bonds picture as proposed by Linus Pauling to explain the electronic structure of benzene rings. Anderson used the RVB model to explain the possible mechanism for superconductivity transition in high T_c cuprate superconductors having an AFM ground state. Later the excitations of the spin liquid were claimed as topological, which are loop-like (in addition to emergent point-like excitations). These quasiparticle excitations have nonlocal statistical interactions such as a charge moving around a magnetic flux and are called anyons. However, spin liquid can have multiple types like different types of magnetic order (Figure 8.2).

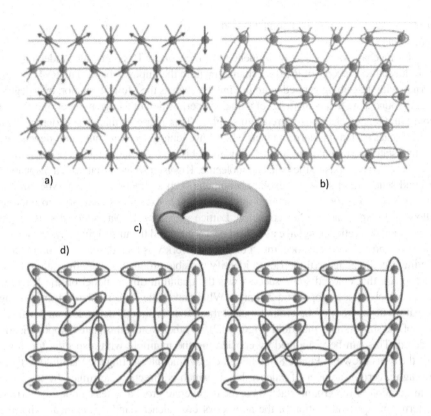

FIGURE 8.2 Schematics showing (a) the Kagome lattice where the antiferromagnetic state is considered as a gapless ground state; this can be contrasted to the (b) RVB ground state which is degenerate and is considered gapped. (c) The RVB ground state is considered topologically nontrivial, as the topology of the spin/charge manifold leading to the singlet pairing determines the state of the system, which is indicated in (d) where two systems of different topological numbers (dimers crossing the cut) constitute two distinct ground states, even though the number of dimers is conserved between them. (Schematics obtained and adapted from Ref. [24].)

8.1.3 ANDERSON'S RVB MODEL

Philip W. Anderson proposed the RVB model based on the linear superposition of spin-singlet pairs spanning different ranges called RVB. It is a possible ground state for the $S = 1/2$ AFM Heisenberg model on a triangular lattice. RVB state is a unique singlet ground state with either a short-range or power-law decay of AFM order. This state of affairs is called spin liquid, where the spin–charge separation takes place. Anderson postulated that spin excitations in an RVB state are $S = 1/2$ fermions, which are called spinons. This is in contrast with excitations in a Neel state, which are $S = 1$ magnon or $S = 0$ gapped singlet excitations. Spinon is similar to 1d spin chain-like domain walls. If the singlet bonds are liquid, two $s = 1/2$ spins formed by breaking a single bond can drift apart with the liquid of singlet bonds filling in the space between them. They behave as free particles and are called spinons. The concept of holons follows naturally as the vacancy is left over by removing a spinon. A holon carrier charge e but no spin. The space is considered the core of a vortex surrounded by spinons. The condensation of vortices creates mass and a gap through repulsion.

This mechanism was heavily influenced by the work of Baskaran and Anderson who introduced the concepts of RVB mechanism within the impurity band. The RVB is concerned with describing gapped quantum matter, such as the Mott's insulator, by a degenerate ground state that is made up of a linear superposition of a large number of different configurations of electrons paired into singlets. This phase of matter is different from paired states such as antiferromagnets which have long-range order but a non-degenerate ground state, as shown in Figure 8.3. This model is very similar to Pauli's original ideas on resonating π-bonds in the benzene molecule. Kivelson showed that such a resonating ground state would allow for exotic phenomena such as fractionalization, whereby the electron wave function is splintered into quasiparticle excitations analogous to photons allowing for spin and charge separation. Lattice gauge theory was adopted as the most convenient theoretical language as it most neatly captured the underlying physics of such systems, and as a consequence, much theoretical work was then developed around identifying specific gauge symmetries to identify possible ground-state configurations. One such phase that enjoyed some success was the quantum dimer model on the Kagome lattice which is an example of a Z_2 phase. What makes the RVB mechanism interesting is that the degeneracy of the ground state of the system is directly dependent on the topology of the spin (dimer) manifold (Figure 8.2). In other words, this state is topologically ordered. This can be visualized by considering the manifold with a specific topology, i.e., the torus shown in Figure 8.4, where the gray line intersecting the manifold will cut through a certain number of valence bonds which is directly determined by the bond configuration of the ground state; thus, the degenerate ground states can be categorized in terms of the conservation of the number of cut valence bands. This means that the topological order of the ground state is a quantum number for that state.

The concepts have been applied to superconducting boron-doped diamond films. The high doping density in the boron-doped diamond system is comparable to the critical density expected for an Anderson–Mott transition, therefore indicating that electron correlations are vital in explaining the transport features of the system. The system was thus described within the RVB impurity band model where diamond, with

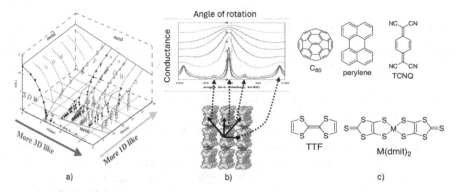

FIGURE 8.3 (a) Superconducting transition of some organic complexes having 1D and layered 3D structures. (b) Conductance as a function of angle shows a central peak and side peaks due to tunneling between layers. (c) Various molecules show superconducting transition due to resonating bonds.

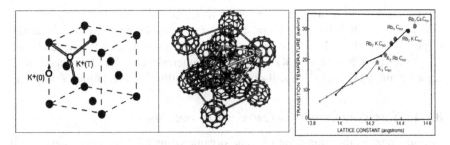

FIGURE 8.4 Superconductivity in doped fullerenes [87]. Lattice structure of solid C_{60} (solid O) showing the tetrahedral (K(T)) and octahedral (K(O)) alkali sites (left). The K(T): K(O) site ratio is 2:1 [88]. (middle) T_c of bulk samples of $A_{3-x} A' \times C_{60}$ where A and A' represent K; R_b or C_s has nearly linear dependence on lattice constant a. Alkali atoms with larger ionic radii produce a greater effective negative pressure and large lattice constant. The lattice constants refer to fcc. (right) Application of pressure causes a lattice contraction and corresponding decrease in T_c. Results determined for potassium and rubidium-doped C_{60} (gray and black lines, respectively) overlap with the zero-pressure data [89].

negligible electron correlations, offers a "vacuum" to the boron subsystem. Within the diamond, the random Coulomb potential would remove the threefold degeneracy of the acceptor states. This is expected to induce a single, narrow band of holes. Such a system would allow for singlet pairing between spins of neighboring neutral acceptors B^0–B^0. Crossing the insulator to metal transition, charged B^+ and B^- states (free carriers induced at the boron sites) get spontaneously generated and are delocalized. This delocalization coupled with the random Coulomb potential causes resonating behavior of the pairing of the singlets resulting in a RVB superconducting state.

To explain the origin of superconductivity in the doped diamond system the well-known RVB model has been proposed. It was proposed that the neutral dopant atoms (D_0) or a pair ($D_0 D_0$) performed a virtual transition to higher energy (U) polar state from the insulating phase. This process is described as $D_0(\downarrow)D_0(\uparrow) \rightarrow D^-(\downarrow\uparrow) D^+_0$, which causes a virtual charge fluctuation leading to the super/kinetic exchange. This leads to an effective Heisenberg coupling between the two dangling spins (D) which becomes AFM (a Kondo state). The RVB state which is a spin-singlet state can also be a triplet state, particularly in a frustrated ferromagnetic system. So when there is competition between inter-site (magnetic and on-site (Kondo) the RVB state emerges to suppress the local magnetism. The formation (depletion) of spin liquid (RVB state) can prevent (enhance) the system from forming a Kondo state which yields a Fermi liquid state. Similarly, an RVB state can be used to stabilize the Kondo state. The instability or fluctuation of the Kondo state can support the re-entrance superconductivity. Although this state can rarely be observed, we explained our results on Kondo-like transition through the formation of an RVB state and the SC transition takes place via an AFM super-exchange process. The spin-triplet effect can also arise from the charge–Kondo effect, which can be seen in a double (or multi-) quantum dot system. It also shows a peak in R–T and R–B behavior. In this article, we would like to extend this study further by introducing the spin-triplet effect where a vortex structure (superconducting and magnetism) can coexist with a superconducting state through a reentrant superconductivity mechanism. The reentrant vortex-bound spin originates from either the GB or from the intra-grain vortex core fluxon. The superconducting transition above 10 K has been claimed in several carbon-based materials. At the time of preparation of the book, we are aware of graphite-intercalated superconductors ($T_c \sim 15$ K), SWCNTs ($T_c \sim 15$ K), alkali-doped fullerenes ($T_c \sim 40$ K), B-doped single crystal diamond ($T_c \sim 11$ K), and B-doped polycrystalline diamond films ($T_c > 10$ K).

8.1.4 SUPERCONDUCTIVITY IN ORGANIC CONDUCTORS

The superconducting transition of some organic complexes has been suggested in the 1960s and 1970s. Notably, $(TMTSF)_2PF_6$ ($T_c = 0.9$ K, at an external pressure of 11 kbar) or Bechgaard salt was synthesized in 1979 (Figure 8.3). The materials can have 1d and layered 3d structures. High hydrostatic pressure was also applied to the materials. A list of complicated materials is given in the table. The transition temperatures can vary from sub-Kelvin to 10 K depending on the structures. In addition to the BCS superconductivity, spin-triplet superconductivity is claimed in these materials. In the grain boundaries of nanocrystalline diamond films, one-dimensional (1D) conduction of filamentary conduction has been found. Hence, the "solitonic" transport model of filamentary conduction can be applied to explain the unconventional superconducting transport properties of boron-doped nanodiamond films.

Fullerenes were discovered by Kroto and Smalley in 1985. C_{60} fullerene molecule composed of 20 hexagons and 12 pentagons has a diameter of 0.7 nm. The C_{60} fullerene has a face-centered cubic (fcc) lattice with a lattice constant of 14.17 Å at room temperature. Fullerene molecule is highly electronegative and forms doped compounds with alkali metals. These doped fullerene molecules are superconductors at temperatures below 20–40 K. K_3C_{60} and Rb_3C_{60} show superconductivity with the onset T_c of 18 and 28 K. Under strong pressure, Cs_3C_{60} compound T_c can even increase to 40K. Although at atmospheric pressure Cs_3C_{60} compound is both insulating and magnetic. Recently hole-doped fullerenes system $C_{60}/CHBr_3$ has exhibited the greatest critical temperature, $T_c = 117$ K at ambient pressure for an organic superconductor. Field-effect doping techniques have been exploited to prepare superconducting fullerenes. The maximum T_c of 52 K for 3 ± 3.5 hole per C_{60}, which is almost five times higher than for electron doping, has been absorbed. Spin-triplet superconductivity can be seen in doped C_{60} which finds a remarkable similarity with BNCD films (Figure 8.5).

The superconductivity of the 4-Å SWCNTs was discovered more than a decade ago and marked the breakthrough of finding superconductivity in pure elemental undoped carbon compounds. The van Hove singularities in the electronic density of states (DOS) at the Fermi level in combination with a large Debye temperature of the SWCNTs are expected to cause an impressively large superconducting gap. The superconducting transition temperature T_c was observed with onset at 15 K. By optimizing the geometry of the carbon nanotubes, a maximum T_c of 60 K is claimed. Unconventional superconductivity in CNT has been shown based on the observation of a BKT transition. This can also be seen in BNCD films, which will be explained later in the chapter. From the electrical transport characteristics, two types of superconducting resistive behaviors were claimed. The first type is the quasi-1D fluctuation superconductivity that exhibits a smooth resistance drop with decreasing temperature,

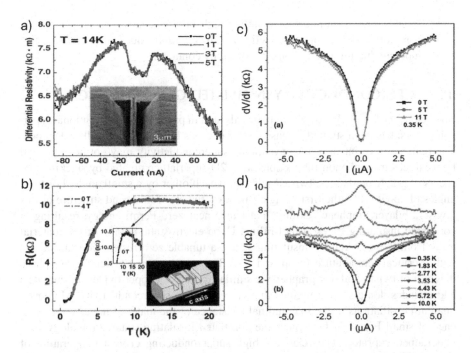

FIGURE 8.5 The superconductivity of the 4-Å single-walled carbon nanotubes (SWCNTs) shows a BKT transition. (a) At low temperatures, the differential resistance also shows a smooth increase with increasing bias current (voltage). (b) The second type is the quasi-1D–3D superconducting crossover transition, which was observed to initiate at 15 K with a slow resistance decrease switching to a sharp order of magnitude drop at 7.5 K [25]. (c) The dip in resistance at low temperature indicates the existence of supercurrent, the zero bias resistance peak above 2.77 K. (d) A competing mechanism (with superconductivity) was suggested in the system. The ground state is singlet superconducting due to the reduction of Coulomb interaction through dielectric screening by the surrounding nanotubes.

initiating at 15 K. At low temperatures the differential resistance also shows a smooth increase with increasing bias current (voltage). These manifestations are shown to be consistent with those of a quasi-1D superconductor with thermally activated phase slips as predicted by the Langer–Ambegaokar–McCumber–Halperin theory. The second type is the quasi-1D–3D superconducting crossover transition, which was observed to initiate at 15 K with a slow resistance decrease switching to a sharp order of magnitude drop at 7.5 K. The latter exhibits anisotropic magnetic field dependence and is attributed to a BKT-like the transition that establishes quasi-long-range order in the plane transverse to the c-axis of the aligned nanotubes, thereby mediating a 1D–3D crossover.

Other cases are as follows:

- Superconductivity in ropes of SWNT is observed in "long" samples ($L > 1$ mm)
- Destruction of superconductivity by normal contacts
- 1D character of superconductivity (phase slips)

- Possible Josephson intertube coupling in disordered ropes
- Proximity-induced superconductivity with very high values of supercurrent!
- Superconducting fluctuations stabilized by contacts?

8.2   SUPERCONDUCTIVITY IN TWISTED GRAPHENE (TBC)

The behavior of strongly correlated materials, and in particular unconventional super-conductors, has been studied extensively for decades but is still not well understood. The realization of intrinsic unconventional superconductivity may not be explained by weak electron–phonon interactions. In a 2D superlattice created by stacking two sheets of graphene that are twisted relative to each other by a small angle. For twist angles of about 1.1° – the first "magic" angles – the electronic band structure of this "twisted bilayer graphene" exhibits flat bands near zero Fermi energy, resulting in correlated insulating states at half-filling. Upon electrostatic doping of the material away from these correlated insulating states, a tunable zero-resistance state occurs with a critical temperature of up to 1.7 K. The temperature–carrier density phase diagram of twisted bilayer graphene is similar to that of copper oxides (or cuprates) and includes dome-shaped regions that correspond to superconductivity. Moreover, quantum oscillations in the longitudinal resistance of the material indicate the presence of small Fermi surfaces near the correlated insulating states, in analogy with underdoped cuprates. The relatively high superconducting critical temperature of twisted bilayer graphene, given such a small Fermi surface (which corresponds to a carrier density of about $10^{11}/cm^2$), puts it among the superconductors with the strongest pairing strength between electrons. Twisted bilayer graphene is a precisely tunable, purely carbon-based, 2D superconductor. It is therefore an ideal material for investigations of strongly correlated phenomena, which could lead to insights into the physics of high-critical-temperature superconductors and quantum spin liquids. The quantum spin liquid behavior of BNCD films will be discussed in the next chapter (Figure 8.6).

SUPERCONDUCTIVITY IN OTHER CARBON

Quantum phases of matter known as superconductors transmit electrical current with zero resistance. Microscopically, this phenomenon arises from the fact that it is energetically favorable for electrons to bind into two-electron states, dubbed Cooper pairs, that move collectively and cooperatively without energy loss. A Cooper pair is said to be spin-singlet, when its two-electron spins (intrinsic angular momenta) point in opposite directions and the pair has a total spin of zero, whereas spin-triplet Cooper pairs have a total spin of 1, and the two-electron spins can be aligned in the same direction. Most experimentally known super-conductors have spin-singlet Cooper pairs; these include metals (such as lead and niobium) that demonstrate conventional superconductivity, and cuprates (layered copper oxide compounds) that exhibit unconventional superconductivity. An unconventional superconductivity associated with spin-triplet Cooper pairs is claimed. This finds a remarkable similarity with the results obtained from BNCD films and will be discussed in this chapter (Figure 8.7).

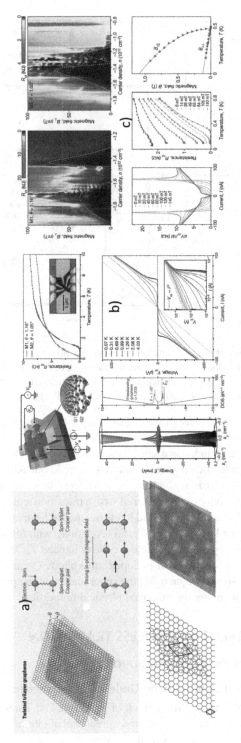

FIGURE 8.6 Superconductivity in twisted trilayer graphene. (a) Magic-angle twisted trilayer graphene (MATTG) is a system of three sheets of hex-agonally arranged carbon atoms. The top and bottom sheets are aligned, and the middle sheet is rotated by an angle θ of approximately 1.6° relative to the other two sheets. (b) Superconductivity results from electrons binding into two-electron states called Cooper pairs. In a spin-singlet Cooper pair, the electron spins (intrinsic angular momenta) point in opposite directions; in a spin-triplet Cooper pair, they can be aligned in the same direction. (c) In the presence of a strong magnetic field in the plane of the material, a spin-singlet Cooper pair breaks apart because a phenomenon known as the Zeeman effect causes the spins to align in the same direction. By contrast, a spin-triplet Cooper pair can survive such a field. Superconductivity in MATTG persists under a strong in-plane magnetic field, and suggest that this observation is evidence for spin-triplet Cooper pairs [26].

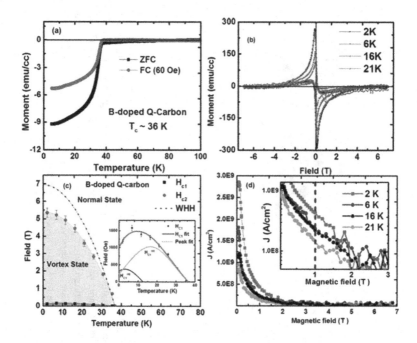

FIGURE 8.7 (a) Magnetic moment vs. temperature plots of the boron-doped Q-carbon films show Tc @ 36 K; (b) The M-H loops below Tc; and (c) the upper and lower critical field are shown with the deconvolution of the lower critical field. (d) The critical current density as a function of the applied magnetic field at various temperatures below Tc is presented [27].

High-temperature superconductivity has been discovered in boron-doped carbon, called Q-carbon having a filamentary network (instead of amorphous carbon). By using a pulsed laser ablation technique, these samples are prepared from nanosecond melting of carbon in a super undercooled state followed by a rapid quenching of carbon. The temperature-dependent resistivity and magnetic susceptibility measurements demonstrate type II superconductivity in this material with a transition temperature of 36 K and an upper critical field of 5.4 T at 0 K, and it follows the BCS superconductivity. It was suggested that a higher DOS near the Fermi level along with higher Debye temperature and phonon frequency are responsible for the enhanced T_c. This discovery of high-temperature superconductivity in B-doped amorphous Q-carbon shows that the nonequilibrium synthesis technique using the super undercooling process can be used to fabricate materials with greatly enhanced physical properties.

8.3 BEREZINSKII–KOSTERLITZ–THOULESS TRANSITION

8.3.1 FIELD-INDUCED SUPERCONDUCTOR–INSULATOR TRANSITION

8.3.1.1 Phase Transitions without Breaking Underlying Symmetries

The pioneering and now Nobel Prize–winning work of Kosterlitz and Thouless (KT) on topological excitations in 2D systems has generated a great deal of interest. Although the transition was first predicted by Berezinskii in his work on superfluid He (hence the B-KT transition), much of the mathematics regarding the physics of this transition was put forth by Kosterlitz and Thouless. The BKT transition describes the metal–superconducting phase transition by invoking the thermal excitation of topological defect pairs called vortex and antivortex. In the metallic or resistive state, the vortices are unbound and proliferate, leading to a finite resistance. At the BKT transition temperature, the vortices bind into pairs, leading to the divergence of the coherence length. The original idea was conceived using the so-called XY-model of a 2D array of spins. Such systems up to the time of KT's discovery had been proven through rigorous theoretical work to not allow phase transitions with long-range order on approaching low temperatures. The transition was very early on and observed experimentally in a range of systems including 2D crystals, thin superconducting films, and even ion-trapped gasses; this provided a great deal of experimental data for the development of a more concrete and refined theory of topological phase transitions that are currently enjoying a great deal of attention. The key feature of the BKT transition is that it occurs due to the unbinding of topological defects that are paired by a logarithmic interaction in 2D, thus breaking the long-range order, as they proliferate without breaking the underlying symmetries of the system. The BKT transition has also been investigated in a range of bulk superconducting systems that possess 2D crystal planes. This is well established in many of the high T_c superconductors where the crystal lattice is composed of 2D CuO planes (ab-plane) which are weakly linked in the third spatial dimension (c-plane). In such layered systems, the presence of other superconducting planes squeezes the field of the pancake vortices into a narrow range along the c-axis. It has been established that Josephson coupling between layers can affect the logarithmic interaction between vortices in the same plane. Because of these finite-size effects, the BKT transition observed in layered bulk superconductors occurs at temperatures slightly higher than purely 2D cases. This has been confirmed in experiments where the number of ab-planes was controlled, thus determining the effective thickness from this analysis to the known thickness of the films. Thus, by measuring thermodynamic transport properties, it is possible to identify the BKT transition.

In any case, even though the BKT effect has been observed in a range of superconducting systems, its occurrence has not yet been investigated in boron-doped diamonds. This is mainly due to the assumption that the system is strictly 3D, despite numerous evidence signifying two dimensionalities. The experimental characteristics of the BKT transition are well documented and generally easily identified through analysis of temperature-dependent transport properties, and in the next section, the occurrence of the BKT transition is investigated in the superconducting boron-doped diamond system. In this study, we present the temperature dependence of both resistance and VI to verify the occurrence of the BKT transition.

In this work, we study a series of BNCD films to investigate the possibility of the BKT transition (vortex–antivortex binding) through the standard resistance vs. temperature (R–T) and current–voltage (I–V) analyses. The identification of the BKT temperature is quite remarkable for this system as the transition is generally limited to 2D films, whereas the samples studied here are approximately 300 nm thick. There are, however, accounts of the BKT transition occurring in bulk samples that have a layered crystal structure. As nanocrystalline diamond films were discussed as a superlattice structure with conduction along grain boundaries, we believe a similar mechanism is at play here. The granularity, or more specifically the alternation of grain and boundary regions, allows for a 2D-like behavior similar to the layered high T_c ceramics. In previous reports, 2D structure in BNCD films has been found from 2D WL transport. Although the further investigation into the nature of the GB will most definitely be required to verify the 2D conduction channels claimed here, there is a growing amount of both experimental evidence and theoretical analysis that suggest the nanodiamond crystal grains are separated by an sp^2-hybridized carbon phase similar to disordered graphene highlighting the possibility of Dirac fermion or chiral superconductivity in this system. The effect of structurally arranged boron in these graphitic boundaries has recently been researched to induce interfacial superconductivity. Although the BKT transition has been reported in carbon nanotubes and predicted for graphene, the vortex structure of BNCD has been studied through scanning tunneling microscopy, and yet, to date, no other report on the BKT transition has been made for this material. Additionally, it is well known that granular superconductors can also exhibit a dual charge–BKT transition, and hence we investigate the magnetic field-induced superconductor–insulator transition where the density of field-induced localized vortices increases up to a quantum critical point above which they condense into an insulating Bose glass state capable of a BKT transition. This transition is expected for disordered superconductors and has been identified in other granular materials through a universal scaling analysis, which expects the collapse of all magnetoresistance isotherms to a single curve described by a critical exponent product (vz) of order unity. The observed features are again related to the temperature-dependent pinning/localization effect arising from the grain boundaries.

These coherence peaks in the voltage–current behavior (Figure 8.8) can be considered a region of quantum criticality, where a BKT-like transition from the

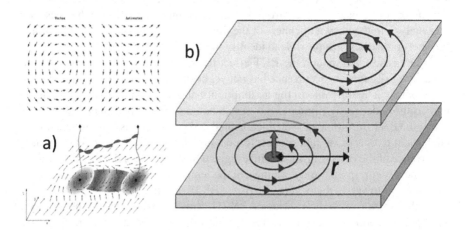

FIGURE 8.8   (a): Topological defects: (top) vortex and antivortex involved the BKT transition once vortex–antivortex biding occurs (bottom). The bound state where vortex–antivortex pairs are locked in position and do not contribute to resistance. (b): The BKT transition has also been experimentally verified in bulk layered superconducting system. This is due to vortex and antivortex pairing between layers. Shown in the schematic above is a depiction of vortex dynamics between superconducting layers and can lead to new novel phases of vortex matter [28].

nonequilibrium singlet-triplet Kondo effects occur. As the current bias increases, the Kondo effect will be destroyed favoring a more coherent BCS-like superconductor–normal metal transition.

The anomalous peak in the $R(T)$ and magnetoresistance data and the coherence peaks in differential conductance vs. T plots can in that case be attributed to regions within charge–BKT and vortex–BKT transitions in which there are phase transitions between metallic phases insulator and superconducting phases as envisaged in theory. However, the current-dependent phase diagrams deviate significantly from the theoretical BKT ones, suggesting a more complex interplay of magnetism, superconductivity, and structure that may require the inclusion of a possibility of Kondo effects. Theoretical studies are seeking to explain the influence of interference of different paths in the granular structure, the evolution of vortex–antivortex interactions, the role of the graphitic phase in the GB, and the influence of level of doping on this unconventional superconductor, as well as providing further evidence for Kondo-like and BKT-like behavior.

One of the hallmarks of the BKT transition is the effect of the universal jump in the superfluid stiffness, which can directly be inferred from the behavior of the exponent a in the I–V characteristics:

$$V \propto I^{a(T)}, \ \alpha(T) = \frac{\pi J_s(T)}{T} + 1. \qquad (8.1)$$

The superliner behavior shown here is due to sufficiently large currents causing the unbinding of vortex–antivortex pairs. According to the KT criterion, the parameter α should jump from $\alpha = 3$ at the BKT temperature to $\alpha = 1$ above the BKT temperature. At temperature below the BKT point, α is expected to increase with decreasing temperature. As shown in Figure 8.8, the voltage–current characteristics are shown on a log-log scale to determine the power-law scaling. The critical parameter $\alpha(T)$ is determined from the slope of the curves for temperatures ranging from 300 mK to 2.1 K. The power-law behavior ($V \propto I^{\alpha(T)}$) is observed in the low current regime, where the $V(I)$ dependence is predicted to arise from the thermal dissociation of the vortex–antivortex pairs at temperatures above the BKT point and from current-induced dissociation at temperatures below the BKT point.

In systems with high tunneling resistance, excess charge can build up in respective arrays or grains due to single-particle tunneling; this in turn leads to charging effects where excess charge in an island/electrode can polarize neighboring electrodes. This has been experimentally related to the formation of charge solitons, which can dominate the transport properties. In 2D arrays when the distance between two charge solitons is smaller than the screening length but larger than the thickness ($d < r << \Lambda$), the interaction between the charge and anticharge pair is essentially logarithmic. This leads to the observation of the charge analog of the BKT transition. As the bound pairs are electrically neutral, they can act as localized entities and thus the system will be in a resistive state. These features have also been observed in granular superconducting systems, where superconducting islands are separated by fine grain boundaries acting as Josephson junctions.

One system of particular interest is the nanocrystalline boron-doped diamond, not only has this system already exhibited a pronounced reentrant resistive phase but also effects of Cooper pair confinement, low dimensionality, and signatures of a BKT-like transition. The boron-doped diamond system is considered a structural granular superconductor, composed of nano-scale crystalline diamond grains. HR-TEM indicates columnar growth, crystal twinning, and pronounced boundary regions between grains; thus the microstructure of the system is more complex than conventional granular superconducting films (usually grown through sputtering or other deposition techniques). It has also been shown that the inter-grain coupling can be tuned by applying pressure, and due to the insulating nature of the intrinsic band structure of the diamond, the boron-doped diamond presents an ideal system to study charging effects between grains as mentioned above. One of the most intriguing aspects of this system is the formation of an anomalous resistive phase, the bosonic insulator.

Magnetoresistance: WL to WAL transition

- We have repeatedly observed temperature-dependent WAL and WL crossover in several samples.
- Features are typical for topological insulators with the strong SOC of surface states although SOC is weak in carbon systems...
- What is the origin of these topological features...?

$$\Delta\sigma(B) = \frac{e^2}{\pi h}\left(F\left(\frac{\tau_B^{-1}}{\tau_\phi^{-1}}\right) - F\left(\frac{\tau_B^{-1}}{\tau_\phi^{-1} + 2\tau_i^{-1}}\right) - 2F\left(\frac{\tau_B^{-1}}{\tau_\phi^{-1} + \tau_i^{-1} + \tau_*^{-1}}\right)\right) \quad (8.2)$$

- Two mechanisms to explain WAL/WL crossover....
- sp^2 phase of GB is similar to graphene.
- WL corresponds to intervalley scattering between Dirac cones and changes to intravalley scattering with variations in temperature.
- Nontrivial surface state due to hole population of band structure.
- HLN formula gives the dephasing rate (due to inelastic scattering at small fields), as well as the intervalley and intravalley scattering rates (elastic scattering).
- Observe an exponential increase in dephasing rate as temperature increases.
- Similar values and trends reported for graphene.

We observe a trend in the data indicating a change from activated behavior to some other phase identified by the drastic decrease in slope and eventual temperature-independent state. This feature has been explained in light of the macroscopic quantum tunneling of vortices and more recently as a quantum or Bose metal phase. In a 2D superconductor, the dissipation is expected to arise from vortex–antivortex dissociation and thus by fitting the activated region, we can extract the vortex binding energy. The dependence of this pairing energy as a function of the applied field is shown in Figure 8.9. From the field-dependent pairing energy we can also extract $H_0 \sim B_{C2}$ found to be 2.46 T for sample B5.

For comparison, the T_c field dependence can also be used to determine the upper critical (B_{C2}), which is shown in Figure 8.9, where B_{C2} (0 K)=2.8 T. This critical field is approximately 12% larger than the value obtained from the activated transport fitting, and this discrepancy (0.34 T) although relatively small indicates that the system does

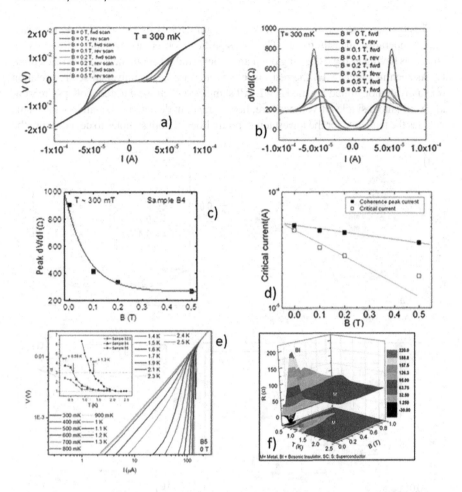

FIGURE 8.9 (a) The voltage–current characteristics at different magnetic field. (b) Evolution of differential resistance (*dV/dI*) peaks (or coherence peaks) with magnetic field variation. (c) Magnetic field scaling of the coherence peaks. The resistance of coherence peak scales in the same manner as the bosonic peaks read from *R(T)* data. (d) Magnetic field scaling of critical current and current of coherence peak (current corresponding to the *dV/dI* peak resistance). (e) Current–voltage response plotted on double log scale. According to the theory, the BKT transition is reached when the $V \sim I^3$. This is indicated by the dotted line and corresponds to a temperature of 1.3 K for the specific sample [29].

not strictly follow BCS theory. The line shows a fit to $B_{C2} = \varphi_0 / \left(2\pi\xi(0)^2\right)(1 - T/T_c)$ from which we extract a zero-field coherence length of 10.84 nm (approximately half the grain size). As suggested previously there is a strong possibility of a field-induced quantum phase, investigated through magnetoresistance measurement. A smeared crossing point can be seen in both samples as indicated by Figure 8.10, which shows the data for B5. However, upon closer examination of the *log-log* plot of the magnetoresistance, we observe that the crossing point of the various isotherms is concentrated at a single point at a low temperature and then starts to drift as the temperatures are increased above 1.3 K.

This temperature was determined to be a significant point from both *I–V* and resistance temperature measurements. This is a significant observation as magnetoresistance isotherms having a single crossing point are widely studied about the charge–vortex duality. As seen in Figure 8.11, the resistance is plotted using the universal scaling relation as usually demonstrated for the H-SIT. All samples, which were tested, collapse reasonably well onto a single curve; there are, however, slight deviations that are notable. This includes the tendency for the lowest temperature lines of both samples to deviate from the

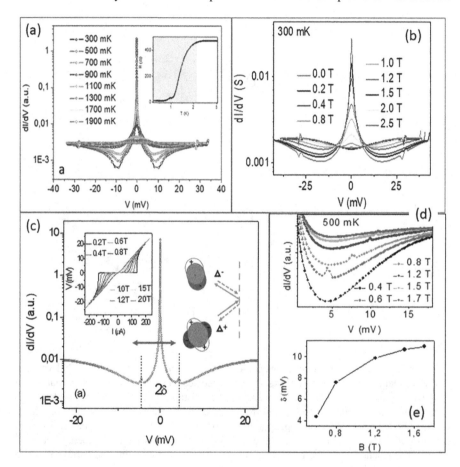

FIGURE 8.10 (a) The **zero-energy peak** is highly dependent on temperature and is only observed at temperatures below the superconductivity transition onset, as indicated in the inset. The peak height decreases exponentially with increasing temperature and FWHM increases exponentially. The ZBCP does not split up until it is completely suppressed and inverted to a V-shape dip, similar to what has been observed in other triplet superconductors. The field strength where this inversion of the peak occurs marks the crossover (B_{C2}) from superconducting to insulating regime as shown in the high-field magnetoresistance data. (c) The ZBCP also shows a distinct field dependence with side peaks at $V = \delta$ mV (see arrow) that split from the zero-energy mode at low fields. Side peaks that follow this field dependence are related to anisotropy of the pairing potential as in the case of *d*- and *p*-wave superconductors. This is a result of the change of sign of the pairing potential, as the Cooper pairs scatter from the boundary region, thus forming the bound states; this is indicated schematically in the inset [11].

collapse, indicated by open circles and triangles in Figure 8.11, compared to all other isotherms which collapse reasonably well to the same line. This may signify a crossover to
alternative scaling behavior at lower temperatures. It can also be observed that the insulating side of the scaling does not follow the steep upturn usually observed for SIT-related
superconductors, and the value of the critical resistance is well below the theoretically
predicted value. Similar scaling behavior has been observed before and was related to
normal-state electrons that contribute significantly to the transport through tunneling;
essentially the localization of charge carriers within the grain is reduced as temperature
increases and thus the insulating state is not as severe. It has also been shown that the
breakdown of the universal scaling could occur due to finite-size effects and the formation of Josephson networks at certain grain sizes. Nevertheless, the scaling behavior here
results in critical exponent products of $z\upsilon = 1.024 \pm 0.3$ and $z\upsilon = 1.028 \pm 0.2$. These values
are well within the theoretical prediction of unity for a 2D system and are confirmed by
the best fit to the dR/dB isotherm data taken at the critical field B_C (Figure 8.11). These
observations are a good indication that this system may very well exhibit the charge glass
state above the H-SIT as predicted for disordered 2D superconductors.

8.4 SUPERCONDUCTING CARBON: A NEW SPIN-TRIPLET SUPERCONDUCTOR?

Hexagonal structures with resonant valence bonds are the basis of the topological
phase in carbon, including diamond, which also hosts a nonplanar hexagonal ring in
each cell. Grain boundaries of polycrystalline and particularly nanocrystalline diamond films show high conductivity, which can be metallic by nitrogen incorporation
and superconducting by heavy boron incorporation. Boron creates a bilayer graphene
structure in diamond films, turning the material into a superconductor. As a result,
a superconducting diamond behaves like a topological superconductor like a spin-
triplet superconductor. Let me explain the spin-triplet superconductivity that has been
claimed in a boron-doped diamond.

8.4.1 ZBCP

As seen in Figure 8.10, the differential conductance obtained at temperatures around
the critical point shows a pronounced ZBCP. The ZBCP in the differential conductance spectrum has before been observed in anisotropic superconductors and
is known to arise from the formation of surface ABS, when a sign change of the
pair potential occurs due to the translational symmetry breaking at the boundary as
shown in Figures 8.9 and 8.10.

$$
\begin{aligned}
J = J_N \int_{-\Delta}^{\Delta} \sum_{\pm} &\frac{D_\uparrow D_\downarrow / D}{1 + R_\uparrow R_\downarrow - 2\sqrt{R_\uparrow R_\downarrow}\cos(2\delta \pm \vartheta)} T(E)\,dE \\
&+ J_N \left(\int_{-\infty}^{-\Delta} + \int_{\Delta}^{\infty} \right) \frac{2\cosh\delta\left[e^{-\delta}D_\uparrow D_\downarrow / D + \sinh\delta\right]}{e^{2\delta} + e^{-2\delta}R_\uparrow R_\downarrow - 2\sqrt{R_\uparrow R_\downarrow}\cos\vartheta} T(E)\,dE,
\end{aligned}
\tag{8.3}
$$

where $T(E) = \tanh\left[\beta(E + eV)/2\right] - \tanh\left[\beta(E - eV)/2\right]$.

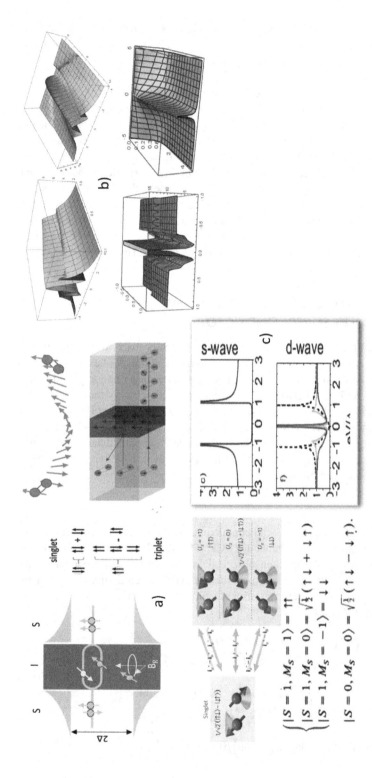

FIGURE 8.11 (a) The robust nature of the ZBCP toward applied filed indicates the possibility of a triplet component to the superfluid, and this mixing is expected to occur due to the spin–orbit interaction at the GB interface due to breaking of structural inversion symmetry and asymmetric confining potential between grains. The resonance ZBCP can be evaluated as a spin-polarized Andreev bound state scenario due to spin precession caused by the Rashba field at the interface [11]. (b) 3D plots of dI/dV with the variation eV/Δ in (a) temperature (T) and (b) superconducting gap (Δ) can explain the experimental data (unpublished). (c) Conductance spectra for an s-wave (gapped) and a d-wave (ungapped) superconductor showing a zero-bias conductance peak.

The peak is found to be sensitive to both the applied magnetic field and temperature. It is well established that this phenomenon occurs in spin-singlet d_{xy}-wave superconductors; however, it is also expected to occur in spin-triplet (p-wave) materials as discussed later in the text. In the case of the p-wave superconductor the ABS forms along the ab-plane. It is thus an indication of the two dimensionality of the system, consistent with the BKT analysis. In turn, the peak presented in the discussed measurements allows one to investigate more deeply the symmetry of the pairing mechanism for the superconductivity of the BNCD. Indeed, the presence of the peak (excess of the DOS at the Fermi Level) can be attributed to the Andreev reflection from the surfaces of the boundaries which are naturally presented in the granular material. The symmetry of the order parameter is crucial for getting a pronounced ZBCP, which is different from the s-wave symmetry. As these ZBCP are a result of scattering from specific particle trajectories, their observation is strongly linked to the crystal orientation of joined grains. For higher angle mismatch a stronger scattering of quasiparticles is expected than in the case of a small lattice mismatch which can influence the symmetry of Δ (however, we do not have angle-dependent ZBCP data).

Having accepted the proposed 2D structure of BNCD films the origin of anisotropic (even frequency) Δ can be established, which is generated at the interference of superconductor (nanodiamond grains) and spin-active boron layer. This ZBCP appearing from the sub-gap (interface) state shows dI/dV peaks at the bias voltages $eV = \pm\Delta_0 \cos\theta/2$ from spin-dependent phase shifts. The two minima between the ZBCP and two satellite peaks appearing for $\theta = 0$ or 2π (Figure 8.12) also suggest two superconducting order parameters present in this system. The peak-to-dip transition is described as $0-\pi$-phase transition in a spin valve, which demonstrates an odd-frequency superconducting order parameter found in other reports. Earlier such results were explained as d-wave Δ but we claim here as a chiral p-wave superconductor, which has not been widely studied, yet our data are very significant.

ZBCP in trilayer superconducting systems can occur in many different superconducting systems, due to a range of underlying physical mechanisms – including

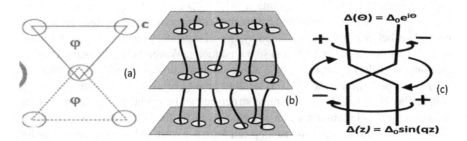

FIGURE 8.12 (a) Multi-rings form two closed loops connected through tunnel junctions and capture an AB phase φ. Shockley model describing a vortex line connecting two triangles which are used to simulate intersecting triangles. (b) Schematics of vortex structures in BNCD samples having layered structures. Cylindrical surface shows the flux lines and vortex cores corresponding to columnar diamond layered structures with the twist [26,27]. (c) AB tube and split AB tube forming core of a vortex.

both WL and WAL, the Kondo effect, ABS, Zeeman field splitting, and topological effects such as Majorana zero-bound states. The origin and nature of the superconducting bias peak tell us about the type of superconductivity present and the effects that occur alongside it. We look at calculating the differential conductance (dI/dV) for a superconductor/ferromagnetic insulator/superconductor junction that exhibits a non-BCS DOS – as a result of the spin-mixing and the proximity effect. In reference a generalized model was derived using Ricatti amplitudes and the S-matrix formalism, which was studied in context with an s-wave singlet superconducting hybrid system; this model was utilized to look at an NbN/GdN/TiN tunnel junction, that exhibited d-wave triplet superconductivity – characterized by the odd-frequency order. As a result of the spin-filter effect – from the different polarization potentials – an asymmetry arises in the spin-channel transmission probability coefficients $D_\uparrow$ and $D_\downarrow$, which leads to a finite spin current in the channel. The tunneling sub-gap appears when the spin-mixing angle $\vartheta \approx \pi$. The differential conductance was calculated from the spectral current density J at the different interfaces, the result of which is closely related to the ratio between the quasiparticle energy E and the superconducting gap Δ. Different boundary conditions were applied for the interface, leading to pronounced satellite peaks and determining the location of the local minima on the dI/dV spectra.

8.4.2 Topological Superconductor

One of the important observations in the non-splitting of the robust features of ZBCP peaks by the magnetic field, which is possible for topologically protected materials is the interplay of spin (Kondo-like) and orbit (RSO-like) properties. The dI/dV shows a strong central ZBCP peak and two satellites like an interference pattern resulting from the RSO effect, which splits Δ into four components, i.e., $\Delta\uparrow\downarrow$, $\Delta\uparrow\uparrow$, $\Delta\downarrow\downarrow$, and $\Delta\downarrow\uparrow$ like in a d-wave superconductor. Since it is already split into two pairs of components like two spin and two orbits, no more splitting of the peaks is possible even if the magnetic field is applied. In a p-wave superconductor, the spins are oriented in the same direction (bosonic), which cannot be split by a magnetic field. Similarly, the triplet pair created by the superposition of $|\downarrow\uparrow+\downarrow\uparrow\rangle$ cannot be split, since the change in either side will be compensated by another component, hence protecting topologically. This behavior can be explained by an odd-frequency (p-wave) order parameter instead of an even parity (p-wave) superconductor. Wavefunction twisting creates a crossing (singular) point ($|\downarrow\uparrow+\downarrow\uparrow\rangle$), where duality cannot be broken by a magnetic field easily and show a ZBCP. The satellite peaks correspond to the turning of the spins either $|\uparrow\uparrow\rangle$ or $|\downarrow\downarrow\rangle$. A complete structure can be explained by an infinity (∞) loop or a p-orbital maintaining an odd frequency. So the GB works like a phase slip junction. The unwinding of the ∞ loop can form a d-wave order parameter consisting of a four-lobed orbital.

The symmetry of the pair wavefunction is given by frequency (time) $\otimes$ spin $\otimes$ parity (even s, d, or odd p-wave) = magnetic field $\otimes$ spin (singlet or triplet) $\otimes$ orbit (q). Even parity is possible for a simple superconductor without a junction. So, the singlet remains as s-wave or even frequency (ESO) and a triplet remains as an odd parity (p-wave) (ETO). So far, we have seen spin-triplet and p-wave superconductivity and also revealed odd-frequency nature from peak features of R–T, R–B, and dI/dV–V data. However, all three odd features cannot be present in a system unless all symmetries are broken by introducing an additional change of the phase of Δ forming a complex chiral phase. A superposition of ABS can give a topological phase. A nontrivial ABS is similar to a Majorana-bound state that has been studied extensively on a theoretical basis. For an odd frequency, reflection from the ABS causes a π flip. So a singlet changes to p-wave or odd parity for OSO as if $|\uparrow\downarrow>$ flips to a $|\uparrow\uparrow>$ or $|\downarrow\downarrow>$ state. For OTO the triplet $|\uparrow\uparrow>$ or $|\downarrow\downarrow>$ can change to a singlet $\uparrow\downarrow$ state as if the twisted p-wave is untwisted to an s-wave and vice versa. ABS is the property of the interface or surface that gives topological protection to the bulk (s) state. The p-wave nature of the interface gives a ZBCP feature, which is also a signature of odd-frequency pairing as the order parameter changes to a d-wave, which also gives the ZBCP feature.

The sign inversion of Δ from the ABS of a p-wave can suggest an odd-frequency and triplet superconductor of an even parity wave. In an odd-frequency and triplet superconductor by breaking the spin rotational symmetry or spatial invariance symmetry, an odd-frequency pairing can be induced. Superconducting diamond having the interface in the form of well-aligned grain boundaries in the form of SL structures has presented SOC-induced spin-triplet superconductivity. The layered microstructure of diamond can accommodate vortices and exhibit FFLO states and odd-frequency superconducting order. The experimentally observed π-phase transition in magnetoresistance from a WL to the WAL state can be explained through the presence of an ABS, which is a doublet state that leads to the spin-triplet state as a superposition. A vortex structure can accommodate the ABS, which is a hallmark of the FFLO phase, which can also be present in the superconducting diamond films exhibiting strong SOC.

Pauli's exclusion principle says that in one state two electrons of opposite spins can be present like ($\uparrow\downarrow$). The half spins have equal mass but opposite polarity. They can have an exchange interaction and bosons (zero or integer spins) as many as possible in one level. From a spin-singlet ($\uparrow\downarrow$) state a spin-triplet state can be created where two electrons of equal polarity can exist. However, it is compensated by two other spins of equal and opposite polarity like $|\uparrow\uparrow>$ and $|\downarrow\downarrow>$. We can have a third type of pairing $|\uparrow\downarrow> + |\downarrow\uparrow>$, which is similar to the subject. To produce a triplet state barrier of ferromagnetic material is needed, since the triplet has an AFM behavior ($\uparrow\downarrow$). This picture becomes interesting with the addition of the second potential barrier. The self-spin picture works very well as a resonance in the well. Reentrant superconductivity can be explained by tunneling between two vortices. Assuming the vortices behave like black holes tunneling between them can be explained by a wormhole. The transition from the BH to wormhole at a temperature can be compared to the transition from a gapless superconductor to a gapped superconductor (vortices), which are connected by a tunnel junction. Let us understand the wormhole state in detail, which can explain the features of spin-triplet superconductors in the next section.

Conventional (s-wave) superconductors are routinely used in fabricating coupled qubits for quantum processing technologies. However, unconventional (spin-triplet) superconductors are expected to offer advantages as basic elements of topological quantum computers for complex operations since the symmetry of the order parameter (Δ) corresponds to a doubly degenerate chiral state.

8.4.3 FFLO and the SSH Model

In the presence of RSOC by applying an electric field, a nonequilibrium magnetization or spin accumulation in the interface takes place with the generation of spin-polarized currents, which can form an effective S-F-S structure. A quasi-classical theory describing the interface of superconductors and ferromagnets in the presence of the SOC has also been developed. All these problems can be solved with the proper understanding of the originally stated FFLO-like state in the 1960s with the modulated order parameter. However, it was based on the d-wave superconductor. Despite numerous efforts, a clear experimental observation of these states which are already found in BEC remains challenging through electrical transport in superconductors. Although the LO state has been rarely observed in some heavy fermion materials and layered organic materials in the presence of vortex phases, the helical state is difficult to see experimentally in non-centrosymmetric superconductors as a combination of $\Delta_1 = \Delta\ (e^{iqr} + \delta e^{-iqr})\}$ and $\Delta_2 = \Delta\ (\delta e^{iqr} + e^{-iqr})\}$, where δ represents the phase shift or the coupling between the layers which is observed from the lattice mismatch at the GB regions. Ultimately a complex-stripe phase, which is intermediate between FF and LO state, was claimed in locally non-centrosymmetric multilayers. The CS phase created from the direct coupling between layers and expressed as $\Delta(r) = \Delta_0 \exp\ (iqr)\ \{1 + \delta \exp(-2iqr)\}$ is presently claimed, which gives a topological insulating phase on the diamond. All these FFLO and vortex features stem from the ABS as a solution of the "soliton" state in the 1D Dirac equation. This is particularly important to match the experimental observations of peaks and oscillatory features of R–T, R–B (with hysteresis), and dI/dV–V spectra which strongly suggest an odd-frequency order pairing in BNCD films.

The ZBCP features are commonly observed in p-wave superconductors where sign change Δ occurs at the ab-planes. In the light of the FFLO state, the order parameter expressed as $\Delta = \Delta_0 \exp(q \cdot r)$ or $2\Delta_1 \cos(q \cdot r)$ which can explain the overall magnetoresistance features and dI/dV features observed in Figure 8.11. Unlike a constant order parameter (Δ) in a BCS superconductor, the FFLO state is defined as $\exp(q\ r)$ and $\cos(q\ r)$, where q stands for the orbital angular momentum equivalent to the magnetic field (works normally to the electron momentum k). This q is responsible for the vortex formation like an AB ring, which is inaccessible to spins like a bound state. In the WL process spins can move randomly; however, in the presence of a bound state (or an impurity center or a doublet state), a spin flipping takes place which makes a transition to a WAL state. WAL state can be created by placing two rings normal to each other. A spin-singlet state when hits an ABS is split into two halves, and after crossing the vortex they rejoin as a superposition of states, hence forming a spin-triplet state. This AB-like barrier forms an ABS at the center Δ which decays exponentially with the increase of energy. Due to the doublet ABS, an oscillatory variation can also be seen. Therefore, FFLO is described as a product of exponential and cosine functions of the orbital momentum. When the singlet function performing a WL process hits the extremely rigid GB, it split into spin triplets which work as another WL orbits perpendicular to the original WL orbit. While a singlet corresponds to the subtraction of two WL processes, the triplet shows the superposition which can travel a long distance since it is unaffected by the local disorder.

8.4.4 VORTEX AND OSCILLATORY Δ: FFLO AND ODD FREQUENCY

Yokoyama et al. showed how Zeeman splitting can create modulation of pair potential and increase odd-frequency pairing at the core centers of vortex lines. Instead of the Zeeman field for GB-interface superconductivity of diamond films, the RSO field split the spin channels confirmed by the non-splitting of the ZBCP by applying a magnetic field. However, the mid-point of the vortex lines was claimed to have even-frequency pairing like an s-wave or d-wave which is the bulk state of a topological superconductor. The surface state is made of a p-wave superconductor; hence, a topological superconductor is constructed. Odd- and even-frequency pairing states explain the central peak and the satellite peaks, respectively, as if this inhomogeneous material is made of two superconductors with p_x and p_y waves. As a combination, i.e., $p_x + ip_y$ (odd d-wave) or $p_x - ip_y$ (odd s-wave), they correspond to parallel and antiparallel vortex and also produce weak and strong ZBCP features, respectively. So the bulk state is ETO (even-frequency spin-triplet odd parity) like a p_x-wave, whereas the interface-induced symmetry is OTE (odd-frequency spin-triplet even parity) like a d-wave superconductor. This p_x-wave is involved with a sign inversion. However, another ETO state is p_y-wave in the bulk where no sign change of MARS takes place, since the interface-induced symmetry is also ETO-type. Therefore, a combination of p_x and p_y as $p_x \pm ip_y$, which is similar to a d_{xy} state, can be revealed. Here two lobes of p_x-wave (or spin-type component) change their sign by the symmetry operation unlike the p_y-wave (or orbit-type component), which remains unchanged by a similar operation. Ultimately, this model resembles the SOC and gives a clear picture of the vortices. This process works through the separation of a spinless charge from a chargeless spinful system. Access spin–charge separation in the $p_x + ip_y$ superconductor within the dual Ginzburg–Landau formulation requires consideration of paired vortex composites. If a vortex in the down-spin boson field pairs with an antivortex in the up-spin boson field and this pair condenses, a transition occurs into a fully gapped superconductor, such as a superconductor with a tightly bound pair. So, our model is consistent with both FFLO and vortex models, since both are derived from the ABS and solitonic state.

8.4.5 TOPOLOGY AND THE SSH MODEL

From the microstructure and electrical transport, we reveal a superlattice-like structure which is probably created from the displacement of (111) plane in diamond induced by boron incorporation. However, a theoretical model explains all these microstructural features and electrical properties have not been available. Due to the reconstruction of the diamond surface described as Pandey chain, the (111) plane is formed, which has some special properties. This has not only a very flat band (close to graphene layers) but also contains a large density of carriers (electrons) resulting in strong electron–electron interactions; hence, superconductivity and magnetism can be possible in this particular (111) layer, which leads to a spin-triplet superconductivity as claimed here. (111) plane forms a triangle, since one bond is absent from the three nearest neighbors on a trigonal (planar) structure instead of a regular tetragonal structure of a diamond crystal. The atoms on the (111) surface form a 2×1 superstructure and a zigzag chain after the (111) plane is displaced by the boron Zak phase from the band structure or dimer state described by the SSH model (including two hopping parameters and two order parameters) can explain the origin of the topological features in superconducting diamond films. Boron atoms further introduce a topological phase to the diamond structure with more deformation along (111) planes yielding an extremely flat energy band and an inversion symmetry.

In Pandey's symmetric chain model, the bond length of the (111) plane was similar to graphite yielding the surface energy band, whereas two surface bands derived from the bonding and antibonding combinations of dangling orbitals along the chain produce the bulk band gap. The calculated Fermi surface was found to be flat and degenerate and nearly degenerate along J–K directions with the presence of electron–hole packets and unstable against Jahn–Teller distortions. Overall, this chain finds a remarkable similarity with polyacetylene with alternating single and double bonds like a dimer which yields a gap at the Fermi surface due to this asymmetry. Such a directionally flat (completely dispersionless) band can produce a straight-line-shaped Fermi surface as observed in type-III Dirac cones and tested from molecular-orbital representation. Starting from a square-lattice model in 2D for spinful fermions, this model has also been extended to a diamond lattice model, where a type-III Weyl semimetal has been constructed. Although a modified 2D SSH model was discussed in this paper, we think that a possible extension in 3D SSH on a diamond lattice structure particularly in the [111] direction can be useful in explaining our observed results.

This chain structure can support the generalized SSH model, which is developed on the alternating single-double (zigzag) bonding arrangements as originally proposed in the Shockley model, however, is treated with a topological phase. Such unique configuration with arbitrary adiabatic deformation can produce a zero-energy edge model which can explain the ZBCP features.

This dimerization picture of the chain can be used to find a nontrivial topological phase (so-called Zak phase) with a zero Berry curvature (a geometric analog of the magnetic field in the momentum space). Earlier Berry phase was quantized and correlated with the Brillouin zone through the addition of an inversion symmetry which takes the value 0 and π only corresponding to the band center having two values 0 and $a/2$. A combination of time-reversal symmetry (an even function) and space inversion symmetry (an odd function) gives a zero Berry curvature, which is equivalent to SOC which eventually produces a Zak phase (π, π) accompanying the fractional wave polarization $(1/2, 1/2)$. All these topological features arise from the difference between hopping between atoms like alternating hopping energy described in the SSH model. Hence, the experimental observation of the π-junction is explained through the Zak phase or zero Berry curvature (Figure 8.13).

Vortex is a bulk boundary correspondence SSH model has been extended from 1D to 2D system and beyond two-band models; however, the application in a truly 3D system such as in a diamond structure has not been realized. Although diamond

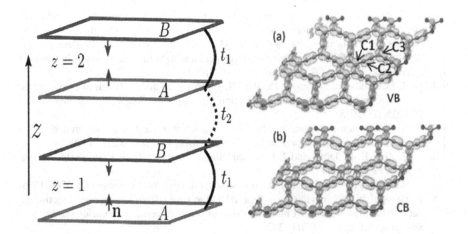

FIGURE 8.13   Sketch of the diamond lattice structure, with the hopping amplitudes. The A (light gray) and B (dark gray) are sublattice planes. The Rashba vector in the SOC is staggered on the sublattices. Two-dimensional carbon topological insulator superior to graphene is shown [90].

(and zinc-blende) structures were studied over decades, the topological properties of the valence band have not been reported. In particular, the spin–orbit split-off bands near the valence band maximum showing strong topological nontrivial behavior have been claimed. A similar feature can be seen in boron-doped diamonds; however, it has not been explained earlier. Although Z_2 gauge field has been applied to a square and a hexagonal lattice, there is a need for its use in the diamond lattice structure.

- **Main problem discussed in this chapter:** The zero-bias conductance peak in superconducting diamond films is explained.
- **What has been achieved?** The unconventional superconducting properties of diamond can compare with other materials; hence, carbon as a strongly correlated system can be established.
- **What has not been achieved?** The origin of recently claimed spin-triplet superconductivity and other magnetic properties of boron-doped carbon remains unresolved. The features of zero-bias conductance peaks cannot be controlled. Hence, carbon-based qubits have remained elusive.

BIBLIOGRAPHY

1. L. Fu, C. L. Kane, and E. J. Mele, Topological insulators in three dimensions, *Physical Review Letters* 98, 106803 (2007); S. Ryu, Three-dimensional topological phase on the diamond lattice, *Physical Review B* 79, 075124 (2009); I. M. Shen and T. L. Hughes, Entanglement of a 3D generalization of the Kitaev model on the diamond lattice, *Journal of Statistical Mechanics* 10, 10022 (2014).
2. A. Pal, J. Ouassou, M. Eschrig, J. Linder, and M. G. Blamire, Spectroscopic evidence of odd frequency superconducting order, *Scientific Reports* 7, 40604 (2017); E. Zhao, T. Löfwander, and J. A. Sauls, Nonequilibrium superconductivity near spin-active interfaces, *Physical Review B* 70, 134510 (2004).

3. M. Eschrig, A. Cottet, and J. Linder, General boundary conditions for quasiclassical theory of superconductivity in the diffusive limit: application to strongly spin-polarized systems, *New Journal of Physics* 17, 083037 (2015); J-i. Inoue, G. E. W. Bauer, and L. W. Molenkamp, Diffuse transport and spin accumulation in a Rashba two-dimensional electron gas, *Physical Review B* 67, 033104 (2003).

4. M. T. Edmonds, *Nano Letters* 15, 16 (2015); J. Narayan and A. Bhaumik, Novel phase of carbon, ferromagnetism, and conversion into diamond, *Journal of Applied Physics* 118, 215303 (2015).

5. G. Baskaran, Resonating valence bond mechanism of impurity band superconductivity in diamond, *Journal of Superconductivity and Novel Magnetism* 21, 45 (2007).

6. Z.-G. Zhu, K.-H. Ding, and J. Berakdar, Single- or multi-flavor Kondo effect in graphene, *EPL* 90, 67001 (2010).

7. J. Jak, Berry's phase for energy bands in solids, *Physical Review Letters* 62, 2747 (1989).

8. T. Mizoghichi and Y. Hatsugai, Type-III Dirac cones from degenerate directionally flat bands: Viewpoint from molecular-orbital representation, *Journal of the Physical Society of Japan* 89, 103704 (2020).

9. F. Liu and K. Wakabayashi, Novel topological phase with a zero Berry curvature, *Physical Review Letters* 118, 076803 (2017).

10. D. Obana, F. Liu, and K. Wakabayashi, Topological edge states in the Su-Schrieffer-Heeger model, *Physical Review B* 100, 075437 (2019).

11. S. Bhattacharyya, D. Mtsuko, C. Allen, and C. Coleman, Effects of Rashba-spin-orbit coupling on superconducting boron-doped nanocrystalline diamond films: Evidence of interfacial triplet superconductivity, *New J. Phys New Journal of Physics* 22, 093039 (2020).

12. B.-H. Chen and D.-W. Chiou, An elementary rigorous proof of bulk-boundary correspondence in the generalized Su-Schrieffer-Heeger model, *Physics Letters A* 384, 126168 (2020).

13. J. C. G. Henriques, T. G. Rappoport, Y. V. Bludov, M. I. Vasilevskiy, and N. M. R. Peres, Topological photonic Tamm states and the Su-Schrieffer-Heeger model, *Physical Review A* 101, 043811 (2020).

14. G.-W. Chern, Three-dimensional topological phases in a layered honeycomb spin-orbital model, *Physical Review B* 81, 125134 (2010).

15. A. J. Willans, J. T. Chalker, and R. Moessner, Disorder in a quantum spin liquid: Flux binding and local moment formation, *Physical Review Letters* 104, 237203 (2010).

16. T. Rauch et al., Nontrivial topological valence bands of common diamond and zinc-blende semiconductors, *Physical Review Materials* 3, 064203 (2019).

17. A. P. Schnyder, S. Ryu, and A. W. W. Ludwig, Lattice model of a three-dimensional topological singlet superconductor with time-reversal symmetry, *Physical Review Letters* 102, 196804 (2009).

18. P. Hosur, S. Ryu, and A. Vishwanath, Chiral topological insulators, superconductors, and other competing orders in three dimensions, *Physical Review B* 81, 045120 (2010).

19. H. Guyot, P. Achatz, A. Nicolaou, P. Le Fevre, F. Bertman, A. Taleb-Ibrahimi, and E. Bustarret, Band structure parameters of metallic diamond from angle-resolved photoemission spectroscopy, *Physical Review B* 92, 045135 (2015).

20. T. Yokoya, T. Nakamura, T. Matsushita, T Muro, Y. Takano, M. Nagao, T. Takenouchi, H. Kawarada, and T. Oguchi, Origin of the metallic properties of heavily boron-doped superconducting diamond, *Nature* 438, 647 (2005).

21. F. D. M. Haldane, Model for a quantum Hall effect without Landau levels: Condensed-matter realization of the "parity anomaly", *Physical Review Letters* 61, 2015 (1988).

22. C. L. Kane and E. J. Mele, Quantum spin Hall effect in graphene, *Physical Review Letters* 95, 226801 (2005).

23. L. Fu and C. L. Kane, Topological insulators with inversion symmetry, *Physical Review B* 76, 045302 (2007).
24. T. I. Baturina, and V. M. Vinokur, Superinsulator–superconductor duality in two dimensions, *Annals of Physics* 331, 236–257 (2013).
25. S. Sachdev, The quantum phases of matter, arXiv:1203.4565 (2012).
26. R. Lortz, Q. Zhang, W. Shi, J. T. Ye, C. Qiu, Z. Wang, H. He, P. Sheng, T. Qian, Z. Tang, N. Wang, X. Zhang, J. Wang, and C. T. Chan, *Proceedings of the National Academy of Sciences* 106, 7299 (2009); Z. Wang, W. Shi, R. Lortz, and P. Sheng, Superconductivity in 4-Angstrom carbon nanotubes—a short review, *Nanoscale* 4, 21 (2012).
27. A. Bhaumik, R. Sachan, J. Narayan, A novel high-temperature carbon-based superconductor: B-doped Q-carbon, *Journal of Applied Physics* 122, 045301 (2017).
28. A. Bhaumik, R. Sachan, J. Narayan, High-temperature superconductivity in boron-doped Q-carbon, *ACS Nano* 11, 5351–5357 (2017).
29. R. Schneider, A. G. Zaitsev, D. Fuchs, and H. von Löhneysen, Excess conductivity and Berezinskii–Kosterlitz–Thouless transition in superconducting FeSe thin films, *Journal of Physics: Condensed Matter* 26, 455701 (2014); L. I. Glazman, and A. E. Koshelev, Thermal fluctuations and phase transitions in the vortex state of a layered superconductor, *Physical Review B* 43, 2835 (1991).
30. C. Coleman, and S. Bhattacharyya, Possible observation of the Berezinskii-Kosterlitz-Thouless transition in boron-doped diamond films, *AIP Advances* 7, 115119 (2017).
31. Y. Iwasa, Superconductivity: Revelations of the fullerenes, *Nature* 466, 191–192 (2010).
32. F. C. Zhang, M. Ogata, and T. M. Rice, Attractive interaction and superconductivity for K_3C_{60}, *Physical Review Letters* 67, 3452 (1991).
33. R. M. Fleming, et al., Relation of structure and superconducting transition temperatures in A_3C_{60}, *Nature* 352, 787 (1991); O. See, et al., Compressibility of M_3C_{60} fullerene superconductors: Relation between T_c and lattice parameter, *Science* 255, 833 (1992); F. Hebard, Superconductivity in doped fullerenes, *Physics Today* 45, 26–32 (1992).
34. M. Zhao, W. Dong, and A. Wang, Two-dimensional carbon topological insulators superior to graphene, *Scientific Reports* 3, 3532 (2013).

9 Hybrid Quantum States
Diamond NV Center and Qubits

9.1 QUANTUM BITS (QUBITS)

Quantum mechanics is based on the uncertainty principle and fundamental particles are nearly invisible. An ideal place to apply quantum mechanics is an atom that consists of negatively charged particles that revolve around a center consisting of positively charged particles leaving the space between these two opposite particles empty. An artificial atom can be made from superconducting (zero electrical resistance) materials, which are popularly known as a qubit in the form of a closed loop of superconductors with a small gap of non-superconducting material (Figure 9.1). The supercurrent flows in the loop in both clockwise and counterclockwise directions, and a superposition of states can be created, which is equivalent to the creation of a void or a vacancy. A spin qubit can be achieved in a NV center hosted by a diamond lattice that has a spin-triplet state. NV centers being the smallest magnets work as spin qubits.

Qubits are the building blocks of quantum computers, and they can exist in multiple states simultaneously. This property, known as superposition, makes quantum computers capable of performing certain computational tasks much faster than classical computers. Just as the fundamental unit of classical information is the bit, there is a fundamental unit of quantum information, the qubit. One distinction between bits and qubits is that a qubit can be placed in a superposition of states, i.e., a bit can be either 0 or 1, but a qubit can be in a superposition state, $|\Psi\rangle = \alpha|0\rangle + \beta|1\rangle$, where the kets are the two "values" that can be taken by the qubit, and α and β represent the (in general complex) coefficients describing the superposition (Figure 9.1).

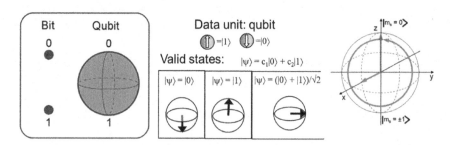

FIGURE 9.1 Unlike a zero and one bit, a qubit can be described as a superposition of two bits or two energy states. The arrows show the states schematically. Spin qubits can be described by a spin-triplet state, e.g., $m_s = 0$ and $m_s = 1$.

 DOI: 10.1201/9781003316411-9

These artificial quantum objects are designed using standard semiconductor fabrication techniques, and they are based on a two-level quantum system. In a quantum system, duality works in the qubit which is an oscillator. Microwave (MW) is the input that gets converted into different frequencies/channels after interaction with qubits. Photons pass through different stages of wheels, which increases the fidelity of the circuit.

One of the most important parameters for quantum information processing is the coherence time, and we may understand this parameter by exploring the meaning of coherence in the quantum setting. The coherence time is the characteristic time taken for the fragile quantum information to leak into the environment or can be seen as the time for a quantum superposition state to become a mixed state, i.e., a state with a classical probability of being found in either state, but without a well-defined quantum phase. In our example above, decoherence takes the superposition state $|\Psi\rangle$ to a mixed state which has the probability of $|\alpha|^2$ being found in $|0\rangle$ and $|\beta|^2$ being found in $|1\rangle$. Decoherence erodes the complex phases and much of the power of quantum computing, which will be discussed in the later part of this chapter.

In a quantum simulator, several qubits can be arranged in different geometric forms (lattice), so that they can be entangled and form a macroscopic state. Using a quantum simulator, we can explain the tunneling problem, coherent backscattering, spin–orbit scattering, Kondo scattering, and other problems (and will be discussed in the next chapter). However, this chapter deals with the quantum simulation of the diamond NV center using a multilevel system that will be useful to explain these quantum many-body problems. An application of these quantum phenomena will be quantum vortices which are observed in superconducting materials followed by the tunneling between the vortices forming a wormhole structure discussed at the end of the last chapter. There are also other kinds of carbon-based qubits (see Chapter 10); however, we focus on the diamond NV center in this chapter.

Recent critical advances show the potential for NV centers in quantum information processing: It has been shown that nuclear spins interacting with the NV electronic spin can be used as qubits, opening the possibility for building local multi-qubit quantum registers. Further, at cryogenic temperatures the electron spin can be interfaced coherently with photons, enabling the linking of such registers to macroscopic networks. Finally, the electron-spin and nuclear-spin states can be determined in a single shot by projective quantum measurements. On these grounds, it has been proved experimentally that the NV center is a promising platform for implementing quantum registers and networks.

9.2 DIAMOND NV CENTERS

9.2.1 NITROGEN-VACANCY CENTER (NVC) IN DIAMOND

With a band gap of around 5.5 eV stretching from the ultraviolet to the far infrared, diamond is the widest band gap semiconductor known. This makes diamonds transparent in the visible region, and any colors observed arise from defects – impurities, dislocations, vacancies, and complexes – which create electronic energy levels within the band gap. Diamonds have several outstanding properties, particularly,

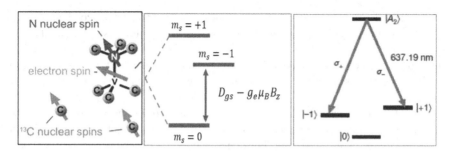

FIGURE 9.2 (left) Nitrogen-vacancy (NV) centers are a point defect dinia mond and consist of a nitrogen and an adjacent vacancy in a diamond lattice. Single nitrogen-vacancy spins and their nuclear spins can be used as spin qubits, which are coupled to their electron spins. (middle) The energy-level structure of the NVC consists of sublevels $m_s = 0$ and $m_s = -1$. (right) The associated (optical) transitions between the sublevels of the NVC ensembles or groups of NV centers are can be used for quantum computing.

mechanical, thermal, and optical properties. However, every natural diamond is not free from defects. One of the commonly observed defects is the NV center which consists of substitutional nitrogen in a diamond lattice next to a vacancy (Figure 9.2). Since nitrogen has five valence electrons, it forms three covalent bonds with three neighboring carbon atoms and one dangling bond with the vacancy oriented along with one of the four crystallographic {111} directions and displays C3v trigonal symmetry. The NV center exists in two charge states, the neutral charge state, known as NV^0, which is a spin-1/2 system, and the negative charge state, known as NV^-, which is a spin-1 system. In the NV^0 state, five electrons interact with the vacancy: two from the nitrogen atom and three from three carbon atoms forming dangling bonds with the vacancy, giving the NV^0 a spin-1/2 character. When an external voltage is applied, one electron is captured by the nitrogen donors. This then makes it have six valence electrons, making it negatively charged. This negatively charged state is denoted as NV^-.

The NV color center is also one of many extrinsic point defects which are introduced in diamonds during synthesis in a laboratory. The remarkable properties of the NV color center of diamond include the following: It is extremely bright, and a single center can be seen at room temperature; it can be spin-polarized by shining white light on it; and its spin state can be read out by using the difference in fluorescence between the bright spin-zero ground state and the relatively dark spin ±1 ground state. The defect behaves much like a single ion or small molecule but comes "pre-packaged" in a robust solid-state environment that it does not strongly interact with. The NV center displays quantum phenomena that can be accessed with relatively simple experimental arrangements, and under moderate environmental conditions.

The NV center shows the emission of light under excitation with visible light and is a stable single-photon emitter (Figure 9.3). It has an electronic spin that can be initialized and read out optically and manipulated by applying a light, electric field, magnetic field, MW radiation, or a mix of either. This produces sharp resonances in the wavelength and the flux of the photoluminescence. These resonances can be

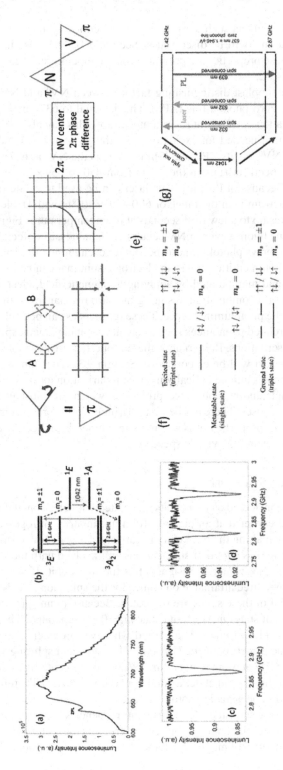

FIGURE 9.3 (a) Photoluminescence spectrum of the nitrogen–vacancy center at room temperature. (b) Simplified electronic structure of an NV center. (c) Zero-field optically detected magnetic resonance (ODMR) spectra. (d) ODMR spectra under the external magnetic field of 2 mT along the NV axis [4]. (e) Proposed NV center model consists of a 2p phase change. A vacancy can be compared to a triangle and is associated with a p phase change. (f) Details of the energy diagram of a spin triplet state. (g) These levels are connected by absorbing of green light and emission of red lights.

justified in terms of phenomena such as spin–orbit interactions, quantum entanglement, and Rabi oscillations (will be discussed in Section 9.3). They also have long coherence times. All these properties are present at room temperature. There are two possible states for the NVC.

Experimentally, the simplest distinguishing factor between NV^0 and NV^- is the spectral position of the zero-phonon line (ZPL). This is equal to 637 nm = 1.945 eV and has a linewidth of a few $\times 10^{12}$ Hz at room temperature. This can also be defined as the energy of transition needed for excitation from the ground state ($^3A_2$) to the excited state (^3E). The NV^0 has its ZPL at 575 nm, whereas the NV^- at 637 nm. The NV center absorption spectrum runs in the range from 450 to 659×10^{-9} m with a maximum of 555 nm. Because of this, a visible laser can be used to excite the NV^- state. The emission spectrum is in the range of 600–850×10^{-9} m with stable photoluminescence. This takes up to a few milliseconds at room temperature (Figure 9.3).

For the imaging of the fluorescence of NV centers in diamonds, a confocal microscope is used. The NV shows photoluminescence under excitation with visible light and is a stable single-photon emitter. It has an electronic spin that can be initialized and read out optically and manipulated by applying a magnetic field, electric field, MW radiation or light, or a combination, resulting in sharp resonances in the intensity and wavelength of the photoluminescence. These resonances can be explained in terms of electron-spin-related phenomena such as quantum entanglement, spin–orbit interaction, and Rabi oscillations. Remarkably, these properties are available even at room temperature (this part will be elaborated in Section 9.3). The spin associated with the NV^- center displays unique optically induced polarization (or initialization), optical readout, and outstanding coherence properties which are most useful for quantum information purposes. There has been surprisingly little work to completely characterize the quantum properties of NV centers in diamonds and comparatively little work that extends beyond the visible spectrum.

9.2.2 THE ELECTRONIC STRUCTURE

The NV center can be considered as two qubits associated with electron and nuclear spins. As we have said earlier that an electron from an impurity nitrogen atom is captured by the vacancy making the NV center negatively charged (NV^-) which also turns it into a spin-1 system as a total of six electrons are involved with the vacancy. The electron spin in the NV center is designated as the first qubit. If we consider the ground state, we have three attainable outcomes for the spin along the NV axis. These are $m_S = 0, \pm 1$. All of these states are separated. Because of anisotropic interactions, the energy level of states $m_S = \pm 1$ and state $m_S = 0$ are separated. The degeneracy of the states $m_S = +1$ and $m_S = -1$ are lifted whenever an external magnetic field is applied along the direction of the NV axis. The zero-field splitting value of the ground state is 2.87 GHz at room temperature. When an NV center is excited and then de-excites back to the ground state, a single photon is emitted. Therefore, the NV center is considered to be a stable photon emitter.

Conversions between the spin states have different frequencies and it turns out that we can manipulate our potential quantum bit with alternating current magnetic-field pulses. If one uses a frequency of magnitude 1.4×10^9 Hz, all the states were affected, except for $m_S = +1$. This state stayed idle because the transition frequency to this state is not the same as the other states due to the shift by the magnetic field. The states $m_S = 0$ is therefore designated as $|0\rangle$ and $m_S = -1$ is thus designated as $|1\rangle$. In this way, the electron spin can be represented as a quantum bit (Figures 9.2 and 9.3).

The nuclear spin of the Nitrogen-14 nucleus is designated as the second quantum bit, and it has a magnetic spin of 1. That means $I = 1$. This means that there are also three attainable outcomes much like the electron-spin energy-level states. They are $m_I = 0, \pm 1$. They are dissimilar to the extent that in the presence of a static magnetic field B_0, the $m_I = -1$ state remains unchanged. This quantum bit is manipulated using alternating current magnetic pulses of frequency 2.9×10^6 Hz. The state of $m_S = 0$ is thus designated as $|\downarrow\rangle$ and $m_S = +1$ is designated as $|\uparrow\rangle$. The difference in frequencies that are used for the manipulation of the electron and nuclear spin enables us to control the qubits independently.

9.2.2.1 Nuclear Manipulation of the Surrounding Carbon-13 Atoms

Whenever the electron spin of the NV center is at $m_s = \pm 1$, a dipolar magnetic field is induced. The nuclear spin has a unique frequency that can be used to selectively control the surrounding spins. If the electron spin is at a state $m_s = 0$, no magnetic field is created. The surrounding nuclear spins thus depend on the state of the electron spin. We have thus created a controlled quantum logic gate between the electron and the nuclear spin. Just as how a classical logic gate allows bits to change between 0 and 1, the quantum gate we created here can influence the outcome of the surrounding nuclear spins. This enables us to control multiple nuclear spins around a single NV center. That means that each NV center becomes a system of six or more qubits.

The desired state is probabilistic. To measure the spin, different optical transitions in the NV center associated with different spin states are recorded. A laser pulse resonant to only the transition for spin-up is applied to detect photons. When the spin is down it stays dark.

9.2.2.2 Spin Decoherence

Spin coherence time is the period a qubit can hold on to its state and thus preserve information. Decoherence of a quantum bit decreases the equality and efficiency of quantum computers and thus it must be suppressed. This decoherence can be caused by noise or nearby electric or magnetic field. In diamond, a problematic source of this magnetic field is the carbon-13 nuclear spin which consists of 1.1% of the carbon atoms. The magnetic field due to the nucleus causes the quantum state of the NV center to be lost and thus shortens the qubit's coherence time which is approximately 5 ns. The error created in information due to decoherence can be corrected using Shor's method. Nine qubits are required to secure or decrease the decoherence of 1 qubit. Up to 5 qubits are required to correct 1 qubit. The quantum gates associated with the NV centers need to be holonomically controlled for error correction as discussed in this chapter.

Active quantum error correction can be used to secure the information for arbitrarily long times at the cost of more qubits if coherence times are longer than some threshold (which depends precisely on the architecture and the type of error correction utilized). A straightforward target level for quantum error correction is the ability to accomplish 104 operations before decoherence. The ground spin states have been mostly utilized in quantum coherence demonstrations up to this point.

The quantity of ensemble work on NV– diamond demonstrates that the ground-state properties are highly repeatable from center to center, unlike the excited states. They are magnetic dipoles in a rigidly defined, nonmagnetic substance, which contributes to their reproducibility. Because phonons do not easily link to the centers, quantum coherence is preserved, with proximal nuclear spins (particularly ^{13}C) being the main cause of decoherence. For testing few-qubit algorithms, the coupling of NV- centers to adjacent spins has produced impressive results. The fact that the NV- offers a clear readout for nuclear and electronic spin dynamics, also known as dark spins, which are often hidden, is one of the most significant characteristics of such coupled systems. NV- and ^{13}C nuclei, as well as NV- and proximal N electronic spins, both exhibit observable spin–spin dynamics. Even though these protocols cannot be scaled, the small (few nanometers) inter-qubit spacing could be a useful feature in larger scalable quantum computers coupled by long-range photonic interactions mediated by color centers. There are two ways to obtain quantum information processing in diamond: (i) The NV- are qubits coupled by photonic interaction and (ii) the photons are qubits and NV- entangles the qubits.

9.3   MANIPULATION OF THE NV CENTER SPIN

External control of the NV center spin by optical and MW fields is a crucial characteristic that makes NV centers useful for solid-state quantum technologies. Using optical excitation, the NV center spin can be brought into the $m_s = 0$ ground state, but an external MW field that resonates with the spin-splitting transition is typically used to coherently manipulate the spin. At low temperatures, pure optical methods have also been used to demonstrate complete control of the single NV center.

9.3.1   METHODS OF DETECTING THE SPIN STATE OF NV CENTERS: SPIN MANIPULATION OF THE NV CENTER IN THE GROUND STATE

The population of NV centers can be shifted from spin sublevel $m_s = 0$ to $m_s = \pm 1$ by applying a resonant MW or radio frequency. For example, a 532 nm laser can increase the $m_s = 0$ population by 90%. At 1042 nm, a singlet state transition has been seen. This permits us to identify a fluorescence change in the magnetic resonance spectra that were optically detected. The electron-spin resonance frequencies under zero magnetic field and a tiny external magnetic field are contained in the optically detected magnetic resonance spectra of a single NV. Although the magnetic resonance of the NV center is frequently employed for strain and pressure measurements, it is also frequently used for sensing thermal, electric, and magnetic fields. By calculating the NV spin Hamiltonian, one may determine the theoretical electron spin resonance (ESR) frequency values.

The ground-state spin Hamiltonian of the NV center under the external magnetic field is $H = hD_{gs} \hat{S}^2 Z + hE(\hat{S}^2 x - \hat{S}^2 y) + \gamma_e B \cdot \hat{S}$. The parameters D_{gs} and E are the ground-state zero-field splitting and strain-induced splitting, respectively. The magnetic field vector is given by B and $\hat{S}_x$, $\hat{S}_y$, $\hat{S}_z$, and $\hat{S}$ are the spin operators. Other constants h and γ_e correspond to Planck's constant and NV electronic gyromagnetic ratio (28 MHz/mT). For investigations requiring a single-photon source, such as quantum entanglement, quantum cryptography, and quantum optics research, single NV centers in diamond crystals are a better option. Due to its reproducibility, ground spin states have been primarily utilized in quantum coherence demonstrations up to this point.

9.3.1.1 Coherent Manipulation of NV Center

The dynamics of the spin state of spin-polarized NV centers can be observed from their interaction with the MW field. Considering the phase angle, ϕ_{ph} to be the projection of the spin state in the x–y plane from the x-axis. When the resonant MW field is applied in this direction such that $\phi=0$, the spin will carry out Larmor precession around the applied field. The frequency of this precision is the Rabi frequency as the spin state oscillates between $m_s=0$ and $m_s=1$ states of $^3A_2$. This can be seen as an oscillatory behavior in the fluorescence, which is called Rabi oscillation. The probability of the spin being in state 0 can be found from its proportionality to the fluorescence. The Rabi oscillation of a single NV center decays exponentially due to decoherence.

Rabi oscillation can be visualized as the spin state on a Bloch sphere. The general representation of a spin state on the Bloch sphere is $|\psi_{sp}>\, = \cos\dfrac{\theta}{2}|0\rangle + e^{i\phi_{ph}}\sin\dfrac{\theta}{2}|1\rangle$, where the angle θ is the angle to the state from the z-axis. The spin state of the system is then represented along the z-axis in the Bloch sphere. At a certain length of the MW pulse, the spin state of the NV center inverts from $m_s=0$ to $m_s=\pm1$, and this is called a π pulse as the argument $\theta=\pi$ in this condition.

9.3.2 RELAXATION TIMES

It is not possible to keep a spin-polarized NV center in a superposition state indefinitely. It eventually loses its coherent state to various factors in the environment and will go back to a thermally distributed system. This process is called relaxation and is classified into two types:

i. Longitudinal electron-spin relaxation time (T_1), which is the lifetime of a spin-polarized state and sets the ultimate limit for the phase coherence time. This happens due to stochastic processes like phonon interaction that irreversibly change the axis spin projection of the state. The T_1 for ground-state spin in NV centers in a diamond can be measured due to the difference in fluorescence intensities as the $m_s=0$ state is bright and the other darker in comparison. It is in the millisecond range at room temperature.

ii. T_2 is the time interval in which the phase coherence of the spin state is lost to the spin bath of the surrounding environment through off-resonant interactions. Through the dipolar interaction with the fluctuating local field produced by the local spin bath, this interaction causes the loss of phase memory of the single spin. For an ensemble of spins, the local fields differ for each spin, and the experimental setup differs for each measurement for an ensemble measurement of a single spin. The trials provide average values of the inhomogeneous dephasing times in both situations, which are typically denoted as T_2^* ($\leq T_2$). At room temperature, T_2 for nanodiamond NV centers are found to be in the range of microseconds.

The surrounding paramagnetic impurities in the lattice, such as N and ^{13}C atoms, which exhibit dipolar coupling to the NVs, are the primary cause of dephasing in diamond NV centers. The decoherence period can be increased either by lowering the temperature, which will significantly slow down the spin-flip process of the spin bath, or by employing samples with fewer P1 centers. The dynamical decoupling approach involves separating the NV spins from their spin bath, e.g., by utilizing various pulse sequences. The longest recorded decoherence time for a single NV center in bulk diamond is 1.8 ms at ambient temperature. The interaction between the ^{13}C nucleus and electron spins is responsible for most of the decoherence in the system.

The electron dephasing period is $T^*_{2,\text{el}} = 3.6$ μs, and the decoherence interaction normally takes place within a few seconds. In contrast, the NV center's nuclear spin scarcely interacts with any spins other than its electron spin. When compared to the decoherence of the electron spin, the nuclear spin's decoherence can be thought of as a second-order process. As a result, $T^*_{2,\text{nu}} = 5.3$ ms is the larger nuclear dephasing period. Therefore, we can disregard the nuclear spin's decoherence. This system is an example of a hybrid spin register. Different dephasing durations combined with various frequencies required to control the two qubits result in various qubit types.

We can link the NV centers into an entire network using quantum entanglement.

9.3.3 USING QUANTUM ENTANGLEMENT TO ENTANGLE NV CENTERS

There are several factors which make NV center ensembles appealing for usage in quantum information processing:

They are a fermionic ensemble that can be considered to be in the bosonic basis when they are in the state of collective excitation.

They can undergo superradiant phase transitions.

They can reach the strong coupling regime when coupled with flux qubits (and even ultrastrong under certain conditions).

They have long coherence times (~1 ms for electron spin, and >1 s for nuclear spin) at room temperature.

Controllable by external means (such as MW and optical fields).

Coherence time can be increased by dynamical decoupling.

They have many energy levels that can be altered for usage as well as dark states, and these transitions can be observable when they fluoresce. By changing the parameters we can utilize their different energy levels, which make them into a qubit, qutrit, qudit, etc.

They can be reduced acting as an effective harmonic oscillator in quantum systems.

They can be utilized in hybrid quantum circuits and can be used as quantum memory.

They can be entangled with each other and other types of qubits.

As we said earlier, the NV center is a stable photon emitter. We can use these photons to achieve entanglement between NV centers, thus linking them. We use 2 NV centers and make them emit a photon that is entangled with a spin state. We initialize the spin states and read out with optical pulses. By employing one of the optical lambda transitions, researchers demonstrated that the polarization of the emission photons preserves good entanglement with the NV spin states. We take the photons and bring them together on a beam splitter, making them indistinguishable. Meaning it is impossible to tell from which NV center it came, and which NV center is in what state. Thus, we have created an entangled state between two separate NV centers. Using quantum computer/simulator NV centers are created and the interactions between three NV centers are established.

9.3.4 SIMULATING ONE NV CENTER (COUPLED WITH SUPERCONDUCTING QUBIT)

The International Business Machines (IBM) Quantum Experience (QE) was used to try to simulate a virtual hybrid system made up of a superconducting (flux) qubit and an NV center, which is a two-qubit system. This was done using the IBM QE (5 qubits) IBMqx4 spin-qubit simulation. It was feasible to substitute the ^{13}C atom with a flux qubit and mimic dynamics similar to those seen in the entangled electron nitrogen atom by adjusting the initial state preparation through MW pulse control of the simulator's logic operations.

This virtual system was shown to achieve a coherence time of 0.35 s for a single NV center qubit and fidelity in the range of 0.82, which are found to be equivalent to similar physical hybrid systems. These results come from calculations of the spin relaxation rate and state coherence. The coupling to various operational modes in the physical system can be altered by adjusting the external magnetic field. In addition, isotopically pure diamond has a decoherence of ~3 ms and a duration of ~14 ms for a single NV center, from NV center experiments.

Their electron-spin coherence period is extended by dynamical decoupling, which also enables each qubit to be individually and uniformly decoupled. With the addition of the sequence, the flux qubit coherence time ranges from 40 to 85 µs.

We should simulate spin relaxation durations (Figure 9.4a and c) by applying pulses to the circuit (Figure 9.4b and d), as the electron spin of the NV center is always connected to the nuclear spin of its nitrogen atom. Within a period of 2.5 µs, the state coherence declines by 0.2 before stabilizing with an increase in pulse

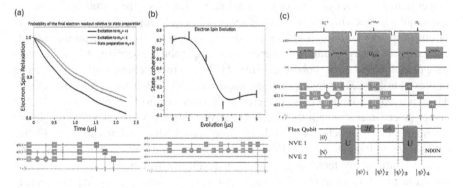

FIGURE 9.4 (a) Probability of excitation with respect to the electron-spin relaxation simulation. (b) Electron-spin evolution state coherence time simulation. (c) (top) The trotterized Hamiltonians used in the qubit circuits, (middle) The quantum circuit after the trotterization, and (bottom) Three-qubit entangle-ment gate circuit between two NVEs and a flux qubit [17]. Applying the time evolution operator a N00N states can be created.

duration. When we examine this in terms of the state excitations to $m_s = +1$ and $m_s = -1$ as well as the state preparation to $m_s = 0$, we observe a consistent relationship in which the relaxation of the state preparation slows down over time and that of the excitation to $m_s = 1$ decreases the fastest.

Figure 9.4d is the simulation with a gate structure of the electron-spin state coherence evolution as in Figure 9.4c which exhibits a steady decrease from 0 to 1 μs and then a rapid decrease from 1 to 3 μs. It then rises before stabilizing at the 4–5 μs interval. The relaxation seen from the electron spin of our virtual system is similar to that seen in physical NV center investigations. In practical tests, the spin center is frequently decohered as a result of disruption by the ^{13}C nuclear spins around it; nevertheless, once the pulse sequence takes effect, the system enters a stable state. Physical experiments have shown that longitudinal relaxation exhibits both temperature and magnetic-field dependency. We link the lower state coherence seen in our quantum gate simulation of the system using the simulator to the decoherence of the virtual system.

It is well known that significant amounts of naturally occurring ^{13}C isotopes can cause the NV center to become decoherent through spin–spin interactions. As a result, it is intriguing to include such components in simulation experiments because their dynamics are similar to those of trials. The NV centers' entanglement associated with the process is maintained by a decoupling method connected to them because of the environment. With an increase in connected NV centers, the coherence time was shown to decrease. This study was expanded to simulate a coupled system made up of three NV centers and a flux qubit and to show effects that were consistent with recently developed theoretical hypotheses. We could use this method in scalable and fault-tolerant information processing as described below.

9.3.5   EVALUATION OF HIGHLY ENTANGLED STATES IN ASYMMETRICALLY COUPLED 3 NV CENTERS BY IBM QE

The decoherence time of the quantum register is very important and should be much longer than the operating time of the quantum circuit. Since the decoherence time of a spin qubit is much higher than that of a single flux qubit (up to 0.5 ms), the accuracy of quantum simulation will be higher, which can operate complex quantum circuits consisting of a large number of superconducting qubits. The diamond NV centers have demonstrated a long coherence time of 0.7 ms at room temperature, which can be improved to 1.8–2.0 ms by suppressing impurities and defects for a single NV center. There is a major source of noise that creates decoherence in NV centers, since the NV center defect interacts with the nuclei of surrounding nitrogen and ^{13}C atoms in the diamond. For ^{12}C enriched diamonds, electron-spin coherence time can reach 3.0 ms.

The entanglement of two NV centers at room temperature has been demonstrated. Until today technologies are not developed to couple a few NV center quantum registers in diamond, which requires very strong coupling between at least three NV centers. One of the major limiting factors of multiple NV center quantum registers is that the coupling required for multiple qubit operations relies on the coupling between the NV centers, which is via the ^{13}C atoms. This inherently shortens coherence times for stronger coupling between NV centers. Nevertheless, researchers claimed the synthesis and analysis of three coupled NV centers using the implantation of $C_5N_4H_n$ ions in the diamond, but until recently the large-scale creation of three coupled NV centers in diamond has proven difficult. Attempts are made using 3 NV centers in different configurations in a triangular form by breaking the symmetry of couplings. Using a quantum simulator, a large improvement of the entanglement in artificially created structures of three NV centers through a cyclic redistribution of couplings has been achieved. This operation demonstrates entanglement after a very short free evolution time in an ordered configuration, which is advantageous for a larger number of operations that can be performed on the quantum register before a loss of accuracy from decoherence.

Unlike an idealized configuration, the demonstration of entanglement in a nonideal system of three coupled NV centers is more useful for simulating a real system such as an extended lattice structure. In such a system the coupling strength between any two NV center defects depends on the size of the quantum dots and the interdot distances. The geometric configurations of the NV centers concerning one another (such as vertical, horizontal, or triangular) would produce the resonance spectra that would affect the luminescence depending on the coupling between the dots. The asymmetry of a given configuration is represented using a coupling constant that depends on the coupling strength v_{AC}, v_{AB}, and v_{CB} between the three NV centers. Initially, an equilateral representation with equal coupling strengths between each NV center $(v_{AB} = v_{AC} = v_{CB})$ was examined. This was compared with isosceles representation with equal coupling strengths between two of the three NV centers $(v_{ik} = v_{ij} \neq v_{jk})$ for $i, j, k \in \{A, B, C\}$. Finally, the scalene representation with different coupling strengths between each NV center $(v_{AB} \neq v_{AC} \neq v_{CB})$ was studied. This is achieved by a cyclic redistribution of the coupling strengths of each coupled NV center pair.

This simulation was performed on the IBM Quantum Experience. A $\pi/2$ MW pulse is simulated using Hadamard gates and π MW pulses are simulated by NOT gates. A strong resonance peak relative to non-resonance can be demonstrated by introducing inequalities in interdot couplings through a distribution of the size and distance. In some irregular configurations, a strong resonant tunneling can be found from the strong entanglement, or a right combination of states as observed by breaking the symmetry of a regular configuration. The unexpected short $2\tau_{ent}$ for the scalene configuration shows the potential of NV center quantum registers for room-temperature operation of quantum computers because the fabrication of disordered configurations of coupled NV centers is far easier to achieve than that of ordered systems. Therefore, this is an important step in the development of accessible, convenient, and efficient quantum computers that are still accurate. These special structures would be useful to develop NV center-based hybrid quantum devices also by adding squeezed states obtained from strong coupling between a resonator and a couple of NV centers. These closed-loop configurations of NV centers can be useful for the development of further understanding of spin-triplet superconductivity and topological qubits. In general, interaction with photons produces sub-radiance and superradiance states, where the resonance peak can be tuned with the symmetry of a three-dot configuration.

Here, the exact coupling mechanism between two NV centers has not been mentioned, which would be interesting for future studies. NV centers created by ion implantation can form many different coupled configurations and an extended lattice structure with distortions that can be simulated by the present technique. Here, we show resonance spectra of three qubits connected by three coupling parameters without considering any geometric phase, which can be extended through holonomic operations of three quantum gates. The holonomic control of three coupled qubits in addition to the replication of some well-known results provides a solid foundation for simulating holonomic control of NV centers using IBM QE. However, there are very few attempts to show the holonomic control of three coupled qubits. A method of performing holonomic control of three-qubit systems (three three-level Rydberg atoms) using a single holonomic gate is proposed; however, NV centers have specific energy levels of transition, which will be studied in the following section. These features can be observed in a spin-triplet superconductor as a strong zero-bias conductance peak with two satellite peaks. Similar results can be obtained from 3 qubits that can be either a set of 3 or 2 NV centers combined with a superconducting flux qubit (Figure 9.5).

9.3.6 GEOMETRIC PHASE

The study of the phase has produced many applications particularly, in holonomic quantum computation by manipulating the circuit parameters where the gates applied on qubits are based on geometric phases. The geometric (or Berry) phase is acquired by the adiabatic evolution of a quantum system around a circuit (path of rotation), which can be Abelian if the energy levels of a system are non-degenerate. Conversely, if a system's energy levels are degenerate, the cyclic evolutions of the degenerate subspaces will produce a path-dependent transformation called the non-Abelian phase. The supposition that non-Abelian geometric phases can be applied to holonomic control of qubits was postulated over two decades ago. Since then, holonomic control of qubits has progressed from speculation to the realization of control of physical qubits using their non-Abelian geometric phases. Such geometric phases have been implemented by a single cycle of non-adiabatic evolution, which is achieved by rotating a single qubit with a holonomic gate.

In a qubit, this geometric phase will have a nonzero solid angle associated with it. The solid angle can be used to store information about the acquired geometric phase.

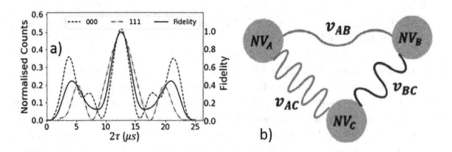

FIGURE 9.5 (a) Return probability of electrons and fidelity plotted as a function of time show three peaks. (b) Three NV centers are simulated by superconducting qubits where the inter qubit (NVC) couplings are asymmetric [15].

The trajectory of the unit vector along the Bloch sphere is equal to half of the solid angle. The pure geometric phase arises from the geometric property of the closed loop or path that can be extremely resilient to the effects of noise since noise does not cause any changes to the area enclosed by the loop. Since the Berry phase is only dependent on the global geometry of the loop and is resistant to small errors quantum gates that rely on the geometric phase have higher performance than other types of gates. Such quantum gates can be protected against the effects of decoherence, which may improve the performance of such gates.

9.3.6.1 Holonomic Operation of Diamond NV Centers by Superconducting Qubits

The holonomic approach to controlling NV center qubits provides an elegant way of theoretically devising universal quantum gates that operate on qubits via calculable MW pulses. The electron return probability can exhibit spin–orbit coupling-like behavior as observed in topological materials based on the extra geometric phase.

Further progression is seen in the development of multi-qubit holonomic gates, both in superconducting qubits and ion traps. NV centers in diamonds show great promise for holonomic control due to having both a stable non-Abelian geometric phase and the means of control via optical excitation. Qubits in NV centers can be rotated by MW, radiofrequency, and Stokes laser pulses. These pulses can be used to implement a set of universal quantum gates, which then operate on the qubits. Such control of NV centers via non-Abelian geometric phases has been demonstrated, with holonomic single-qubit gates implemented on NV centers demonstrating high fidelity, even at room temperature.

A three-qubit system produces a similar effect to RSOC and vortex structures in materials that are highlighted in the paper. There have been attempts to acquire the geometric phase in a qubit circuit of IBM QE. Also, the improved return probability (P_r) of electrons in the asymmetric arrangements of three entangled qubits has been demonstrated. However, a set of three qubits accumulating a geometric phase dependent on time has not been achieved. A holonomic gate can be defined as an operation or gate that causes a qubit rotation. The qubit acquires an Abelian phase that can be implemented on qubits in optical systems such as NV centers or superconductors. For the three-qubit system in a trigonal arrangement that occupies a closed space, we can generate a non-Abelian geometric phase and show the superiority of off-resonant holonomic gates over the on-resonant gates. This development shows particularly interesting potential, as NV centers are viable qubits at room temperature, and the operation of universal gates on such qubits is a step forward in the development of universal quantum computers. Another very important property of a holonomic gate in an NV center is that it also has a dark and bright state.

The circuit for the implementation of a **one-qubit** holonomic gate consists of three main processes, namely the initialization of the qubit, the holonomic gate implementation, and the measurement (Figures 9.6 and 9.7). We start by initializing a Bloch sphere to the states $|z\rangle$ and $|-z\rangle$, since the final states of the qubit are measured on the standard $|z\rangle$ basis. We also initialize the standard states $|x\rangle$ and $|y\rangle$. Unlike an ordinary qubit where $|-z\rangle = 0$ and $|-z\rangle = 1$, in this Bloch sphere the state $|z\rangle = 1\rangle$ and the state $|-z\rangle = |2\rangle$. A set of pulses is used to implement a single-qubit unitary holonomic

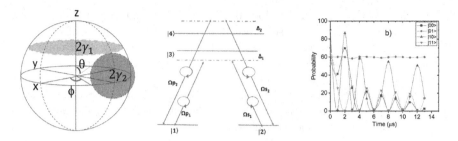

FIGURE 9.6 (left) The two rotation paths of the qubits implemented in the NV center. These paths start and end at the same position on each qubit. The first path is implemented on the first qubit, while the second path is implemented on the second qubit. (middle) The effective energy diagram of the spin triplets (^{3}A and ^{3}E) for the 4-level NV center including the selection rule. The Rabi frequencies are Ω_{pi} for the applied pump and Ω_{si} for the Stokes field with $i = 1$ and 2. The detuning is $\Delta_1 = \Delta_2 = \Delta$. (right) The return probability of the state $|1\rangle$ after evolution in the rotation path for varying degrees of the degeneracy of the ground states over time [18].

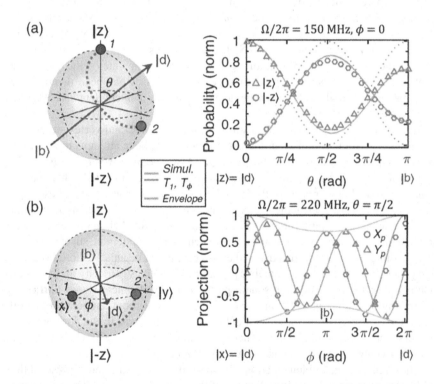

FIGURE 9.7 Simulation of holonomic control on a single qubit. (a) The rotation path is implemented by a single holonomic gate. Since the path forms a closed loop, a geometric phase can be created via a single iteration. Schematic of the Bloch sphere representation of the $\pi/2$ gate transformations. (b) The two rotation paths of the qubits are implemented in the NV center. These paths start and end at the same position on each qubit. The first path is implemented on the first qubit, while the second path is implemented on the second qubit [20].

gate U with parameters θ, φ, and λ. Here θ and φ are used to denote the rotation angles in the Bloch sphere and $\lambda = \Delta / \Omega$, where Δ and Ω correspond to the detuning frequency of the pulse and Rabi frequency of the qubit, respectively. The holonomic gate on IBM QE θ and φ is implemented using R_X, R_Y, and R_Z gates, respectively. The pulse with the frequency Δ is appended into the circuit to achieve the detuning. To this end, we reproduce some of the previous work through emulations on IBM QE. The discrepancy between the emulated and simulated results can be used to calculate the accumulated phase.

We use the GHZ frame of reference when mentioning any qubit states in the three-qubit holonomic control procedure. It is impossible to demonstrate all the nine levels of an NV center using qubits because, for N number of qubits, the total number of levels is 2^N. To get around this problem most holonomic simulations omit the $|0\rangle$ state and replace the poles of the first qubit with the states $|1\rangle$ and $|2\rangle$ (this corresponds with most holonomic simulations). These states can degenerate. The system for the three-qubit system in the rotating frame is given as

$$H = \sum_{i=1}^{8} |\omega_i|i\rangle\langle i| + \frac{i}{2}\left(\Omega_{P_1}e^{-iV_{P_1}t} - \Omega_{P_2}e^{-iV_{P_2}t}\right)\left(|1\rangle\langle 7| - |1\rangle\langle 8|\right)$$

$$\left(-\frac{i}{2}\left(\Omega_{S_1}e^{-iV_{S_1}t} - \Omega_{S_2}e^{-iV_{S_2}t}\right)\left(|2\rangle\langle 7| + |2\rangle\langle 8|\right) + \frac{i}{2}\left(\Omega_{P_3}e^{-iV_{P_3}t}\right) \quad (9.1)$$

$$\left(|1\rangle\langle 3|\right) - \frac{i}{2}\left(\Omega_{S_3}e^{-iV_{S_3}t}\right)\left(|2\rangle\langle 4|\right).$$

The Hamiltonian in the interaction picture is given by

$$H_I = \begin{pmatrix} 0 & 0 & i\Omega_1/2 & 0 & 0 & 0 & i\Omega_2/2 & i\Omega_3/2 \\ 0 & 0 & 0 & 0 & 0 & -i\Omega_6/2 & -i\Omega_5/2 & i\Omega_4/2 \\ -i\Omega_1^*/2 & 0 & \Delta_1 & 0 & 0 & \Delta_1 & 0 & 0 \\ 0 & 0 & 0 & 0 & 0 & 0 & 0 & 0 \\ 0 & 0 & 0 & 0 & 0 & 0 & 0 & 0 \\ 0 & i\Omega_6^*/2 & \Delta_1 & 0 & 0 & \Delta_1 & 0 & 0 \\ -i\Omega_2^*/2 & i\Omega_5^*/2 & 0 & 0 & 0 & 0 & \Delta_2 & 0 \\ -i\Omega_3^*/2 & i\Omega_4^*/2 & 0 & 0 & 0 & 0 & 0 & \Delta_3 \end{pmatrix}. \quad (9.2)$$

Here Δ_1, Δ_2, and Δ_3 are the detuning frequencies and Ω is maintained at 15 MHz. The coupling between the ground states and the E_x, E_y states (in Figure 9.8) was not considered because such a coupling would require that all the nine states of the NV center be used, which is not possible given the fact that 3 qubits can only encode a maximum of 8 states ($|1\rangle = |000\rangle$, $|2\rangle = |001\rangle$, $|3\rangle = |010\rangle$, $|4\rangle = |011\rangle$, $|5\rangle = |100\rangle$, $|6\rangle = |101\rangle$, $|7\rangle = |110\rangle$, $|8\rangle = |111\rangle$). We set different values for Δ_i and solve the Schrödinger equation with the system in the $m_s = 1$ states. In our simulation, we focus on the representation using the states $|1\rangle$ and $|2\rangle$ since it corresponds with the dark and bright

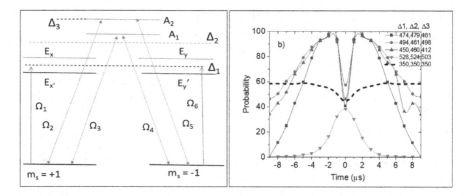

FIGURE 9.8 (left) Energy diagram of the spin triplets ^3A and ^3E for the complete 9-level NV center including the selection rule. The Rabi frequencies are Ω_{pi} for the applied pump and Ω_{si} for the Stokes field with $i = 1, 2$, and 3. The detuning is $\Delta_1 = \Delta_2 = \Delta_3 = \Delta$. Here Δ_1, Δ_2, and Δ_3 are the detuning frequencies and Ω is the Rabi frequency. (right) Evolution of the state |1> over pulse sequence duration for varying values of detuning frequencies Δ_1, Δ_2, and Δ_3. The solid dashed line represents the on-resonant case [18].

states of an NV center. Since there are three coupled qubits and each has a dark and bright state, we can represent the triplet nature of the ground states (or nuclear spins) of the NV center. By encoding the nuclear spins in one qubit, the electron spin states in the second, and intermediate states in the third qubit, we propose a method for holonomic control of three qubits.

We consider three rotation paths. For the first qubit, we implement a holonomic gate with $\varphi = 0$ and $\theta = 2\pi$, which is a complete rotation around the y-axis. The Rabi frequency for the first qubit Ω_1 is equal to the qubit frequency stated in IBM QE at 4966 MHz. A similar holonomic gate was implemented on the second qubit. For this holonomic gate $\theta = -2\pi$, which is a complete rotation in the opposite direction from the first qubit. The third qubit is rotated around a circular path around the y-axis with θ ranging between $\pi/6$ and $-\pi/6$. The same is true for φ. We keep the detuning frequency for the first holonomic gate Δ_1 variable for varying values of the detuning frequencies for the other holonomic gates Δ_2 and Δ_3. In qubits, a dark state is a state where a qubit undergoes trivial dynamics (i.e., the qubit can be described by the qubit Hamiltonian). For the creation of geometric phases, the rotational path traversed by the qubit is usually a closed loop and can be implemented on an arbitrary axis such as the one shown in Figure 9.6a.

Dark states can be created by the strong overlapping of two orthogonal states like the spin–orbit interaction. In condensed matter physics, it is well known that the geometric phase arises from the RSOC. This, however, has not been experimentally demonstrated using a quantum simulator that requires at least three qubits. Earlier, a three-qubit system was employed to generate a synthetic magnetic field, a chiral ground-state current, or a chiral spin state. However, the geometric phase was not captured, which was attempted in the present work. The purpose of the holonomic operation is to create a geometric phase through a closed loop through multiple iterations. Dark states can be described as the bound states or the vortex core that arises from the strong spin–orbit coupling, which results in weak (anti-) localization phenomena without breaking the time-reversal symmetry as observed in Figure 9.8b. This picture becomes very interesting for three vortices that can be controlled by a single operation. We see fine control of the P_r with the Rabi frequency (or RSOC strength). At zero frequency, P_r has a peak at the origin or zero detune time (similar to the weak localization phenomena observed in the magnetic field–dependent resistance of metal with the effect of RSOC). This becomes a minimum by the application of the Rabi frequency, which is similar to weak anti-localization. This is a hallmark feature of RSOC in a topologically protected material. While two holonomic operations perform a weak localization-like feature, the third holonomic gate breaks the time-reversal symmetry. Hence, the operation of a 3-qubit system works more efficiently than a 2-qubit system.

For the 3-qubit system, since there are more dark states, transitions between the dark states can be observed just through the holonomic control of one of the 3 qubits in the system. The transition is also quite interesting because it causes the amplitudes of the states other than the initial states to increase for a short duration. Also with 3-qubit systems, the system can exhibit geometric dependence, which is not possible for one and two-qubit systems, since only one configuration exists for these systems while there are two configurations for the qubits in the three-qubit system.

9.3.6.1.1 Holonomic Quantum Computation

Using two qubits we can create a four-level system that can be extended by adding a third qubit. A set of three qubits delivering an eight-level system can create a geometric phase. The holonomic gate operation can run these qubits, which can also be implemented in NV centers. This can be used for color images.

Holonomic computation can be performed on qubits implemented in NV centers or superconductors. However, the NV center is seen as a more promising candidate for the implementation of holonomic computation because (i) its quantum state can be manipulated by laser pulses or MW fields, (ii) it has a high fidelity at room temperature, and (iii) has a long-lived spin triplet. The lower levels of NV centers can function as qubits. Dark states can also control the level of decoherence: The phase difference between the dark state of a holonomic gate and the initial state of the qubit determines the decoherence of the system.

A dark state is a state of a three-level atom that cannot emit or absorb photons. Dark states have an eigenvalue of zero. Dark states of Λ-three-level systems can be used for quantum memory storage. Dark states are a special case of dressed superpositioned states as such states do not interact. It is like a bound state.

Dark states are needed in gate operation to remove dynamic phase shifts. This allows the proper measurement of the geometric phase. In the Rydberg atom the Rydberg state $|r\rangle$ is coupled to the state $|0\rangle$ by a pair of lasers with blue detuning $(-\Delta)$ and red detuning (Δ), respectively, but with the same Rabi frequency (Ω_0) and with the state $|1\rangle$ with the same Rabi frequency (Ω_1).

In the three-qubit holonomic control procedure, we use the GHZ frame of reference when mentioning any qubit states. Now in reality it is impossible to demonstrate all the nine levels of an NV center using qubits because for N number of qubits, the total number of levels is 2^N (Figure 9.9).

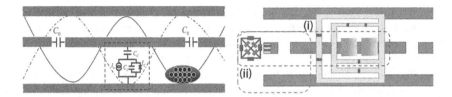

FIGURE 9.9   (left) Combining superconducting flux qubits and NV centers [1]. (right) Proposed quantum processor of superconducting loops and diamond consists of NV centers [19].

9.4   HYBRID SUPERCONDUCTING AND PHOTONIC QUBITS

One-qubit system can only be encoded with the ground states of the NV centers (or with one dark and one bright state). This means that only one dark state can be measured. However, since the ground states are degenerate no transitions are possible. For the 3-qubit system, since there are more dark states, transitions between the dark states can be observed just through the holonomic control of one of the 3 qubits in the system. The transition is also quite interesting because it causes the amplitudes of the states other than the initial states to increase for a short duration. Also, with 3-qubit systems, the system can exhibit geometric dependence, which is not possible for one and two-qubit systems since only one configuration exists for these systems while there are two configurations for the qubits in the three-qubit system.

9.4.1   COUPLING DIAMOND NV CENTERS TO A SUPERCONDUCTING FLUX QUBIT

One of the key drivers in quantum technology research is the development of qubit systems that can coherently process quantum information and store it for the desired amount of time. The NV^- center is an intriguing quantum system with some of the most properties suited to make qubits. The NV^- is exceptional due to its weak interaction with the environment, which results in exceptional coherence time. Experiments on coherent coupling in bulk diamonds have used around 10^7 spins. However, if the NV centers are very close to the flux qubit, the coherent coupling regime can be attained with fewer spins. A SQUID placed in a resonator can be used to tune the frequency of the resonator. Thus, quantum information storage, readout, and transmission can be accomplished by adjusting the frequency of the resonator into resonance with the SC qubit and the spin ensemble.

It is possible to obtain direct coupling between a flux qubit and a spin ensemble via cavity-mediated coupling. However, due to its short coherence period, the flux qubit's application in quantum computing is limited. NV center is used as quantum memory in hybrid systems as it has a very long coherence time.

To create hybrid quantum systems, the NV center can be connected with other quantum systems, such as superconducting flux qubits. Since hybrid quantum systems can combine two separate characteristics of the subsystems—the tunability of artificial atoms like quantum circuits and the lengthy coherence durations of atoms or spins—they have been actively advocated for quantum computation and quantum simulation. Additionally, either direct or indirect coupling strategies can be used to realize strong and controllable connections between two subsystems. One particular hybrid quantum system that combines NV centers in a diamond crystal with superconducting qubits has recently attracted attention. Long coherence periods and stable energy levels in NV centers, as well as excellent tunability and scalability in superconducting qubits, are advantages of this hybrid quantum system. Additionally, the magnetic coupling between superconducting qubits and NV centers has the potential to be three orders of magnitude stronger than the coupling between NV centers and a transmission line resonator. This hybrid system is a strong contender for replicating the numerous aspects of many-body systems due to its particular benefits. To replicate a Jaynes–Cummings (JC) lattice, for example, a hybrid quantum architecture made of inductively linked flux qubits with an NV center ensemble atop each qubit loop was proposed. It is feasible to explore the quantum phase transition between the localized and delocalized phases in a Bose–Hubbard-like model using this hybrid quantum system. The JC lattice simulated using them is simpler and more controllable than those realized using linked cavities because the flux qubits and the NV centers can be controlled by external magnetic fields.

A diamond NV center is effectively linked to the flux qubit. The coupling strength between flux qubits and spins can be modified to achieve strong and even ultrastrong coupling regimes by either constructing the hybrid structure in advance or regulating spin excitation frequencies via external magnetic fields. These hybrid quantum systems can be utilized to simulate coupled boson systems, including the demonstration of bilinear coupling. Furthermore, their scalability allows us to build hybrid arrays that simulate 1D bosonic lattices with tunable coupling strengths. The hybrid array is simplified to the tight-binding model of a 1D bosonic lattice in the strong coupling regime. The hybrid array exhibits additional interesting features in the ultrastrong coupling regime. Because of the bilinear coupling in this regime, for example, it can display quasi-particle excitations with an energy gap from the ground state. Furthermore, it is discovered that these quasi-particle excitations and the ground state are stable under certain conditions that may be tuned using an external magnetic field. This gives an experimentally accessible way for probing the system's instability. Here, we seek to explore the direct coupling of the flux qubit with a spin center.

9.4.2 Implementation of a Hybrid Qubit System

The magnetic field applied to operate the flux qubit, together with the strain field in the diamond crystal, lift the degeneracy of the ground state $m_s = \pm 1$ of the NV center. Thus, the $m_s = 0$ and $m_s = 1$ form an effective two-level system to which one can couple the resonant flux qubit. The Hamiltonian for this coupled system can be written as $H = H_{\mathrm{NV}} + H_{\mathrm{FQ}} + H_{\mathrm{int}}$, where H_{NV} is the ground-state Hamiltonian of the electronic spin of the NV center and H_{FQ} is the Hamiltonian of the flux qubit. The interaction Hamiltonian is written as $H_{\mathrm{int}} = g_e \mu_B \sigma_Z B_{FQ} \cdot S$, where B_{FQ} is the magnetic field generated by the flux qubit.

Because the coherent coupling is not possible along the quantization axis of the NV center, which is considered as the z-axis in the present study, the interaction Hamiltonian can be rewritten as $H_{\mathrm{int}} = g_e \mu_B B_{FQ}^{(x,y)} \sigma_Z \left(S_x \cos \cos \theta_{xy} - S_y \sin \sin \theta_{xy} \right)$, where $B_{FQ}^{(x,y)}$ is the magnetic field generated by the flux qubit in the x–y plane. θ_{xy} is the angle from the x-axis to $B_{FQ}^{(x,y)}$ in the x–y plane, which depends on the position of the NV center. The strength of the coupling of a single NV to a flux qubit is $g = g_e \mu_B B_{FQ}^{(x,y)}$. On the other hand, if an ensemble of NV centers with the same crystallographic orientation is coupled to the flux qubit, each of the NV centers in the ensemble would have slightly different coupling strengths. One can show that the individual coupling strengths couple to give a total coupling strength for the ensemble of $G = \sqrt{\left(\sum n g_n^2 \right)}$.

Previous studies of coupling NV centers in diamond to flux qubits used approximately spins in bulk diamond to achieve strong coupling. Instead of using many spins, the coupling strength can also be increased by increasing the magnetic field at the point where the NV center is located. This can be achieved by placing the individual NV centers closer to the superconducting line of the flux qubit. The NV centers are randomly distributed inside the diamond, which is assumed to be spherical for simplicity. When we place the diamond crystal in the vicinity of the flux qubit, each NV center would experience a (slightly) different coupling strength. The interaction between the NV centers and the magnetic field produced by the flux qubit leads to a coupling between the two systems.

9.4.3 Hybrid Quantum States: Superconducting Diamond and NV Center

Although we have already established the fabrication techniques for diamond nanowires and resonators, there are a number of outstanding tasks:

- Capacitive properties of boron-doped diamond
- Coupling diamond to other superconducting materials (aluminum)
- Measuring resonator properties of diamond
- Fabricating coupled diamond/aluminum flux qubits
- Controlling properties of diamond nanowire through capacity or inductive coupling
- Deeper investigations into bound states of boron-doped diamond and their potential for topological computation.

In the boron-doped superconducting diamond film, the insulating peak observed shows the presence of a quantum phase-slip effect, which is not observed in any other carbon system (Figure 9.10). Also, diamond nanowires show the quantum phase-slip effect, which can be operated at high frequencies. This shows the presence of a new qubit system in carbon. Carbon creates quantum states, namely the NV centers with nitrogen and superconducting diamond with boron. Both quantum states have a spin-triplet configuration, although, for a boron-doped diamond, it was not firmly established. Both triplet states can undergo a superposition and the result is not seen experimentally. Such kind of doping effect cannot be seen in other forms of carbon or other materials. Other low-dimensional carbon, e.g., graphene and nanotubes, can also be useful to make qubits in association with standard superconductors, which will be discussed in Chapter 10.

One of the key challenges for the implementation of scalable quantum information processing is the design of scalable architectures that support coherent interaction and entanglement generation between distant quantum systems. A nanotube double quantum dot (DQD) spin transducer that allows achieving steady-state entanglement

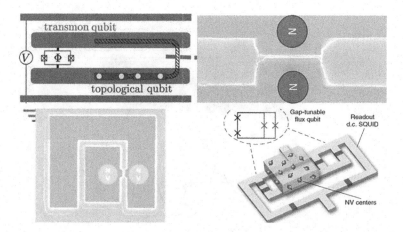

FIGURE 9.10 (Left) Diamond flux qubit with micro-bridge junctions, with embedded NV centers for N00N state entanglement (unpublished). (bottom, left) A loop of superconducting diamond and a phase-slip bridge are constructed. (top, right) A superconducting nanowire can be coupled with two diamond NV centers. This is compared with coherent coupling of a superconducting flux qubit to NV centers [2].

between NV center spins in diamonds with spatial separations up to micrometers has been proposed. The distant spin entanglement further enables us to design a scalable architecture for solid-state quantum information processing based on a hybrid platform consisting of NV centers and carbon–nanotube DQDs. In this work, we propose an efficient strategy for steady-state entanglement generation between NV center electron spins at micrometer distances, which is mediated by the leakage current of a carbon–nanotube DQD. Each quantum dot interacts locally with a single NV center in the proposed hybrid platform. Due to the Pauli's exclusion principle, we find that the NV center electron spins will be driven into a maximally entangled state along with the electrons being blocked in the DQD. The scheme requires only voltage control of the nanotubes and MW driving of the NV center electron spins, which is feasible within current state-of-the-art experimental capabilities. In addition, the steady-state entanglement of the electron spins can be exploited to realize an entangling gate between the nuclear spins associated with the NV centers via the hyperfine coupling. Therefore, the hybrid platform allows the generation of nuclear-spin cluster states for universal measurement-based quantum computation with excellent scalability and provides a new route toward solid-state quantum information processing (Figure 9.11).

9.4.4 NANODIAMOND QUANTUM SENSOR FOR RAPID DIAGNOSIS OF VIRAL DISEASES

A medical test kit for the early detection of viruses can help the respective governing bodies to control the infection rates. An affordable and economical rapid diagnosis for the early detection of viral diseases by making use of nanodiamond quantum properties can be developed. Further, this quantum sensor is a thousand times higher sensitive to viruses compared to traditionally used gold nanoparticle-based lateral flow tests, which means it can detect a disease at an early stage and can become crucial for reducing the transmission risk of infected individuals. NV center is a structural

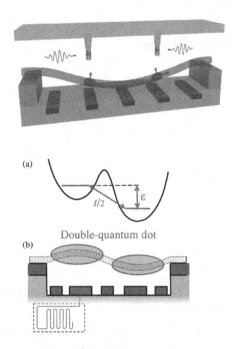

FIGURE 9.11 (a) Schematic of the hybrid building block: diamond pillars (transparent gray) containing single NV centers placed above carbon nanotubes bridging on the source and drain contacts. The gates that create the DQD (gray) are below the nanotube. The DQD confined in the nanotube is used to mediate the generation of steady-state entanglement between the electron spins (black) of two driven NV centers, where diamond nanopillars can be located close to the DQD using the technology developed for diamond scanning probes [21]. (b) A suspended carbon nanotube hosting a double quantum dot, whose one-electron charged state is coupled to the second flexural mode. Sketch of the electronic confinement potential and of the two main parameters, the hopping amplitude t and the energy difference ϵ between the two single-charge states (top). Physical realization (bottom). One of the gate electrodes is connected to a MW cavity for dispersive qubit readout [22].

defect in a highly ordered diamond, which makes it unique for its quantum optical properties. The nanodiamond quantum properties can improve the paper-based lateral flow tests for the early detection of viruses ranging from SARS-CoV-2 to HIV and can provide a rapid diagnosis (Figure 9.12).

The NV centers can signal the presence of an antigen or other target molecule by emitting a bright fluorescent light. In the past, fluorescent markers have been limited by background fluorescence, either from the sample or the test strip, making it harder to detect low concentrations of virus proteins or DNA that would indicate a positive test. However, the quantum properties of fluorescent nanodiamonds allow their emission to be selectively modulated, meaning the signal can be fixed at a set frequency using a MW field and can be efficiently separated from the background fluorescence, addressing this limitation. The fluorescent properties of quantum diamonds could be used for ultrasensitive diagnostic tests. In a blood sample, nanodiamonds studded with antibodies could be used to capture a virus. This would then attach to DNA that sticks to a test strip over a MW resonator. MWs would change the spin in the captive diamonds, enhancing their fluorescent signal and making the technique 100,000 times more sensitive than other methods (Figure 9.13).

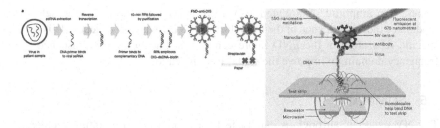

FIGURE 9.12　(left) Single-copy detection of HIV-1 RNA on Lateral Flow Assays (LFAs) using (right) Reverse Transcription and Recombinase Polymerase Isothermal Amplification (RT-RPA), and FNDs. Schematic illustration of the use of FNDs in LFAs [23].

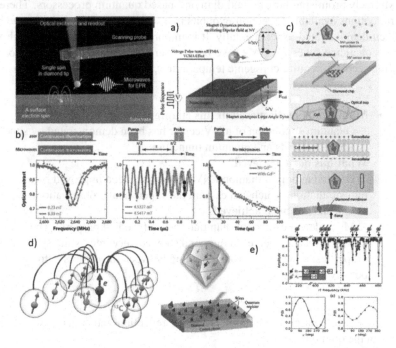

FIGURE 9.13　Basic principle of scanning magnetometry: (top left) (a) A sharp tip with a NV center at the apex is used to map out the three-dimensional magnetic field vector above a magnetic nanostructure, such as an isolated electron spin. (middle) (b) Sensing techniques and protocols, showing (top) pulse-timing diagrams and (bottom) example measurements. (right) (c) The integration of nanodiamonds into cell membranes. See details in Ref. 24 (d) The electron spin of a single NV center in diamond acts as a central qubit and is connected by two-qubit gates to the intrinsic ^{14}N nuclear spin and a further eight ^{13}C nuclear spins surrounding the NV center. (e) This technique uses an electron spin qubit to selectively control many individual nuclear spin qubits, while simultaneously decoupling them and thus protecting them from unwanted interactions in the system. 10-spin-qubit quantum system that can store quantum information for very long times up to 75 seconds [25].

Florescent nanodiamonds (FNDs) are immobilized at the test line in a sandwich structure in the presence of double-stranded DNA amplicons. Exciting at 550 nm (green) produces fluorescence emission centered at 675 nm (red), imaged with a camera. An amplitude-modulated MW field, applied by the resonator, selectively modulates the fluorescence of the immobilized FNDs at a set frequency. This enables specific separation of the FND fluorescence from background fluorescence in the frequency domain, to improve the signal-to-noise ratio.

9.5 SUMMARY

The NV center is an obvious candidate for a matter-based quantum architecture. Proven single-qubit operations and readout, long coherence times, and Stark shift control centers coupled with transform-limited single-photon generation for flying qubit generation and the potential for photon-mediated qubit–qubit interactions, are all extremely promising for eventual diamond-based quantum processors. There are several options available for coupling qubits in a scalable fashion, though none have yet been demonstrated experimentally. It is important to realize that although very long decoherence times for NV– have been observed at room temperature, optical coupling seems to require cryogenic temperatures.

- **Main problem discussed in this chapter.**
  Quantum simulation of a NV center has been demonstrated by super-conducting qubits and quantum tunneling.
- **What has been achieved?**
  Understanding of diamond NV centers has been improved by quantum simulation, which can explain the experimental observation.
- **What has not been achieved?**
  Many defect centers of diamonds are not studied properly. The nitrogen-vacancy center in diamond showed very interesting spin-qubit properties; however, a strong coupling between the centers is found to be very challenging.

Q. Calculate the scattering parameters for the reflection and transmission of MW signals in the quantum wells.

BIBLIOGRAPHY

1. A. Blais, A. L. Grimsmo, S. M. Girvin, and A. Wallraff, Circuit quantum electrodynamics, *Reviews of Modern Physics* 93, 025005 (2021); Z. L. Xiang, S. Ashhab, J. Q. You, and F. Nori, Hybrid quantum circuits: Superconducting circuits interacting with other quantum systems, *Reviews of Modern Physics* 85, 623 (2013).
2. X. Zhu, S. Saito, A. Kemp et al. Coherent coupling of a superconducting flux qubit to an electron spin ensemble in diamond. *Nature* 478, 221–224 (2011).
3. Z.-R. Lin, G.-P. Guo, T. Tu, Q. Ma, and G.-C. Guo, Quantum computation with graphene nanostructure, In *Physics and Applications of Graphene - Theory*, S. Mikhailov (Ed.), InTech (2011), ISBN: 978-953-307-152-7.
4. H. Aman, Applications of nitrogen-vacancy centers in nanodiamonds for nanoscale thermometry and magnetometry, M. Phil. Thesis, The University of Queensland (2019).
5. M. Ban, Photon-echo technique for reducing the decoherence of a quantum bit, *Journal of Modern Optics* 45, 2315–2325 (1998).

6. I. B. Djordjevic. Chapter 1- Introduction. In *Quantum Information Processing, Quantum Computing, and Quantum Error Correction* (Second Edition), I. B. Djordjevic (Ed.) pages 1–29. Academic Press (2021).

7. A. D. Greentree, B. A. Fairchild, F. M. Hossain, and S. Prawer, Diamond integrated quantum photonics, *Materials Today* 11, 22–31 (2008).

8. H. Haffner, C. F. Roos, and R. Blatt, Quantum computing with trapped ions, *Physics Reports* 469, 155–203, 2008.

9. X. Hu and S. D. Sarma, Hilbert-space structure of a solid-state quantum computer: Two-electron states of a double-quantum-dot artificial molecule, *Physical Review A* 61, 062301 (2000); J. A. Jones, Quantum computing and nuclear magnetic resonance, *PhysChemComm* 4, 49–56 (2001); Z. Ju, J. Lin, S. Shen, B. Wu, and E. Wu, Preparations and applications of single color centers in diamond, *Advances in Physics X* 6, 1858721 (2021).

10. G.-Q. Liu and X.-Y. Pan, Quantum information processing with nitrogen vacancy centers in diamond, *Chinese Physics B* 27, 020304 (2018).

11. P. W. May. CVD diamond: A new technology for the future? *Endeavour* 19, 101–106 (1995).

12. A. Mutiara, R. Refianti, and J. S. K. Karamoy, On a testing and implementation of quantum gate and measurement emulator (qgame), *International Journal of Engineering & Technology* 5, 2186 (2013).

13. A. Trabesinger, Quantum simulation, *Nature Physics* 8, 263–263 (2012).

14. J. Yun, K. Kim, and D. Kim, Strong polarization of individual nuclear spins weakly coupled to nitrogen-vacancy color centers in diamond, *New Journal of Physics* 21, 093065 (2019).

15. D. Mahony and S. Bhattacharyya, Evaluation of highly entangled states in asymmetrically coupled three NV centers by quantum simulator, *Applied Physics Letters* 118, 204004 (2021).

16. S. Bhattacharyya and S. Bhattacharyya, Demonstrating geometric phase acquisition in multi-path tunnel systems using a near-term quantum computer, *Journal of Applied Physics* 130, 034901 (2021).

17. F. Mazhandu, K. Mathieson, C. Coleman, and S. Bhattacharyya, Experimental simulation of hybrid quantum systems and entanglement on a quantum computer, *Applied Physics Letters* 115, 233501 (2019).

18. S. Bhattacharyya and S. Bhattacharyya, Holonomic control of a three-qubits system in an NV center using a near-term quantum computer, *Entropy* 22, 1593 (2022).

19. K. Mathieson and S. Bhattacharyya, Hybrid spin-superconducting quantum circuit mediated by deterministically prepared entangled photonic states, *AIP Advances* 9, 115111 (2019).

20. B. B. Zhou, P. C. Jerger, V. O. Shkolnikov, F. J. Heremans, G. Burkard, and D. D. Awschalom, Holonomic quantum control by coherent optical excitation in diamond, *Physical Review Letters* 119, 140503 (2017).

21. W. Song, T. Du, H. RalfBetzholz, and J. Cai, Nanotube double quantum dot spin transducer for scalable quantum information processing, *New Journal of Physics* 22, 063029 (2020).

22. F. Pistolesi, A. N. Cleland, and A. Bachtold, Proposal for a nanomechanical qubit, *Physical Review X* 11, 031027 (2021).

23. B. S. Miller, L. Bezinge, H. D. Gliddon, et al., Spin-enhanced nanodiamond biosensing for ultrasensitive diagnostics, *Nature* 587, 588–593 (2020); I. Lovchinsky, A. O. Sushkov, E. Urbach, et al., Nuclear magnetic resonance detection and spectroscopy of single proteins using quantum logic, *Science* 351, 836–841 (2016).

24. R. Schirhagl, K. Chang, M. Loretz, and C. L. Degen, Nitrogen-vacancy centers in diamond: Nanoscale sensors for physics and biology, *Annual Review of Physical Chemistry* 65, 83–105 (2014); S. Sotoma, H. Okita, S. Chuma, and Y. Harada, Quantum nanodiamonds for sensing of biological quantities: Angle, temperature, and thermal conductivity, *Biophysics and Physicobiology* 19, e190034 (2022).

25. C. E. Bradley, J. Randall, M. H. Abobeih, et al., A ten-qubit solid-state spin register with quantum memory up to one minute, *Physical Review X* 9, 031045 (2019).

10 Quantum Simulation of Carbon Structures

Close and Open Quantum Systems

10.1 SUPERCONDUCTING QUBITS

- Consists of a thin layer of insulating material between two superconducting materials.
- Insulator acts as a barrier to the flow of Cooper (electron) pairs and the superconducting phase.
- When voltage applied, current flows between superconductors by tunneling effect.

10.1.1 JOSEPHSON JUNCTION AND COOPER-PAIR TUNNELING

Superconducting Josephson junction (JJ) shows many interesting properties. JJ arrays have been used in a variety of devices like superconducting magnetometers, qubits, and bilevel systems. This is possible because of the new physics discovered, when an insulator is sandwiched between two superconducting materials called a Josephson junction.

In classical electrodynamics, when an insulator is sandwiched between two metals, the resistivity of the insulator blocks any current flow. The electron wavefunction decays exponentially in the metal–insulator boundary and never reaches the other side. However, if the insulator is thin enough, the wave function can tunnel through it and reach the other side to conduct current (see Figure 10.1).

10.1.2 JOSEPHSON RELATION

In a JJ the Cooper pair from one side can coherently tunnel to the other side unlike in the case of a metal insulator where the electron wave function's coherence is lost in the tunneling process. The relations for voltage and current in a JJ was given by Brian Josephson in 1962 as $I(t) = I_c \sin[\delta(t)]$, $V(t) = \dfrac{h}{2e} \dfrac{d\delta(t)}{dt}$, where I_c is the critical current of the system and $\delta(t)$ is the difference in phase between the superconducting electrodes. The voltage in a JJ exists only when the phase difference changes with time, a current can still flow when it is static. However, a phase difference is necessary for the current to flow.

DOI: 10.1201/9781003316411-10

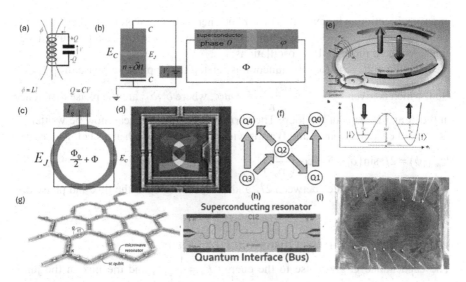

FIGURE 10.1 (a) A qubit can be defined as a combination of a coil (producing a magnetic field) and a capacitor with (+) and (−) polarity. (b) Hence, charge and phase qubits are constructed. The charging energy the capacitor and the Josephson junction controls the property of the qubits and the quantum circuits. (c) The superconducting qubit is made based on a superconducting loop (which confines magnetic flux) using the properties of Josephson junction. (d) In the superconducting loops the supercurrents rotate both in clock-wise and anti-clockwise directions. (e) The superconducting component acts as a qubit state that can be written through excitation or read from using microwave pulses through transmission lines, thus allowing us to manipulate the qubit and perform logic operations. (f) Five qubits can be connected in a star shape. (g) Qubits can form an array of hexagonal ring (graphene-like) structures [27]. (h) Qubits are designed in association with microwave resonators. (i) A superconducting diamond resonator is shown.

The energy related to this junction is known as the Josephson energy given by
$$U = \int I(t)V(t)\,dt = \frac{\phi_o\, I_C}{2\pi}\cos(\delta) = E_j\cos(\delta), \text{ where } E_J = \frac{\phi_o\, I_C}{2\pi} \text{ and } \phi_o = \frac{h}{2e} \text{ is}$$
the flux quantum. Here the energy again varies as the cosine of the phase difference between the electrodes.

10.2 SUPERCONDUCTING QUANTUM INTERFERENCE DEVICES (SQUID)

SQUID is a device that consists of a superconducting loop with JJs placed in each of the two arms A and B of the loop. This superconducting loop can only allow flux that is an integral multiple of the flux quanta $\phi_o = \dfrac{h}{2e}$ through it. As an external flux is applied to the loop, the current flows in the loop to make the flux passing through the loop an integral multiple of ϕ_o. When the flux $\phi < \dfrac{n}{2}\phi_o$, the current moves in a direction to counter this new flux, and when $\phi > \dfrac{n}{2}\phi_o$, the current moves in the opposite direction to increase this flux to the energetically favorable higher flux state. This phenomenon creates an oscillation in the voltage of the loop and the flux change can be counted through these oscillations.

Since the voltage across the junction is related to the phase, a phase drop is created, which is given by $\phi = \dfrac{\phi_o\,(\delta_2-\delta_1)}{2\,\pi}+m\phi_o$, where δ_1,δ_2 are the phase difference in the A and B arms of the loop. The current out of the loop can now be written as the sum of currents from A and B arms with the current in both arms being identical:

$$I_{\text{squid}}\left(\phi\right)=2I_c\sin\left(\delta_1+\delta_2\right)\cos\left(\frac{\pi\phi}{\phi_o}\right).$$

The current oscillates between $2I_c$ and 0 depending upon the flux and phase difference in the junctions.

10.2.1 Intuitive Charge Qubits and Flux Qubits

The capacitance C gives rise to the energy $E_c=\dfrac{1}{2}\dfrac{e^2}{C}$, and the flux in the junction gives rise to $E_J=\dfrac{\phi_o I_c}{2\pi}$. The ratio of these two energies determines whether we will have a charge qubit or a flux qubit. If $\dfrac{E_J}{E_c}=1$, then we have a charge qubit. The superconducting island is then capacitively coupled to Cooper-pair reservoirs. The eigenstate of this qubit is determined by the number of Cooper pairs localized in the island. On the other hand, if $\dfrac{E_j}{E_c}=50$, we get a flux qubit. The flux qubit usually has three JJs present in a loop. As per the discussion in the earlier section, the external flux creates a two-level system of currents moving counter to each other, thus forming a two-level system. The superposition of these two currents gives us the logical qubit. Higher energy levels are far away from these two lowest eigenstates and hence can be neglected while considering this system.

The relative magnitudes of E_c and E_j determine the properties of the system to a huge extent. If E_c is much smaller than E_j , then any fluctuation in the capacitive energy will have little effect on the Larmor frequency ω. Larmor frequency is related to the difference in energy between the ground state, E_0 and the excited state, E_1 as: $\hbar\omega=E_1-E_0$ of the qubit. If $E_j \gg E_c$ in the flux qubit, then the system will be extremely sensitive to flux fluctuation. Therefore, we can manipulate the ratio between the flux and capacitive energies to manipulate the resistive regime of the system. The duality that emerges is quite interesting. In the superconducting regime, the capacitive energy is much lesser than flux energy, and thus the JJ tunnel is dominant over the Cooper pair, but the Cooper pair is free to move and conduct. In the insulating regime, the Cooper pairs are localized and form a bosonic insulator, while $E_j=E_c$. The idea of making qubits was applied to superconducting diamonds as discussed later in the chapter. However, carbon nanotubes and graphene ribbons were used as interconnects in the superconducting circuits to construct hybrid circuits which were tested at microwave or radio frequencies (RF) (see Figures 10.2–10.5).

10.2.2 Nanotube-Based Qubits

In Chapter 3 we introduced fascinating transport properties of CNTs which might be useful in quantum technologies in the future although this material suffers from high defect density.

A recent review [56] explained device prospects of CNT devices which is used in the book.

- CNT charge qubits have previously shown DC magnetic field sensitivity similar to NV centers and electric field sensitivity superior to a single-electron transistor [1].
- It has been possible to create circuit QED spin-photon interface and extended coherence periods for CNT spin qubits [2]. The spin coherence period in CNTs has a theoretical limit that can be more than tens of seconds [3]. Furthermore, industry has taken over since then with the aim of achieving quantum computers based on CNT spin [4].
- Nanomechanical resonators made of suspended carbon nanotubes have quality factors as high as 5 million [5]. Quantum sensing and computing will benefit from the potential for a CNT-based nanomechanical qubit to couple to a broad variety of modalities for external fields and display a lengthy qubit decoherence period [6].
- In the telecom bands, single-photon emitters housed in carbon nanotubes (CNTs) are very desirable due to functionalized SWCNTs' several distinct benefits over other materials [7].
- Highly aligned films of SWCNTs with exciton-polaritons create a novel system with on-demand USC [8], which has applications in quantum information processing and sensing.
- CNTs' mechanical and electronic characteristics allow them to be used in hybrid quantum devices, which enhance superconducting circuits [9].

Nonetheless, a number of significant obstacles that impede the commercial use of carbon nanotubes (CNTs) in quantum technology remain unresolved. These include the following: (1) producing ultrapure CNTs sorted by chirality and (2) assembling CNTs on a macroscopic scale while maintaining their separation and orientation.

These quantum devices make use of CNTs' many degrees of freedom (spin, charge, orbital, and lattice). As a result, CNTs are used in a wide range of quantum technology applications, including sensing, communication, and computation. Achieving simultaneous macroscopic alignment and chirality sorting to produce single-crystal SWCNTs with the same chirality or (n,m) indices is an intriguing next step in this area. Until now, high-quality CNT qubits are not achieved. Developing strong coupling and entanglement techniques for CNT-based qubits is another crucial path.

For spin qubits, pure ^{12}C should be used to manufacture carbon nanotubes in order to eliminate nuclear spin noise. Suspended CNTs are sought after to better isolate qubits from the surrounding environment [10].

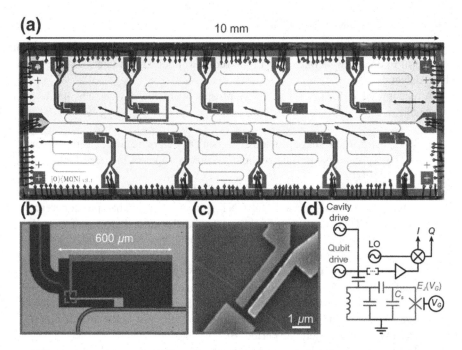

FIGURE 10.2 Device with a quantum circuit made of carbon nanotubes. (a) The device chip's optical picture. Ten multiplexed $\lambda/4$ resonators with various resonance frequencies are addressed by a coplanar microwave transmission line in the middle. For the manufacturing of qubits, each resonator has a cut out in the ground plane near the electric field antinode. Moreover, each resonator includes a single DC line that enables voltage tuning of the qubit frequency. (b) An optical picture of a single qubit with its island capacitively connected to its resonator and its opposite side shorted to the ground plane. The CNT JJ is capacitively shunted to the surrounding ground plane by the island. To adjust the frequency of the circuit, a side gate (red) is employed. (c) A CNT is in contact with two superconducting contacts that are spaced 300 nm apart and a side gate. (d) A schematic of the device's electrical circuitry and a sketch of its readout and control circuits. The resonator is capacitively connected to the qubit. A side gate that receives voltage (V_G) adjusts the qubit's Josephson energy E_J. A typical heterodyne detection method is employed to measure the response of the qubit, which is capacitively connected to the transmission line that is used to transfer microwave tones to it [9].

The primary issues with CNT-based single photon emitters (SPE) include producing indistinguishable photons, managing the spectrum variety of the SPE, and boosting the source brightness.

10.2.3 GRAPHENE-BASED QUBITS

The key component of modern superconducting (transmon-like) qubits is a Josephson junction (JJ) made of aluminum oxide (AlO_x) which can suffer from charge noise. Researchers attempted to develop new materials especially epitaxial superconducting-semiconductor devices. There were attempts to incorporate graphene-based materials into superconducting quantum computing devices since graphene enable

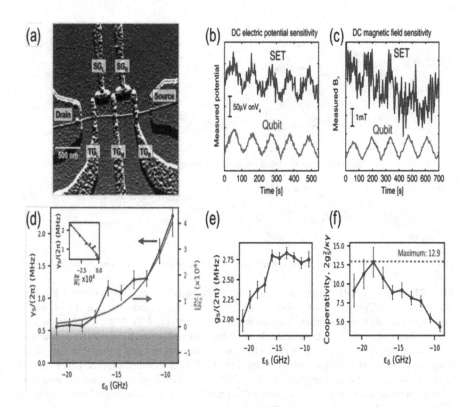

FIGURE 10.3 Hybrid circuit via their interactions with superconducting microwave circuits, solid-state systems are used in quantum electrodynamics (QED) to probe coherent quantum physics. A critical step in the creation of a hybrid superconducting qubit that makes use of a Josephson junction made of a carbon nanotube. It is demonstrated that a local electrostatic gate may be used to tune the frequency of a quantum circuit by connecting a carbon nanotube with a superconducting Pd/Al bilayer. By examining the gate-tunable resonator frequency, a substantial dispersive coupling to a coplanar waveguide resonator is established [56].

the qubit to change states through voltage, much like transistors in today's traditional computer chips.

In earlier days (2018) researchers from MIT and elsewhere recorded the "temporal coherence" (more than 50 nanoseconds) of a graphene qubit which represented a critical step forward for practical quantum computing [11]. The researchers sandwiched a sheet of graphene in between the two layers of a van der Waals insulator called hexagonal boron nitride (hBN) for the Josephson junction. A weak link constructed from a graphene heterostructure was employed by some researchers to create a completely operational superconducting circuit. When voltage gets applied to the qubit, electrons bounce back and forth between two superconducting leads connected by graphene, changing the qubit from ground (0) to excited or superposition state (1). Because the materials are so pristine, the traveling electrons never interact with defects. This represents the ideal "ballistic transport" for qubits, where a majority of electrons move from one superconducting lead to another without scattering with impurities, making a quick, precise change of states. The enhancement that stands out is the coherence

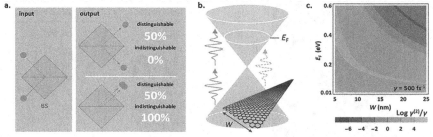

Basic operating principles of nanoplasmonic quantum logic gate

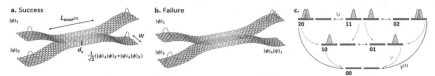

Surface-plasmon-based SWAP$^{1/2}$ gate comprised of nonlinear graphene nanoribbons.

FIGURE 10.4 SWAP1/2 gate based on surface plasmons and made of nonlinear graphene nanoribbons. A Coulomb interaction is used to connect nanoribbons so that the plasmonic modes can couple to one another. After an interaction length $L_{1/4}$ LSWAP$^{1/2}$, the plasmon has a 50–50 chance of staying in the same mode or having switched between ribbons for a separation dz between the ribbons. Because of this, when one plasmon is input into each mode, $|1\rangle_1|1\rangle_2$, the output state is found with one plasmon in each mode, $|1\rangle_1|1\rangle_2$, in which case the gate succeeds, or with both plasmons in one of the modes, $|2\rangle_1|0\rangle_2$ or $|0\rangle_1|2\rangle_2$, in which case the gate fails. When a separable single qubit is input into each mode ($|\phi\rangle$, $|\psi\rangle$), an entangled state is created. The Hong-Ou-Mandel (HOM) effect causes the plasmons to depart the gate in the same output mode when there is no nonlinearity in the waveguide and the plasmons are assumed to be indistinguishable, which results in the gate failing for $|1\rangle_1|1\rangle_2$ at all times. Yet, the significant nonlinearity of the graphene waveguides, induced by the Zeno effect, lowers the likelihood that two plasmons would be discovered in the same nanoribbon and raises the likelihood of success [57].

time and device quality that are appropriate for executing temporal manipulation, or quantum gates. Although the performance graphene JJ qubit is still lacking, its unique mechanics and control are too compelling to pass up [12], however alternative ways are suggested such as quantum dots in double layer graphene [12–16].

In Chapter 3 we have discussed quantum dot prospects of graphene. In conventional semiconductors like silicon quantum dots can produce quantum bits in future quantum computers. Some researchers have recently demonstrated that bilayer graphene is even more advantageous than other materials in this regard. One of the prerequisites for quantum computing is a robust read-out mechanism, which is enabled by the virtually perfect electron-hole symmetry of the double quantum dots they have developed [12–16].

In terms of semiconductors, bilayer graphene is special because of its low band gap, which makes it useful for confining charge carriers in discrete regions, or so-called quantum dots, which are highly intriguing for quantum technologies. An external electric field may be used to adjust this band gap between zero and around 120 milli-electronvolts. These have the ability to trap either a single electron or its opposite, a hole—basically, an electron that is absent from the solid-state structure—depending

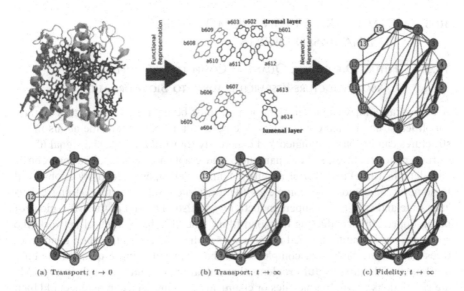

FIGURE 10.5 A crucial process in biology, the binding of gene regulatory proteins to the DNA, could be predicted using a quantum processor. On a D-Wave Two X device, the study was conducted. Nonetheless, 10 qubits can be utilized to build the DNA structures, which can then be simulated on an International Business Machines Quantum Experience (IBM QE) if a sufficient mathematical model is available. Future carbon research will use quantum computers to investigate some of the most crucial issues: Examples of nanostructure growth mechanisms at the atomic level include comprehending chirality, functionalizing carbon nanostructures, storing data in nanostructures, and transporting (or transferring) charges between nanostructures as shown in the bottom part of the figure (a) to (c) as transport changes from t = 0 to infinity [58].

on the applied voltage. One property that is absent from traditional semiconductors is the ability to capture both electrons and holes using the same gate configuration.

It is possible to generate a double quantum dot in this material where the spin characteristics of each opposing quantum dot, each containing an electron and a hole, nearly precisely mirror each other.

Because of the symmetry, an extremely reliable blockade mechanism is produced, which may be utilized to accurately determine the dot's spin state. This is beyond the capabilities of any other two-dimensional electron system or ordinary semiconductors.

The almost flawless symmetry and robust selection procedures are highly appealing for single-particle terahertz detector realization as well as for qubit operation. Furthermore, it is well suited for linking superconductors and quantum dots of bilayer graphene, two systems where electron-hole symmetry is crucial.

From a physics point of view, the characteristics of unusual materials may be investigated using the superconducting circuits surrounding the 2D material stack and the infrastructure of qubit management and readout already established for quantum computing. We still have a long way to go in the materials science of quantum computing before we have more, better, and varied qubits. Future large-scale quantum computer implementation may benefit from the development of durable semiconductor spin qubits. However, we are still in the early stages of developing quantum dot-based (graphene) qubit devices. It is still not possible to realize the first quantum bit in graphene.

10.3 MOLECULAR CARBON, BIOMOLECULES, AND DNA-BASED QUBITS

10.3.1 FUTURE DIRECTIONS: QUANTUM COMPUTER IN COMPLEX CARBON STRUCTURE RESEARCH LEADING TO BIOINFORMATICS

A biological complex molecule's basic unit must be understood through a slightly simplified structure and connectivity. We need to understand how the atoms in the structures can become entangled and coherently transmit an electrical signal (data). Carbon nanostructures, such as nanotubes and graphene, demonstrated long coherence lengths (time) and carrier mobility. Spin-triplet superconductivity is observed in some organic molecule complexes (with a protein-like structure). Long coherence time and spin-triplet superconductivity are also observed in the GB regions of boron-doped superconducting diamond films. The GBs have a neural network-like structure, according to detailed microscopic studies. As a result, carbon nanostructures that mimic biological complexes can be analyzed using a quantum simulator. Some of the first useful problems that quantum computers are likely to tackle are simulations of small molecules or chemical reactions. The computers could then be used to accelerate the search for new drugs or to kickstart the development of energy-saving catalysts to accelerate chemical reactions (Figure 10.5). By delving into the secrets of proteins in DNA, we can continue our journey of developing targeted cancer therapies. Quantum computing would enable us to map proteins in their entirety, like how we map genes.

10.3.2 SUPERCONDUCTING PHASE-SLIP QUBITS

In the previous subsection the possibility of graphene-based quantum devices is discussed since it can be used for a large area. In this subsection we show the prospect of diamond-based quantum devices. It is known that the majority of semiconductor producers typically use 300 mm silicon wafers to supply state-of-the-art solutions to high-tech sectors including consumer electronics, telecommunications, aerospace, and military and defence. Industrialists attempted to develop large-area electronics of carbon-based materials which is an alternative to silicon. However, deposition of diamond thin films on a large substrate for developing next-generation processors was found to be difficult. Recently, AKHAN Semiconductor announced the first-ever 300 mm complementary metal-oxide-semiconductor (CMOS) diamond wafers which can also be useful for quantum device applications.

In the previous chapters, we have described nanocrystalline diamonds as unique granular superconductors where the long order and coherence are mediated by phase coupling between individual grains. Such materials can therefore be used as platforms to develop hybrid systems that intrinsically combine local GB properties such as SOC, with long-range transport of superconducting Cooper pairs, a combination that ensures the interdependency of phase, charge, and spin. In this chapter, we describe some exotic phenomena including the observation of anomalous phase shift (APS), a key property being exploited for superconducting spintronics. Superconducting diamond substrates were used as SQUID devices to show current oscillations as a function of magnetic fields. However, these structures can be tested at microwave frequencies to observe resonant features (Figure 10.6). In addition to the JJ another type of junction, e.g., phase-slip junction can be made in diamond. In one-dimensional (1D) superconductors, quantum phase slip (QPS) arises from the quantum tunneling of the superconducting phase with a sudden change of 2π by interacting with localized charge (quantum dots) or defects (disorder) which can create a π-type JJ. Recently, QPS has restored interest not only in fundamental physics but also in applications in quantum circuits. Additionally, in the presence of strong SOC, APS (φ_0) can be seen in a superconducting (SQUID) device which opens up a fast-growing research direction in future qubit technology (Figure 10.7). QPS describes the suppression of Josephson current through a backflow as found in SOC-related phenomena such as anomalous Hall effect. In a closed ring, it can capture a geometric phase, which yields a topological insulating character with modulation in the transport properties, most importantly the Aharonov–Casher effect. Generally, it is understood that a non-s-wave and odd-frequency superconducting order parameter can be responsible for this phase change in addition to microstructure and thermal effect. However, a model unifying the QPS (or APS) and SOC in the presence of inhomogeneity of the microstructure is yet to be developed in granular superconducting materials. This requires analysis of bias dependence conductance peak measured at different magnetic fields and microwave frequencies in a device structure as presented herein in boron-doped nanocrystalline diamond (BNCD) films.

Nowadays, JJs have been fabricated from superconducting diamond and BNCD films by creating a sharp potential step through deposition that can show QPS. From the microstructure of the BNCD films signature of a phase, a slip junction is shown in this material, which can be verified firmly by a detailed analysis of the magnetic field-dependent ZBCP features and the frequency-dependent measurement of QPS (resonator) devices. For spin mediated interactions, a spin-triplet superconductor can be an ideal platform to study such extraordinary phenomena created by (locally)

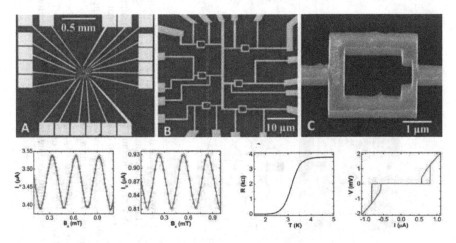

FIGURE 10.6 Monolithic device structures fabricated from boron-doped superconducting diamond: (a) Contact pad geometry, (b) array of SQUID devices connected to electrode contacts, and (c) image of single SQUID structure. (bottom) IV characteristics show clear Josephson junction features with the field-dependent critical current oscillatory behavior. (c) The weak links are observed on the right of the image and have a thickness of 500 nm [59].

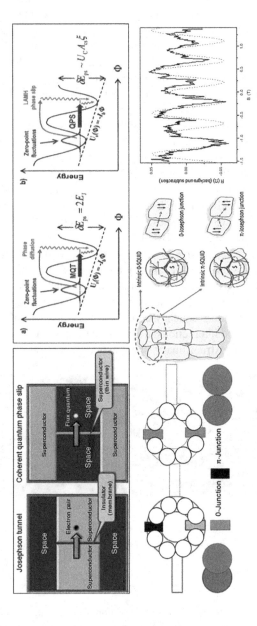

FIGURE 10.7   In low-dimensional macroscopic phase, coherence breaks down due to topological phase fluctuations [60].

- In quasi-1D, these fluctuations are called **quantum phase slips**: tunneling of order parameter between states that differ by 2π is shown (Top left).
- To date, theory cannot account for all observations in superconducting nanowires.
- (Top right) **Duality** between quantum phase slip of vortices and Josephson tunneling of Cooper pairs.

Microstructure shows steps in diamond crystals, and hence supports the origin of QPS and also the charge accumulation. Based on the overlapping of the layers, we define a 0 (left) and a π (right) junction.

(Bottom left) Andreev bound states giving phase π on a junction between superconducting grains. An array of JJ and QPSJ in a square and ring structure [61]. (bottom, right) Magnetoresistance oscillations observed below T_c. Area of oscillations match determined grain size at low temperature. Oscillations can only be fit when taking into account dominant p-Junction (Bottom,middle).

charge accumulation in an s-wave superconductor, which is presently shown through the (nice) arrangement of diamond nanocrystals (and GBs), as well as in the fabricated diamond nanowires. In this way, JJs with pronounced SOC can exhibit π-junction behavior and signature of zero-point fluctuation in the diamond QPS devices through a pronounced zero-bias conductance peak. Although most experimental data on π-junctions involves engineered superconducting/ferromagnetic interfaces, there is a growing body of literature on how complex order parameters in (SOC-induced) unconventional superconducting materials can also lead to π-junction behavior. We present BNCD films as a platform for the study of QPS junction features having well-defined GBs (Figure 10.7) accommodating bilayers of charge and exhibiting spin-triplet superconductivity, as well as features of the APS and phase modulation which can be useful for constructing diamond-based resonator circuits.

Diamond nanowires with a very narrow constriction behave as a phase-slip junction (in addition to the JJ) and a SQUID (or a flux qubit) coupled with a transmission line. In the weak-coupling limit, the SQUID stays a bit far from the transmission line but absorbs the microwave power. When the SQUID loop stays on the transmission line, it resonates with the incident radio frequency (RF) and the absorption line splits which resembles Zeeman splitting as the loop creates a magnetic field. The splitting signifies the creation of a bilevel system consisting of a forward and a reverse supercurrent in the loop, hence a superposition of states. This work can find new qubit applications which were suggested in graphene. Further, the superconducting diamond nanowires are associated with the SOI, which can introduce an additional phase shift as seen in a topological superconductor. The presence of spin-triplet superconductivity in this system can show promise for a new kind of superconducting spintronics controlled by the GB structures like 1D superconductors.

Based on the analysis of data we propose an equivalent circuit consisting of four quantum dots, two are connected in series and the other two in parallel representing a JJ and a quantum phase slip junction (QPSJ) element in the circuit. The equivalent circuit shows two JJs in series and two other JJs in parallel which is equivalent to a QPSJ. The JJ and QPSJ connected in parallel can be found consistent with the recent theoretical descriptions. Theoretical predictions of such materials explain how current loops can form from the spontaneous current and how the phase coupling between grains leads to exotic states such as chiral and orbital glass phases. Furthermore, these systems are expected to make ideal candidates for investigating the XY model or a double XY model. The study of materials that intrinsically give rise to π-junctions is thus important for not only fundamental research but also potentially for quantum technology development, particularly topological qubits.

The RF measurements have been carried out using a vector network analyzer. The sample is located on the mixing chamber plate of a dilution refrigerator cryostat at a 25 mK temperature. Isolation from room-temperature thermal microwave radiation is performed with cryogenic microwave attenuators and an isolator. On the output side, the signal has been amplified using a high-electron mobility transistor amplifier. The current bias has been separately applied through DC lines using bias-Tees. The transmission amplitude has not shown any significant dependence on the applied RF power in the range of, available in this setup, applied powers (up to -60 dBm at the sample). The experimental spectrum is measured as a function of applied current bias I_b. Several features in the wide-frequency range measurement of the sample can be attributed to the finite bandwidth of the cryogenic microwave components, specifically, the microwave isolator, which has a passband from 2 to 12 GHz, and the amplifier, which operates in the 0.3–14 GHz range (Figure 10.8).

In the boron-doped superconducting diamond film, the insulating peak observed shows the presence of a QPS effect, which is not observed in any other carbon system. Also, diamond nanowires show the QPS effect, which can be operated at high frequencies. This shows the presence of a new qubit system in carbon. Carbon creates quantum states, namely the NV centers with nitrogen and superconducting diamond with boron. Both quantum states have a spin-triplet configuration, although, for a boron-doped diamond, it was not firmly established. Both triplet states can undergo a superposition and the result is not seen experimentally. In the transparent diamond structure, both boron and nitrogen would add color, which will make the strong diamond more attractive through multi-functionality. Such kind of doping effect cannot be seen in other forms of carbon or other materials. We propose how a combination of NV center and diamond superconductivity qubits will bring a change in qubit research.

10.3.3 Spin Qubits Based on Spin-Triplet Properties

10.3.3.1 Qubit and Pauli Spin Matrices

Pauli spin matrices represent the spin angular momentum operator for ½ integer spin particles. In three-dimensional (3D) space, they are σ_x, σ_y, σ_z, and I. They obey several relationships or commutation rules as follows. The SOC can be expressed by the spinor as follows: So far, we have explained two types of particles, e.g., bosons and fermions having integer and half-integer spins, and there is always competition between these two particles. How do we combine them in a system or how do we make a conversion between these two particles? A spin chain like a magnet does not have an energy gap that can resist the superconducting system which has a small

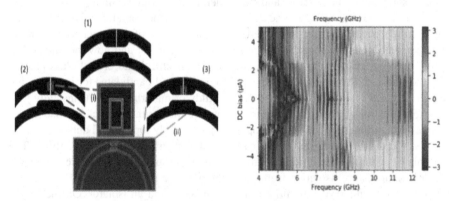

FIGURE 10.8 An 18 nm diamond nanowire device used as a QPSJ, the RF characteristics. Bias and microwave frequency-dependent variation of S21 phase recorded from the BNCD devices. QPS configurations with (1) consisting of the resonator, (2) consisting of the resonator and QPS, and (3) consisting of the resonator and QPS. Figures (i) and (ii) show the experimental diamond system composed of the nanowire and resonator, respectively (unpublished).

energy gap. The spin-half particles can be split into two halves or spinors. A pair of spinors that are duals can open up a band gap. In other words, the spin half particles can be described by $\sigma_x + i\sigma_y$ for 1D or two-dimensional (2D) chains. For the third dimension, another spinor σ_z can be added to the system. These three spinors σ_x, σ_y, and σ_z can be associated with another spinor, call the identity spinor 1. The combination of spinors that are opposite to each other (duals) can be arranged in a cyclic order, e.g., $[\sigma_x, \sigma_y] = 2i\sigma_z$, $[\sigma_x, \sigma_z] = -2i\sigma_y$, and $[\sigma_y, \sigma_z] = 2i\sigma_x$. A combination of two of the three spinors produces the third spinor as σ_x, σ_y, and σ_z. A beam of light can be split into two halves by using a polarizer that can rejoin. This two-level operation forms a qubit and can be described as the rotation of a circle. In general, a sphere (called a Bloch sphere) having three axes of rotations can describe the full operation of a qubit by using three spinors σ_x, σ_y, and σ_z. A qubit operates based on the superposition of two states, e.g., 0 and 1. They can also be compared to Orbit (or charge) and Spin, respectively. The superposition of orbit and spin form a triplet state where two spins are aligned in the same direction or combined in such a manner as to produce an integer or $S = 1$ spin. The spin can be expressed by two spinors as $\sigma_x + i\sigma_y$. The fermionic operator can be converted into a tensor product of the spinors. Similarly, bosonic operators can be mapped by spinors $\sigma\pm$ states so the whole condensate with the magnetic impurities can be described by the three spinors. The total energy or the Hamiltonian can be expressed by the three spinors. As a result, the quantum mechanical problem can be described by the spinors.

10.4 BASICS OF QUANTUM SIMULATIONS APPLIED TO HYBRID MATERIALS

We understood the dual nature of carbon bonds, which yields entirely two different properties, particularly in electrical transport (superconductivity and spin-related property). Now, we propose a quantum hybrid structure based on the arrangement of artificial atoms or superconducting qubits. The proposed hybrid structure of superconducting qubits arranged in hexagonal rings to mimic bilayer graphene can potentially offer new insights into quantum mechanics and its applications. By mimicking the properties of graphene, the system can be used to explore the transition between weak localization (WL) and weak anti-localization (WAL), the Kondo effect, and topological features. Additionally, the combination of superconducting qubits and spinors can provide a platform to study the relationship between spin and orbit, as well as bosonic and fermionic operators. The construction of NV centers using superconducting qubits and the creation of an artificial carbon structure that can store maximum information can also be explored. By creating a three-level spin-triplet system, the possibilities of quantum information processing and quantum communication can be further investigated.

10.4.1 QUANTUM TUNNELING AND COHERENCE BACKSCATTERING BY SUPERCONDUCTING QUBITS

Quantum tunneling has been simulated in a quantum well, as well as in a multilayer superlattice system. This simulation can explain the experimentally observed tunnel spectra seen from semiconductor quantum wells as discussed before. However, we would like to see how to apply this simulation in a real system such as a granular array of metals and superconductors. We are interested to see the quantum interference effect in this system.

Such a model was proposed where tunneling occurs between metallic grains through weakly conducting channels. We also consider a two-dot system with tunneling between them $H_T = \sum_{\alpha\beta\sigma} t_{\alpha\beta} e^{iV_{mn}t} a_{\alpha\sigma}^+ a_{\beta\sigma} + h.c.$, where V_{mn} is the voltage applied to the barrier (α, β being the single-particle states). The coordinate y runs along the interface S. The transverse coordinates z in the adjacent grains are defined in such a way that at the interface $z = z' = 0$. The tunnel amplitude is expressed with the eigenfunctions X_α, X_β of an electron within dots $t_{\alpha\beta} = a \int dy dz_z\ X_\alpha^*(y,z)\ d_{z'}\ X_\beta(y,z')\ |_{z=z'=0}$, for a system with two potential wells separated by a barrier. This can be reduced via the Suzuki–Trotter decomposition.

Now, in tunneling, the Hamiltonian is given as $H = K + V$, where K is the kinetic operator represented as $\dfrac{p^2}{2m} = (F^{-1})\dfrac{P_p^2}{2m}F$, while V is the potential and represented as $V = \sum\limits_{i1,i2,in=3}^{4} C_{i_i,i_2...i_n} \otimes {}_{k=1}^{n}\sigma_{i_k}$. The kinetic energy operator can be expressed as $K = FDF^{-1}$, where F and F^{-1} are the quantum Fourier and inverse quantum Fourier transformations, respectively.

The diagonalization operator is given by the following expression: $D = \Phi_\pi Z_0 Z_1$, where

$$\Phi_\pi = e^{-i\left(\frac{\pi}{8}\right)^2 \left(\frac{R\pi}{2}\right)^2 \Delta t} \text{ and } Z_j = e^{-i\left(\frac{\pi}{8}\right)^2 c_j \sigma_z^j \Delta t}. \tag{10.1}$$

The tunnel junction tunneling takes place in the two-qubit system in a Bell-state superposition. The protocol for simulating the tunnel of a single path in a double potential well on IBM QE requires two qubits and a set of Hadamard and CPHASE gates. In this case the 4 basis states 00, 01, 10, and 11 register 4 lattice sites as position variables for the particle (like 4 positions of a 2×2 image). We borrow this mechanism to simulate tunneling in our circuit for WL. There is a potential V, which is implemented using a single gate. This represents the double potential well with 2 peaks at 00 and 10 with 2 troughs at 01 and 11. The time interval is set at 0.1 and the amplitude at 20. The situation is that a particle is trapped inside one of the two wells by the preparation of the 01 states. In our simulation, we make use of Hadamard, CNOT, and U1 gates. Tunneling here means the entanglement of the two states in which it takes place between. Now the only gate that allows for such an entanglement is the CNOT gate where the target and control qubits are entangled. Now, this entanglement allows the two states to become level and may resonate together. Such entanglement creates an unobstructed path that allows the particle to cross through the barrier. Tunneling often requires the formation of a Bell state, as such a state is both the superposition and entanglement of two basic states, which allow for such a path to form between the qubits where two states are encoded on. In this chapter, we combine tunneling with the coherent backscattering or so-called weak localization (WL) process. In Figure 10.9, we see a two-path WL with inter-path tunneling. The tunneling is shown by a quantum Fourier transformation (QFT), diagonalization, and

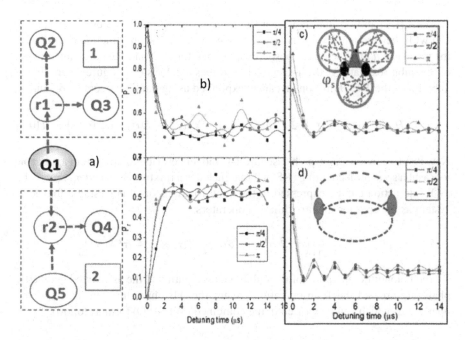

FIGURE 10.9   a) Qubits arrangements Q1-Q5 and r1, r2 on IBM QE for emulation of WL (and SOI) for a two-path system (1,2). Emulation of b) WL and weak anti-localization (WAL) without SOI effect shows a sharp fall and increase of the return probability, respectively measured at ϕ ($\pi/4$ to π). c) Emulation of periodic oscillations (Φ_0) and a supressed WL effect for an A–B ring. Inset: multipath WL model with tunnelling applied to a multi-quantum dot system forms an A–B phase φ_s. d) Emulation of $\Phi_0/2$ oscillations for a tube. Inset (c): multi-path tunnel junction forms an A–B tube.

the inverse QFT which are used to convert potential energy to a kinetic operator and vice versa. This process takes place on the two-qubit register (Figure 10.9).

10.4.2   WL TO WAL GEOMETRIC (AB) PHASE

10.4.2.1   Coherent Backscattering

We start with a multipath configuration on IBM Quantum Experience by creating circuits to emulate WL and incorporate pseudospin (Figure 10.9a). The double-path WL model shows multiple backscattering (wavevector changes from $-k$ to k) in paths 1 and 2. The qubits can be symmetrically connected to the register qubit (r) (and also symmetrically connected to the input qubit) and undergo a detuning process, which is effectively described by the Tavis–Cummings (T-C) Hamiltonian:

$$H_{WL} = \sum_{r=1}^{2} \omega_r a^\dagger a + \sum_{i=1}^{5} \hbar\omega_i \sigma_i^+ \sigma_i^- + \sum_{i=1}^{5} \hbar g(a^\dagger \sigma_i^- + a\sigma_i^+). \tag{10.2}$$

Here $r \in [1;2]$ and $i \in [1;5]$ represent the register qubits and phase qubits (including Q1) with the frequencies ω_r and ω_i, respectively. With the annihilation operator for the resonator a and the qubit (bosonic) spin operator $\sigma_i^{-,+}$ and the coupling strength of the qubit-resonator g, the T-C model describes a 2D system (Figure 10.9b). In the qubit form, the T-C Hamiltonian can be expressed using Pauli's spinors σ_x, σ_y, and σ_z.

$$H_{WL} = \hbar\omega_r\left(1+\sigma_z\right)/2 + \hbar\omega_i\left(\sigma_z/2\right) + \sum_{i=1}^{5} \hbar g \left(\sigma_x \otimes \sigma_x + \sigma_y \otimes \sigma_y\right). \quad (10.3)$$

Now just like in the single path case, the T-C model describes a 2D system applicable to Rashba spin-orbit coupling (RSOC) that includes a mixture of σ_x and σ_y. Spin-orbit (SO) interaction can be expressed in the qubit rotations or spinor using a four-level system as the creation and annihilation operators:

$$H_R = \frac{\omega\sigma_z}{2} + \frac{i}{2}\left[\left(\sigma_x^1 \otimes \sigma_y^1 - \sigma_x^1 \otimes \sigma_y^2\right) - \left(\sigma_x^2 \otimes \sigma_y^1 - \sigma_x^2 \otimes \sigma_y^2\right)\right]. \quad (10.4)$$

By the holonomic operation of the qubit, a closed path and the associated geometric phase are created. Such an operation creates the AB phase through the $(\sigma_x \otimes \sigma_y)$ operations, which can be written in the combined (WL + SO) form as follows:

$$H_{WL+SO} = \hbar\omega_r\left(1+\sigma_z\right)/2 + \hbar\omega_i\left(\sigma_z/2\right) + \sum_{i=1}^{5} \hbar g \left(\sigma_x \otimes \sigma_x + \sigma_y \otimes \sigma_y\right)$$

$$+ \lambda_{SO}\left(\sigma_x \otimes \sigma_y - \sigma_y \otimes \sigma_x\right). \quad (10.5)$$

All these models were implemented on the IBM quantum computer. To emulate the WL process P_R decreases exponentially until it reaches a value of approximately 0.5. The variation of P_R with angles (ϕ) $\pi/4$-π for WL can be found to be opposite to the SOI effect due to destructive interference from the phase shift. In the presence of strong SOI, the MR changes sign to positive compared to negative MR observed for pure WL effect. This can be executed by a 2π rotation, which reverses the sign of the pseudospin state (since for spin 1/2 particles a 4π rotation transfers the pseudospin function into itself).

As stated previously, we would like to see the effect of WL via the A-B rings. Such loops were used in interacting quantum circuits to show the generation of magnetic flux; however, there was no quantum interference effect as shown previously. In our work, we are only able to simulate linear chains of superconducting square loops due to the architecture of the IBM Melbourne device. Such loops allow for the implementation of superconducting circuits which require multi-loop quantum interference. For the A-B ring with the two-path WL simulation, we measure the P_r for the ground state vs. t that was used to generate and sustain the A-B ring (in μs) at different values of ϕ. These oscillations are a result of the $\pi/4$ pulses (see Figure 10.9c and d), which continuously drive the qubit in a Φ_0 periodic manner. These oscillations happen in conjunction with the exponential decay of the P_r of the ground state, and the amplitude of the oscillations decreases until they completely disappear. For the accumulated data we notice that the P_r reaches its maximum at $t = 0$. Then, as in normal WL, P_r decays until it reaches a stable value of around 0.5 with increasing time. However, unlike in WL, it oscillates with a period of Φ_0. A tube is constructed using two rings and two tunnel barriers, which produces oscillations with a period of $\Phi_0 / 2$. This can be shown through the presence of periodic $h / 2e$ oscillations. The effect of changing the ring into a cylinder can be described as the flux-dependent part of the WL correction originating from the phase coherence between multiple scattering paths, which can be obtained in multi-qubit arrangements. By maintaining the time-reversal symmetry in the paths, the relative phase becomes $4\pi\Phi / \Phi_0$. Due to the noise, the oscillations decrease with time (t).

10.4.3 Simulation of Spin–Orbit Coupling and Spin–Flip Effect

The matrix element for a transition from the momentum state k to k' describing the SOI has the form: $V_{k-k'}[1 - i \in k \times k' * \sigma]$. This equation describes the rotation of the electron spin (σ) by the angle between k and k' (i.e., ϕ and θ) around the z- and x-axes. We extend the discussion about a spin ½ atom with axial orbital symmetry which is absorbed (attached) on top of a graphene lattice having sublattices A and B. The spin projection is $\pm\frac{1}{2}$ for the sublattice pseudospin degrees of freedom s $(s = \pm 1)$ with spinors a and b. The effective interaction is expressed as $hg(q) = \tau_0 \otimes s_0 \otimes (v_F q \cdot \sigma - \mu\tau_0)$, where $v = k - k_\tau$, $\tau = \pm$, and $\sigma = (\sigma_x, \sigma_y)$. The interaction represents a solid angle with the unit vector n normal to the area traversed on a qubit or Bloch sphere. For a relative wavevector q and RSO strength (λ), other parameters τ_0, σ_0 $(\pm 1/2)$, and s_0 are defined as identity matrices in the valley, spin, and pseudospin spaces, respectively. The RSO term is given as

$$H_R = i\lambda \ \tau_0 \sum_q a_\sigma^\dagger (q) (\sigma \times s) \cdot \hat{z} \ b_\sigma (q) = i\lambda \ \tau_0 \sum_q a_\sigma^\dagger (q) (\sigma_x s_y - s_y\sigma_x) b_\sigma (q).$$ Spin–

orbit scattering has been described as $\Gamma_{\alpha\beta,\gamma\beta} = \dfrac{\hbar}{2\pi N (E_F)} \left[\dfrac{1}{\tau_0} S_{\alpha\beta}S_{\gamma\beta} - \sum \dfrac{1}{\tau_{so}^i} \sigma_{\alpha\beta}^i \sigma_{\gamma\beta}^i \right]$.

Magnetic impurity (Kondo) scattering can be described (as s–d interactions): $= \dfrac{\hbar}{2\pi N (E_F)} \left[\dfrac{1}{\tau_0} S_{\alpha\beta}S_{\gamma\beta} + \sum \dfrac{1}{\tau_s^i} \sigma_{\alpha\beta}^i \sigma_{\gamma\beta}^i \right]$, which has a close resemblance to the spin–

orbit scattering (s–p) scattering by adding a set of four levels to the SOC diagram.

The Kondo scattering is described by the spin–flip scattering on feedback effect:

$$H_{sd} = -\frac{J}{\Omega} \sum_{R_n k, k', \sigma, \sigma'} \sigma_{\sigma, \sigma'} \cdot S_n a_{k\sigma}^\dagger a_{k'\sigma'}. \tag{10.6}$$

The Kondo effect is generally described as an s–d interaction through the magnetic impurity level: $H_{imp} = \int_d \sum n_{d_s} + U_{nd\uparrow nd\downarrow}$.

The operator d_s in the term $n_{d_s} = d_s^\dagger d_s$ explains the destruction of an electron of spin s in the impurity level. In a previous report, the tunneling of an electron to or from the impurity level has been considered through the nearest carbon p_z orbital as shown in the energy diagram. The mixing between the impurity and the host has been described by a hybridized Hamiltonian term as $H_{mix} = V \sum_s d_s^+ a_s (0)$.

This is explained as the interaction between the as $s=\pm\frac{1}{2}$ as $|1\rangle$, $|2\rangle$, and the d-levels ($|5\rangle$ – $|8\rangle$) levels through the p-levels $|3\rangle$ and $|4\rangle$. In practice, we make a Kagome lattice by fusing two triangles with a common center treated as the impurity (magnetic) atom (Figure 10.10a). At the central spin qubit (qubit 3) a π-pulse is applied, which signifies the Kondo (spin) flip. Other qubits 1, 2, 4, and 5 have a trivial rotation. Therefore, an s-d-s-like configuration is observed for one such triangle. Unlike the SO interaction, the operations are not holonomic for the Kondo lattice.

10.4.4 SPIN–ORBIT COUPLING AND FOUR-LEVEL SYSTEM

Instead of a two-level system in a spin qubit, a four-level system is employed, which can describe a spin-triplet system. In the previous chapter, we have described the spin–orbit and spin–flip interactions. The qubits can perform a holonomic rotation which produced a geometric phase. These points will be addressed again in the next chapter.

We start with a linear (or a bi-linear) chain consisting of atoms in the ground state. An excited state is created by applying an electric or magnetic field. The ground state is split into two states, and they interact with each other through the exchange of fields. The ground state and the excited states represent an "s" wave and a "p" wave, respectively, which are symmetric and antisymmetric. Since the "p" level is orthogonal to the "s" level, it can be split into two orthogonal states which are aligned with the plane of the lattice.

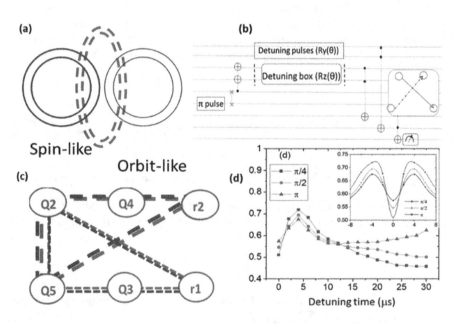

FIGURE 10.10 (a) Two spins and two orbits in a double-path system. (b) Qubit arrangements show a cross-linking of qubits assigned to two spins and one orbit. (c) SOI in a circuit consists of qubits: Two spins and two orbits. (d) Emulating SOI in a circuit consists of two spins and one orbit in one path shows a dip at the zero-detuning time for rotational angles π-$\pi/4$. The complete spectrum matches with the experimental observation of the magnetoresistance showing a minimum and two satellite peaks [39].

This is called the 2D projection of the excited state onto the lattice plane. As a result, both the ground and excited states can be split into two levels, each of which gives a total of four levels. Two ground levels and two excited levels can interact as spin–spin interactions or orbit–orbit interactions, respectively. Mixing all these levels yields SOIs. This interaction can be described by two quantum wells and each well has two levels.

The same principle must be applied to both spin qubits as well, but the difference will be that the detuning pulse must drive the qubits parallel to the plane. First, the orbit qubits interfere with the first register. This enables coupling to form between both qubits. The same is done to couple the spin qubits. Now the registers will interfere together onto the third register; this allows the coupled spins to couple with the orbit to form the complete SOC. This whole process is to be repeated for the other path. We then interfere with both registers encoding the SOC with each other. This allows interaction and energy transfer between both SOCs. The final register is measured to acquire the resistance in the system. This resistance is equivalent to the density of states in the configuration.

The atom is described as a dipole when the field interacts and resonates with the natural frequency of the atom. The atom is split into σ^+ and σ^-, and can interact with annihilation and creation operators, respectively. They can be dissociated into $\sigma_x + i\sigma_y$ and $\sigma_x - i\sigma_y$, respectively. The interaction can be expressed as $\frac{\pi}{2}(\sigma_+ b + \sigma_- b^\dagger)$, where γ is the coupling constant. In the four-level system, the spin–spin or orbit–orbit interactions, i.e., the $|gg|$ and $|ee|$ interactions, do not produce any loop or phase since this interaction is expressed as $2\sqrt{2}(\sigma_x \otimes \sigma_x + \sigma_y \otimes \sigma_y)$ where the qubits have successive rotations along the x- and y-axes. However, in the SOI, a cross-term is introduced as $\begin{pmatrix} \sigma_x^1 \otimes \sigma_y^2 - \sigma_x^2 \otimes \sigma_y^2 \\ \sigma_x^1 \otimes \sigma_y^1 - \sigma_x^2 \otimes \sigma_y^2 \end{pmatrix}$ in addition to the creation and annihilation operators. SOI creates an extra split of the energy level or a new qubit to the atom-cavity system. This operation is creating a close loop or an area as observed in holonomic operations. Such operations can produce an AB phase.

Based on the excited and ground state $|e\geq\begin{pmatrix} 0 \\ 1 \end{pmatrix}$ and $|g\geq\begin{pmatrix} 1 \\ 0 \end{pmatrix}$ for a spin-half system $\sigma_- = |ge|$ and $\sigma_+ = |eg|$. The atomic operator is given by $\sigma_z = |ee| - |gg|$. Now $\sigma_\pm = \frac{\sigma_x + i\sigma_y}{2}$ can be expressed in terms of the Pauli spin matrices σ_x and σ_y.

One of the methods of preparing a pseudospin or a spin-half involves the coupling of two spins via a capacitor that leads to the formation of a spin qubit (or a spin-triplet). Instead, we use a method that involves applying a Hadamard transformation to the spin which provides a basis for implementing pseudospins in quantum systems where the physical components cannot be manipulated. The pseudospins described as $(|1> \pm |2>)/\sqrt{2}$ (in Figure 10.10a) create a Coulomb potential by separating two charges as described in the RSO for a planar configuration. To simulate the SOI, we propose a circuit requiring 3($\times$2) qubits, 2($\times$2) phase qubits (Q2/Q3 and Q4/Q5), and a readout (r1/r2) qubit for one (or a two-path) WL system (Figure 10.10b and c). These four qubits corresponding to four levels perform SOIs. One phase qubit is declared as the fermionic "pseudospin" qubit, while the other qubit is declared as an "orbit" or bosonic spin qubit (Figure 10.10a–c). For RSO, the momentum states associated with the orbits (k, k) and spins (σ) are acting perpendicular to each other (Figure 10.10c). Through the rotation of the qubits, this configuration effectively describes magnetic and electric fields with bosonic spins (|0>, |1> states) and fermionic pseudospins ($\pm$ |1/2> states) as |3>/|4> and |2>/|1>. To create the cross-product or the momenta, a cross-link between the two scalene triangles by connecting r2 → Q4, Q2, Q5, and r1→Q3, Q5, Q2 is constructed (Figure 10.10b and c). Both closed paths are associated with the quantum backscattering or WL process; however, they are placed normal to each other like an electric and magnetic field by assigning two qubits as pseudospins. Figure 10.3c and the inset show SOI in 2-path WL that includes two spins and two orbits in each path.

To include the geometric phase in the four-level system created by four qubits we perform a holonomic operation for each qubit (Figure 10.10). In Figure 10.10b, we see that the circuit initializes the qubit and then implements the holonomic gate using a pulse schedule. Then the system is measured on the $|z\rangle$ basis. Two of them are rotating in two opposite directions and creating a closed path or area. Hence, two closed paths are created that can be combined to yield the RSO in a two-path system. An A-B phase can be created by the holonomic operation of a qubit or an $\sigma_x \otimes \sigma_y$ operation. This is equivalent to an A-B ring (or cylinder), where spin and charge are confined and a spinful excitation orbits an $hc/2e$ vortex (Figure 10.10d). An A-B ring collects the geometric phase from the closed-loop or the holonomic interactions. These are presented schematically in Figure 10.10d.

10.4.5 SIMULATING VORTEX AND TOPOLOGICAL INSULATORS BY SUPERCONDUCTING QUBITS

Conventional (s-wave) superconductors are routinely used in fabricating coupled qubits for quantum processing technologies. However, unconventional (spin-triplet) superconductors are expected to offer advantages as basic elements of topological quantum computers for complex operations since the symmetry of the order parameter (Δ) corresponds to a doubly degenerate chiral state. So far topological superconductors are made based on heavy fermions by creating layered structures; however, such a complicated multi-element hybrid system finds difficulties in device applications today. Therefore, a need for using a single system such as diamond (or other carbon structures) is revealed by doping light elements in a gas phase where the symmetry of the lattice at the GB can be broken intrinsically without using a heavy magnetic material so that all superconducting transition features can be easily understood in a simple hybrid system. Incorporation of layers or interfaces in BCS or s-wave superconductors of the bulk diamond can lead to a new class of topological superconductors which can be achieved using superconducting spin valves such as those developed by combining superconducting diamond grains separated by the well-aligned GBs. Since the microstructure of nanocrystalline diamond can be complicated, unconventional superconductivity mixed with magnetic phase has been observed. There are several possible mechanisms that can lead to ferromagnetic or antiferromagnetic states in nanostructured carbon systems and hence can create spin-triplet superconductivity channels in BNCD films; however, as of yet no experimental demonstration of an FFLO state has been reported.

10.5   SHOCKLEY MODEL SIMULATION

At the temperature where the effect of RSOC is not dominant due to short spin scattering length, the effect of WL dominates, where the forward and backward scattering paths are overlapping without breaking the symmetry in twin GBs as observed from the ultra-high resolution transmission electron microscopy (UHRTEM) microstructures (in Chapters 7 and 8 and Figure 10.7). The rings can be triangular like Kagome lattice which can also match with the microstructure of the films; however, the connectivity of the edges of the triangles is important to define the vortices. When the triangles are placed in a line and connected by tunnel barriers, the structure can be described by an A-B ring or a vortex, since the two WL orbits are overlapping and produce a Φ_0 oscillation. Due to the SOC, the rings can be separated and form a tube-like structure. In that case, the edges of the triangles can be connected to a short tube; however, the structure will appear as a bilayer.

This configuration is equivalent to the splitting of a vortex, which produces a $\Phi_0/2$ oscillation with the increase of the magnetic field. Now, if all the end points of the triangles are interconnected, RSOC configuration is developed. This configuration is developed from the enhanced spin tunneling between the adjacent lattice points, as the spin scattering length increases at lower temperatures yielding a WAL-like picture in the samples with a $\Phi_0/2$ oscillation. All these features can be simulated with a quantum simulator.

In BNCD films, we have achieved features of spin-triplet superconductivity from the ZBCP and odd parity (p-wave) of the order parameter, although most of the beyond superconductors' d-wave superconductivity has been achieved. We have also seen intrinsic Andreev π-phase from the geometric structure of the GB where two superconducting layers are connected by a JJ at an angle $\pi/2$. This structure looks very similar to layers connected by an interlayer path. However, independent of the intrinsic Andreev π reflections, an orbital phase is associated with the spin-triplet Cooper pairs and the p-wave properties of Δ. Competition between these two phases has not been well studied but can be applied to the present system to achieve an odd-frequency pairing, since a π-phase changes the parity of Δ.

Recently a model showing spontaneous Josephson π-junctions with topological superconductors has been suggested, which can be applied to the present material based on the rotation of the electron's spin caused by the SOC. From the understanding of the nature of the magnetic properties (singlet to triplet Cooper pairs) and the orbital parity of the order parameter (even function s- and d-wave to the odd function p-wave), we can understand the core structure. If the s-wave parity corresponds to the evenly distributed space, then the p-wave can match the idea of twisted space. A vortex core made of p-wave is one twister or two, which is a combination of p_x and p_y twisted space which are normal to each other and form a $p_x + ip_y$.

10.5.1   ABS AND VORTEX

Vortex works as a repulsive force at the interface (associated with chiral Δ or a p-wave Δ) that repels the spins from a certain region to some extent (like one pole of a ferromagnetic layer repels the opposite poles). The vortex structure can be explained by the Shockley model describing the surface states in the topological insulator which has already been applied to diamond lattice. Diamond structures consist of two sets of atoms, each having three nearest neighboring atoms forming a triangle on each side of atoms A and B. So, two triangles can be connected by RSOC and form a vortex line. Based on RSOC, the layers are connected by 1D vortices and created by modulating the interlayer coupling or hopping parameters. This model finds a remarkable similarity with the FFLO model since they originate from the ABS as a "soliton" state and describe a 3D topological phase on the diamond lattice. A bound state can be described by a triangle created on a curved space with the change of a π-phase which can often be seen in the diamond microstructure as triangular sections created by two different lattice planes converging toward a point, although do not belong to a plane. This mismatch of the two planes does not form a twin boundary (with a zero-phase difference) but a π shift Δ. This mismatch of planes gives rise to the helical structure or a topological phase in diamond as observed from the twisted ribbon-like structure of the GBs. This also creates a modulation of the projected width of the ribbons, which is observed from the microstructure analysis. ABS provides thermodynamic stability to the FFLO state which is associated with the modulation of Δ. The origin of the ABS is linked to the "solitonic" state which is similar to resonating valence bonds with the alternating single and double bonds. In the GB region, such a structure has already been observed in the form of a polyacetylene chain but more importantly in the hexagonal ring structures (or even triangular structures) formed by the alternation of single and double bonds. Spin and (parity) orbit are perpendicular to each other, and they have opposite or complementary properties (like singlet – odd or triplet – even) odd frequency. Odd frequency corresponds to an ABS from the Andreev reflections. This also allows the transmission of Cooper pairs through the ABS. So, the transmissions ($|t>$) and the reflections ($|r>$) can form a superposition of states ($(|t> + |r>)/\sqrt{2}$) like a beam splitter, which can split the photon followed by recombination. ABS works like a beam splitter, which can have both reflection or transmission and works as a Hadamard gate.

Earlier, the spin centers were defined through tunneling and superposition of states, which can be applied to small molecules and NV centers. The energy-level splitting of an NV center will be ideal for describing the SOI. An NV-center qubit has not only a spin-triplet ground state but also orbital-degenerate Zeeman sublevels, which can be explained by an excited-doublet form level system. A boron atom also forms a spin-triplet structure in a diamond lattice as observed from spectroscopic analysis; hence, such a structure can be simulated here.

10.5.2 Topological Superconductor

From the unusual electronic transport properties, we reveal that they can be correlated with the microstructure of HBDDF and are also different from single crystal diamonds (see Chapters 7 and 8, and Figure 10.7). Hence we propose a model to explain the topological features which are developed based on the Fu–Kane model and the Schockley model of a topological insulator (Figure 10.11).

According to the Fu–Kane model, the deformation of a diamond lattice through the incorporation of boron atoms in the (111) direction may lead to a topological insulator state of the diamond lattice. It is exactly (111) direction reconstruction of

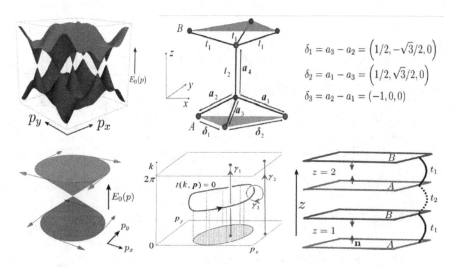

$$\delta_1 = a_3 - a_2 = \left(1/2, -\sqrt{3}/2, 0\right)$$

$$\delta_2 = a_1 - a_3 = \left(1/2, \sqrt{3}/2, 0\right)$$

$$\delta_3 = a_2 - a_1 = (-1, 0, 0)$$

FIGURE 10.11 The Rashba vector in the spin–orbit coupling is staggered on the sublattices. The bold curve in the middle corresponds to $t(k; p) = 0$, while γ_1, γ_2, and γ_3 are contours around which one can calculate the phase winding. The shaded area in the (p_x, p_y)-plane is where the Shockley criterion is satisfied; hence, surface states exist in this region. Sketch of the spin structure of the surface state wavefunctions superpositioned on the energy spectrum. The starting point of the arrows are the momenta, while the arrow indicates the direction of the spin. Sketch of the diamond lattice structure, with the hopping amplitudes. The A and B are sublattice planes. Particle–hole symmetric energy spectrum for the in-plane Hamiltonian as a function of in-plane momenta p_x and p_y [6].

a diamond lattice by boron doping. Based on those observations we assume that the type of mentioned interface might be s-wave/topological insulator state. If it is a reasonable hypothesis then, as it was shown theoretically, such types of interfaces can contribute to the overall superconductivity but with a p-wave order parameter due to a specific interface scattering. In other words, the boron-doped diamond should demonstrate p-wave superconductor phenomena in addition to the common s-wave superconductor observations.

We explain the topological transformations by assuming that the topology of the initial pure diamond Fermi surface is strongly affected by the boron doping process and that finally leads us to a modified Fermi surface, corresponding to the 3D topological insulator state. The problem is to establish a particular type of boron-induced deformation and, subsequently, make a connection with the corresponding topological phases. However, on a pure diamond surface, a dimerized atomic structure of the diamond (111) direction was proposed as follows (Figure 10.11).

10.5.3 GEOMETRIC PHASE ACQUIRED BY THE GB

From the microstructure and electrical transport, we reveal a superlattice-like structure which is probably created from the displacement of (111) plane in diamond induced by boron incorporation; however, a theoretical model explains all these microstructural features and electrical properties have not been available. Due to the reconstruction of the diamond surface described as Pandey chain, the (111) plane is

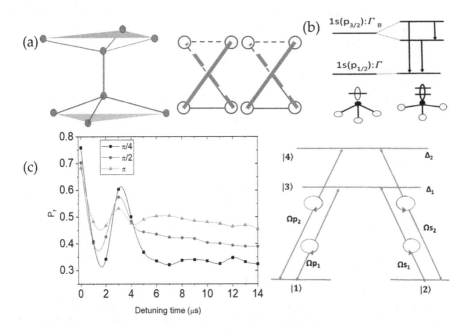

FIGURE 10.12 (a) Shockley model of topological phase. Qubit arrangements show two connected triangles as closed loops and describe spin–orbit interactions. (b) The boron acceptor can be present in the GB, which removes the degeneracy of the $1s(p_{3/2})$ spin–orbit split level, hence creating a strong splitting of the energy levels. A four-level energy ($|1\rangle$ – $|4\rangle$) diagram of the spin triplets of a boron center showing the Rabi frequencies ($\Omega_{s,p}$) and the detuning is $\Delta_1 = \Delta_2 = \Delta$. (c) Variation of P_r (or magnetoresistance) with detuning time (magnetic field) when spin–orbit is included. The qubits are also rotated in different angles from $\pi/4$ to π to show the intensity variation of the side peak, which is a signature of RSOC [54].

formed which has some special properties. This has not only a very flat band (close to graphene layers) but also contains a large density of carriers (electrons) resulting in strong electron–electron interactions; hence superconductivity and magnetism can be possible in this particular (111) layer which leads to spin-triplet superconductivity as claimed here. The (111) plane forms a triangle since one bond is absent from the three nearest neighbors on a trigonal (planar) structure (Figure 10.12) instead of a regular tetragonal structure of a diamond crystal. The atoms on the (111) surface form a 2×1 superstructure and a zigzag chain after the (111) plane is displaced by the boron. Zak phase from band structure or dimer state (including two hopping parameters and two order parameters) can explain the origin of the topological features in superconducting diamond films. Boron atoms further introduce a topological phase to the diamond structure with more deformation along (111) planes yielding an extremely flat energy band and an inversion symmetry (in Chapters 7 and 8 and Figure 10.7). Based on the previous model we propose a structure that can explain the MR due to SOC. The carbon atoms are arranged as a set of qubits on a quantum simulator (Figure 10.12a–c). In diamonds, boron atoms form a spin–orbit state verified by Raman spectroscopy (see Chapters 7 and 8). Also, the formation of a new acceptor level (of approximately 37 meV) associated with the B-C sheets was suggested which can be mixed with the spin–orbit split diamond valence band. This can be shown in a four-level energy diagram in Figure 10.11. At the temperature where the effect of RSOC is not dominant due to short spin scattering length, the effect of WL dominates where the forward and the backward scattering paths are overlapping without breaking the symmetry in twin GBs as observed from the UHRTEM microstructures (in Chapters 7 and 8 and Figure 10.7). The rings can be triangular like Kagome lattice which can also match the microstructure of the films; however, the connectivity of the edges of the triangles is important to define the vortices (Figure 10.7). This model consists of two complementary planes connected by a line that represents a vortex. This is formed by alternating bond lengths like a polyacetylene chain. However, this model can be applied to a 3D lattice, like a diamond structure. When the triangles are placed in a line and connected by tunnel barriers, the structure can be described by an Aharonov–Bohm ring or a vortex since the two WL orbits are overlapping and produce a MR oscillation. Due to the SOC, the rings can be separated and form a tube-like structure. In that case, the edges of the triangles can be connected to a tube; however, remain almost flat like a bilayer. In Figure 10.11, the return probability (P_r) of the electrons with detuning time can be compared to the MR with the **B** field described in Chapters 7 and 8. The variation of P_r shows an initial decrease at low fields up to a certain **B** field followed by a side peak. This is a signature of WAL and oscillations due to the RSOC. We have performed a holonomic operation of the qubit rotation to capture a geometric phase and simulate RSOC which possesses topological features. Details of the quantum simulation are given in references 32 and 34.

- **Main problem discussed in this chapter.**
  Quantum simulation of simple lattice structure has been demonstrated. The quantum wells are filled with disordered carbon, which can be maximally entangled. The maximally entangled structure can be compared to the theoretical model of a black hole.
- **What has been achieved?**
  Understanding of quantum transport properties has been improved. Tunneling between two disordered carbon quantum wells can be explained by a wormhole model.
- **What has not been achieved?**
  The model structure of amorphous carbon has not been simulated yet. Although quantum simulators are used for ordered structures and maximally entangled ones, it is very difficult to handle structurally disordered systems.

BIBLIOGRAPHY

1. I. Khivrich and S. Ilani, Atomic-like charge qubit in a carbon nanotube enabling electric and magnetic field nano-sensing, *Nature Communications* 11, 2299 (2020).
2. T. Cubaynes, M. R. Delbecq, M. C. Dartiailh, R. Assouly, M. M. Desjardins, L. C. Contamin, L. E. Bruhat, Z. Leghtas, F. Mallet, A. Cottet, et al., Highly coherent spin states in carbon nanotubes coupled to cavity photons, *npj Quantum Information* 5, 47 (2019).
3. D. V. Bulaev, B. Trauzettel, and D. Loss, Spin-orbit interaction and anomalous spin relaxation in carbon nanotube quantum dots, *Physical Review B* 77, 235301 (2008).

4. C12 Quantum Electronics. Available online: https://www.c12qe.com

5. J. Moser, A. Eichler, J. Güttinger, M. I. Dykman, and A. Bachtold, Nanotube mechanical resonators with quality factors of up to 5 million, *Nature Nanotechnology* 9, 1007–1011 (2014).

6. F. Pistolesi, A. N. Cleland, and A. Bachtold, Proposal for a nanomechanical qubit, *Physical Review X* 11, 031027 (2021).

7. X. He, N. F. Hartmann, X. Ma, Y. Kim, R. Ihly, J. L. Blackburn, W. Gao, J. Kono, Y. Yomogida, A. Hirano, et al., Tunable room-temperature single-photon emission at telecom wavelengths from sp3 defects in carbon nanotubes, *Nature Photonics* 11, 577–582 (2017).

8. W. Gao, X. Li, M. Bamba, and J. Kono, Continuous transition between weak and ultrastrong coupling through exceptional points in carbon nanotube microcavity exciton-polaritons, *Nature Photonics* 12, 362–367 (2018).

9. M. Mergenthaler, A. Nersisyan, A. Patterson, et al., Circuit quantum electrodynamics with carbon-nanotube-based superconducting quantum circuits, *Physical Review Applied* 15, 064050 (2021).

10. J.S. Chen, K.J. Trerayapiwat, L. Sun, et al. Long-lived electronic spin qubits in single-walled carbon nanotubes. *Nature Communications* 14, 848 (2023).

11. J. I.-J. Wang, D. Rodan-Legrain, L. Bretheau, et al., Coherent control of a hybrid superconducting circuit made with graphene-based van der Waals heterostructures, *Nature Nanotechnology* 14, 120–125 (2019).

12. L. Banszerus, S. Möller, K. Hecker, et al., Particle–hole symmetry protects spin-valley blockade in graphene quantum dots, *Nature* 618, 51–56 (2023).

13. K. Hecker, L. Banszerus, A. Schäpers, et al. Coherent charge oscillations in a bilayer graphene double quantum dot, *Nature Communications* 14, 7911 (2023).

14. L. Banszerus, Tunable interdot coupling in few-electron bilayer graphene double quantum dots, *Applied Physics Letters* 118, 103101 (2021).

15. C. Tong et al., Pauli blockade of tunable two-electron spin and valley states in graphene quantum dots, *Physical Review Letters* 128, 067702 (2022).

16. H. C. Park et al., A strain-engineered graphene qubit in a nanobubble, *Quantum Science and Technology* 8, 025012 (2023).

17. M. A. Nielsen and I. L. Chuang, *Quantum Computation and Quantum Information*, Cambridge University Press, Cambridge, (2011).

18. R. Loredo, *Learn Quantum Computing with Python and IBM Quantum Experience: A Hands-on Introduction to Quantum Computing and Writing Your Own Quantum Programs with Python*, Packt Publishing, Birmingham (2020).

19. S. Russell and P. Norvig, *Artificial Intelligence: A Modern Approach*, Pearson, London, (2020).

20. S. Ansari, *Building Computer Vision Applications Using Artificial Neural Networks: With Step-by-Step Examples in OpenCV and TensorFlow with Python*, Apress, Springer Science+Business Media LLC, New York City, (2020).

21. L. Fu and C. L. Kane, Topological insulators with inversion symmetry, *Physical Review B* 76, 045302 (2007); L. Fu, C. L. Kane, and E. J. Mele, Topological insulators in three dimensions, *Physical Review Letters* 98, 106803 (2007).

22. S. S. Pershoguba and V. M. Yakovenko, Shockley model description of surface states in topological insulators, *Physical Review B* 86, 075304 (2012).

23. K. C. Pandey, New dimerized-chain model for the reconstruction of the diamond (111)-(2x1) surface, *Physical Review B*, 25, 4338 (1982); B. Pamuk and M. Calandra, Competition between exchange-driven dimerization and magnetism in diamond (111), *Physical Review B* 99, 155303 (2019); B. Pamuk, J. Baima, F. Mauri, and M. Calandra, Magnetic gap opening in rhombohedral-stacked multilayer graphene from first principles, *Physical Review B* 95, 075422 (2019).

24. J. Jak, Berry's phase for energy bands in solids, *Physical Review Letters* 62, 2747 (1989).

25. T. Mizoghichi and Y. Hatsugai, Type-III Dirac cones from degenerate directionally flat bands: Viewpoint from molecular-orbital representation, *Journal of the Physical Society of Japan* **89**, 103704 (2020).
26. Y. Chen, P. Roushan, Sank, D., Neill, C., Lucero, E., Mariantoni, M., Barends, R., Chairo, B., Kelly, J., Megrant, A., Mutus, J. Y. P., O'Malley, J. J., Vainsencher, A., Wenner, J., White, T. C., Yin, Y., Cleland, and A. N., J. M. Martinis, Emulating weak localization using a solid-state quantum circuit, *Nature Communications* **5**, 5184 (2014).
27. D. Litinski, M. S. Kesselring, J. Eisert, and F. von Oppen, Combining topological hardware and topological software: Color-code quantum computing with topological superconductor networks, *Physical Review X* **7**, 031048 (2017).
28. M. Bunruangses, A. E. Arumona, P. Youplao, N. Pornsuwancharoen, K. Ray, and P. Yupapin, Modeling of a superconducting sensor with microring-embedded gold-island space-time control, *Journal of Computational Electronics* **19**, 1678–1684 (2020).
29. S. Sachdev, Strange metals and the AdS/CFT correspondence, *Journal of Statistical Mechanics* **101**, P11022 (2010); S. Sachdev, Holographic metals and the fractionalized fermi liquid, *Physical Review Letters* **105**, 151602 (2010).
30. A. Altland, D. Bagrets, and A. Kamenev, Sachdev-ye-Kitaev non-fermi-liquid correlations in nanoscopic quantum transport, *Physical Review Letters* **123**, 226801 (2019).
31. J. Maldacena and X. L. Qi, Eternal traversable wormhole, *arXiv:1804.00491v3*; J. Maldacena, D. Stanford, and Z. Yang, Diving into traversable wormholes, *Fortschritte der Physik* **65**, 1700034 (2017).
32. I. Danshita, M. Hanada, and M. Tezuka, Creating and probing the Sachdev-Ye-Kitaev model with ultracold gases: Towards experimental studies of quantum gravity, *Progress of Theoretical and Experimental Physics* **2017**, 083I01 (2017); I. Danshita, M. Hanada, and M. Tezuka, How to make a quantum black hole with ultra-cold gases, *arXiv:1709.07189v1*.
33. Z. Luo, Y. Z. You, J. Li, C. M. Jian, D. Lu, C. Xu, B. Zeng, and R. Laflamme, Quantum simulation of the non-fermi-liquid state of Sachdev-Ye-Kitaev model, *npj Quantum Information* **5**, 53 (2019).
34. M. Franz and M. Rozali, Mimicking black hole event horizons in atomic and solid-state systems, *Nature Reviews Materials* **3**, 491–501 (2018).
35. A. Chen, R. Ilan, F. de Juan, D. I. Pikulin, and M. Franz, Quantum holography in a graphene flake with an irregular boundary, *Physical Review Letters* **121**, 036403 (2018).
36. Dharmaraj, B. K. Behera, and P. K. Panigrahi, Simulation model for complexity in black holes and demonstration of power of one clean qubit using IBM QX, *Quantum Studies: Mathematics and Foundations* **8**, 167–178 (2021).
37. D. Jafferis, A. Zlokapa, J. D. Lykken, D. K. Kolchmeyer, S. I. Davis, N. Lauk, H. Neven, and M. Spiropulu, Traversable wormhole dynamics on a quantum processor, *Nature* **612**, 51–55 (2022) DOI: 10.1038/s41586-022-05424-3.
38. D. Mahony and S. Bhattacharyya, Realizing highly entangled states in asymmetrically coupled three NV centers at room temperature, *Applied Physics Letters* **118**, 204004 (2021) DOI: 10.1063/5.0043334.
39. S. Bhattacharyya and S. Bhattacharyya, Demonstrating geometric phase acquisition in multi-path tunnel systems using a near-term quantum computer, *Journal of Applied Physics* **130**, 034901 (2021). DOI: 10.1063/5.0049728.
40. F. Mazhandu, K. Mathieson, C. Coleman, and S. Bhattacharyya, Experimental simulation of hybrid quantum systems and entanglement on a quantum computer, *Applied Physics Letters* **115**, 233501 (2019).
41. S. Bhattacharyya and S. Bhattacharyya, Holonomic control of a three-qubits system in an NV center using a near-term quantum computer, *Entropy* **22**, 1593 (2022).
42. K. Mathieson and S. Bhattacharyya, Hybrid spin-superconducting quantum circuit mediated by deterministically prepared entangled photonic states, *AIP Advances* **9**, 115111 (2019).

43. S. Bhattacharyya and S. Bhattacharyya, Quantum interference correction to coupled SYK model: Simulation of a traversable wormhole, (unpublished).

44. I. M. Georgescu, S. Ashhab, and F. Nori, Quantum Simulation, *Reviews of Modern Physics* **86**, 153 (2014); I. Buluta and F. Nori, Quantum Simulators, *Science* **236**, 108 (2009).

45. Z. L. Xiang, S. Ashhab, J. Q. You, and F. Nori, Hybrid quantum circuits: Superconducting circuits interacting with other quantum systems, *Reviews of Modern Physics* **85**, 623 (2013); J. Q. You and F. Nori, Atomic physics and quantum optics using superconducting circuits, *Nature* **474**, 589 (2011); J. Q. You and F. Nori, Superconducting circuits and quantum information, *Physics Today* **58**, 42 (2005); X. Gu, A. F. Kockum, A. Miranowicz, Y. X. Liu, and F. Nori, Microwave photonics with superconducting quantum circuits, *Physics Reports* **718–719**, 1 (2017).

46. G. B. Lesovik, I. A. Sadovskyy, M. V. Suslov, A. V. Lebedev, and V. M. Vinokur, Arrow of time and its reversal on the IBM quantum computer, *Scientific Reports* **9**, 4396 (2019).

47. L. Lamata, A. Parra-Rodriguez, M. Sanz, and E. Solano, Digital-analog quantum simulations with superconducting circuits, *Advances in Physics: X* 3, 1457981 (2018).

48. K. Xu, J. J. Chen, Y. Zeng, Y. R. Zhang, C. Song, W. Liu, Q. Guo, P. Zhang, D. Xu, H. Deng, K. Huang, H. Wang, X. Zhu, D. Zheng, and H. Fan, Emulating many-body localization with a superconducting quantum processor, *Physical Review Letters* **120**, 050507 (2018).

49. P. Roushan et al., Chiral ground-state currents of interacting photons in a synthetic magnetic field, *Nature Physics* **13**, 146 (2017); P. Roushan et al., Observation of topological transitions in interacting quantum circuits, *Nature* **515**, 241 (2014).

50. U. L. Heras, A. Mezzacapo, L. Lamata, S. Filipp, A. Wallraff, and E. Solano, Digital quantum simulation of spin systems in superconducting circuits, *Physical Review Letters* **112**, 200501 (2014).

51. K. Seo and L. Tian, Quantum phase transition in a multiconnected superconducting Jaynes-Cummings lattice, *Physical Review B* **91**, 195439 (2015).

52. J. M. Reiner, M. Marthaler, J. Braumüller, M. Weides, and G. Schön, Emulating the one-dimensional Fermi-Hubbard model by a double chain of qubits, *Physical Review A* **94**, 032338 (2016).

53. L. Lamata, A. Parra-Rodriguez, M. Sanz, and E. Solano, Digital-analog quantum simulations with superconducting circuits, *Advances in Physics: X* **3**, 1457981 (2018).

54. B. Rost, B. Jones, M. Vyushkova, A. Ali, C. Cullip, A. Vyushkov, and J. Nabrzyski, Noisy simulation of quantum beats in radical pairs on a quantum computer, *arXiv:2001.00794v1*.

55. S. Endo, I. Kurata, and Y. O. Nakagawa, Calculation of the Green's function on near-term quantum computers, *Physical Review Research* **2**, 033281 (2020).

56. A. Baydin, F. Tay, J. Fan, M. Manjappa, W. Gao, and J. Kono, Carbon nanotube devices for quantum technology, *Materials* **15**, 1535 (2022).

57. A. Calafell, J. D. Cox, M. Radonjić, et al., Quantum computing with graphene plasmons, *npj Quantum Information* **5**, 37 (2019).

58. M. Faccin et al., Community detection in quantum complex networks, *Physical Review X* **4**, 041012 (2014).

59. S. Mandal et al. The diamond superconducting quantum interference device, *ACS Nano* **5.9**, 7144–7148 (2011).

60. A. J. Kerman, Flux–charge duality and topological quantum phase fluctuations in quasi-one-dimensional superconductors, *New Journal of Physics* **15**, 105017 (2013).

61. S. Bhattacharyya, D. Mtsuko, C. Allen, and C. Coleman, Effects of Rashba-spin-orbit coupling on superconducting boron-doped nanocrystalline diamond films: Evidence of interfacial triplet superconductivity, *New Journal of Physics* **22**, 093039 (2020).

62. S. Bhattacharyya, Microstructure and anisotropic order parameter of boron-doped nanocrystalline diamond films, *Crystals* **12**, 1031 (2022).

11 Conclusion
New Directions

In this book, we explain the basics of quantum transport, magnetic properties, and superconductivity observed in various carbon systems and compare them with advanced quantum features of carbon as a unique material. We discover the multi-functional nature of carbon as a synthesis of beauty, strength, and intelligence particularly in the topological phase of this material. The topological features are seen as a battle between a localized state and several delocalized states (attractive and repulsive forces). Philosophically, this is described as the interactions between two zeros. The central one is fixed as an absolute (or a great) zero which is surrounded by many small zeros. In materials, they are described as a vortex state in the spin liquids. The orbit-like zero state interacts with two (or many) spin-like states and forms a spin-triplet structure. Therefore, interactions of two (spin) qubits can describe the properties of the localized and delocalized states. It can also be described as the interactions between two quantum dots that create space-time distortion. The space between the localized states is distorted, which plays the most crucial role in determining the electronic structures of a material.

An atom consisting of an electron spin and a nuclear spin can be considered a qubit having a two-level system. Two atoms can form a four-level system. The electronic structure has a symmetric (ground) s-state and an antisymmetric part or an excited p-state which can be considered a two-level (a qubit) system. Two s electrons and two p electrons in a carbon atom consisting of can hybridize which is similar to the superposition of quantum states in the qubits. The superposition of 2s and 2p electrons can also yield a planar structure of graphite where one carbon atom is linked with three neighboring carbon atoms. In **Chapter 1**, we see a linear chain, one single bond (p-state, orbit-like, spin 1) alternates with a double bond (s-like, spin zero). A unit of the single and double bond can be considered a two-level system or a qubit. Therefore, a resonating structure can be described as a multi-qubit system.

In addition to the resonating structure of carbon, this unit of ring structure is considered to be the building block of any carbon structure (as discussed in **Chapter 2**). Space-time distortion has been discussed in **Chapter 3** in the form of gate bias, quantum dots, and quantum wells. The wells are separated by a barrier that resembles a spin-triplet state or the vortices are connected through a tunnel barrier. It is a four-level system of qubits. In **Chapter 4**, we introduce spins to the defect centers and study spin–spin (electron–electron) interactions. The spins associated with the magnetic impurity centers interact with the spins of the conduction electrons. Hence, we consider a spin-triplet state responsible for the transition of the conductivity (with temperatures).

DOI: 10.1201/9781003316411-11

Regarding the sp^2-bonded ring structure, the pi-orbitals stick out of the plane of the ring and act as a magnetic dipole although the magnetic moment is very weak. Carbon being a light element cannot bend space, and spin-orbit coupling is found to be significantly weak in a planar hexagonal structure. Through the breaking of the symmetry, topological defects can be introduced in the planar structures particularly, in the form of pentagons and heptagons. In the five-fold symmetry, a curved structure is produced. The bending of space can distort space time and acts as an orbit surrounded by hexagons as spins. The unit consists of a pentagon and (two) hexagons can be considered as a spin qubit (one orbit and two spins).

In **Chapters 1–4**, we have not considered disorder (or randomness) in the structure which are introduced in **Chapter 5** both in the diamond and graphite phases. The quantum wells are filled with disordered carbon which can be maximally entangled. **Chapter 6** shows quantum transport in disordered carbon films and nanostructures. The weak localization effect can be created with two orbits (rotating in opposite directions), whereas the weak antilocalization can be established with three orbits (associated with one orbit and two spins). The interactions between zeros can explain spin-orbit couplings and the Kondo effect.

By introducing a dopant, the symmetry of a structure can be broken and a spin-qubit or a spin-triplet state evolves. A resonating valence band structure was described as the SSH model (particularly, in conducting polymers) and the FFLO-type superconductivity, as seen in **Chapters 7 and 8**. Following the discussion in the previous chapters, superposition can be obtained in several ways. Four electrons can be distributed equally in space and connect the neighboring carbon atoms and form an sp^3-bonded (diamond) structure. The electric fields originating from the strain of the bonds are distributed all over the space equally. However, a vacancy center in the diamond structure can break the symmetry and create a spin-triplet state in association with a foreign atom such as nitrogen. This structure creates a magnetic dipole and a spin qubit which are discussed in **Chapter 9**. In addition, space-time distortion in the maximally entangled structure can be compared with the theoretical model of a black hole. Tunneling between two disordered carbon quantum wells can be explained by a wormhole model in **Chapter 10**.

I attempted to include some transport features of carbon materials in this book however, I believe this is a very little contribution to the field which is truly very vast like an ocean. The world is changing fast both technologically and culturally. During the time of compiling the book, our daily life was severely affected by the COVID-19 pandemic. Researchers tried to develop carbon-based detectors to find the virus followed by the discovery of the COVID-19 vaccine. In search of the origin of life, James Webb telescope was launched and we received some fascinating images of deep space. Plants were grown in lunar soil. The first complete (gapless) human genome sequence was published.

Scientists found phosphorous in Saturn's moons suggesting that it may be capable of supporting life.

In cosmology, following the detection of gravitational waves, images of Blackhole were published. Quantum computer research has progressed significantly and improved with the claim of more than one thousand qubits in a chip which can be useful to solve cosmological and biological problems.

After many years of graphene research, there was a claim to produce the first functional semiconductor made from graphene. This will improve the bio-functionality of carbon materials and human brain research. Over many years, researchers have been trying to demonstrate room temperature superconductivity in materials having complex structures and recently this attempt became phenomenal. Unconventional superconductors can be very useful in making topological qubits. Can carbon complement different areas of research from cosmology to bioscience? Our interest in learning new things has increased hence several unsolved problems are attracting us. I think we have to review our progress and failures to improve our thought processes in addition to the extensive use of artificial intelligence.

The following problems in carbon should be discussed:

Chapter 1. Why carbon has so many allotropes in all dimensions from 0D to 3D? We know carbon cycles and the transformation of energy associated with carbon structures. We don't know why and how to control the cycle and the transformation of energy. For example, the mechanism for photosynthesis (classical or quantum) is not known. A quantum processor will be useful to simulate the process.

Chapter 2. In general, carbon nanostructures can be synthesized, but it is difficult to control the quality of the samples that complies with microelectronic devices. In most cases, we do not know the growth mechanism. Moreover, the (substitutional) doping mechanism of carbon nanostructures remains unknown since it is dominated by defects.

Chapter 3. Understanding dimensionality in carbon is challenging due to the folding and bending of bonds which introduces an extra dimension to the system. For nanotubes, we have one dimension and the chirality of the tubes. We do not know how this extra dimension can control the electronic properties of nanotubes. Quantum simulations of this complicated structure should be performed.

Chapter 4. One of the most important questions arises whether pure carbon structures can be ferromagnetic. Defects in carbon structures can change the properties of carbon, but we do not know the structure of the defective regions. We cannot control the defect structures as well as ferromagnetic doping.

Chapter 5. Multi-layer superlattice structures can be prepared artificially which can show prominent resonant tunnel conductance. In many cases, we cannot control the resonance peaks (both position and width) through synthesis and engineering. Raman spectra of disordered carbon have not been understood fully since a proper theory is unavailable.

Chapter 6. Nitrogen-incorporated nanocrystalline diamonds showed high metallic transport through the atomically thin grain boundaries. The conduction mechanism is not clearly understood since the structure of the grain boundaries has not been established experimentally.

Chapter 7. The observed superconducting properties in boron-doped diamonds are mixed with unconventional features due to defects or grain boundaries. The superconducting pairing mechanism is unknown.

Chapter 8. The origin of recently claimed spin-triplet superconductivity and other magnetic properties of boron-doped carbon remains unresolved. The features of zero-bias conductance peaks cannot be controlled. Hence, carbon-based qubits have remained elusive.

Chapter 9. Many defect centers of diamonds are not studied properly. The nitrogen-vacancy center in diamond showed very interesting spin-qubit properties; however, strong coupling between the centers is found to be very challenging.

Chapter 10. Although quantum simulators are used for ordered structures and maximally entangled ones, it is very difficult to handle structurally disordered systems.

Index

1d system 76, 170
2DEG 78, 86, 88, 170, 179, 204
3DSL 168, 169, 171, 172, 174

ABS 202, 203, 206, 209, 211, 233, 236–239, 291, 292
absorption 34, 41, 137, 141, 180, 248, 281
activated carbon 22
AFM 19, 152, 218, 219, 221, 237
Agni 3
Aharonov–Bohm 100, 124, 185, 295
allotropes 11, 24, 115, 130
amorphous carbon 11, 22, 26, 49, 58, 60, 101, 135, 137, 138, 139, 141, 143, 157, 175, 176, 180, 198, 226, 295
Andreev reflection 202, 235
angular momentum 96, 120, 196, 200, 238, 282
anomalous 76, 84, 114, 117, 118, 122, 192, 207, 211, 229, 230, 279
antiparallel 96, 97, 117, 200, 239
antivortex 188, 203, 207, 211, 217, 227, 228–230, 239
arc discharge 23, 31, 35, 36, 41, 46
artificial atoms 181, 262, 283
artificial intelligence 126, 301
astronomy 18, 126
asymmetry 19, 113, 114, 133, 194, 204, 208, 236, 240, 255

backscattering 70, 98, 158–164, 179, 190, 200, 209, 245, 284, 285, 290
ballistic transport 74, 76–80, 275
band(s) 9, 60, 71, 83, 103, 117, 121, 139, 152, 191, 194, 194, 204, 208, 221, 224, 240, 241, 273
band gap 8–10, 13, 24, 35, 60, 67, 74, 80, 91, 110, 130, 136, 141, 145, 147, 151, 163, 164, 194, 240, 245, 276, 283
barriers 35, 43, 70, 74, 79, 82, 90, 109–112, 132, 139, 145, 179, 216, 287, 291, 295
BCS theory 116, 188, 189, 231
Berry phase 85, 160, 161, 208, 240, 256, 257
biomolecules 20, 278
BKT transition 190, 192, 203, 207, 211, 217, 218, 222, 223, 227, 228, 229, 231
blackhole 5, 300
Bloch sphere 251, 257, 259, 283, 287
BNCD 116, 196, 202, 203, 207, 209, 211, 222, 224, 228, 235, 238, 279, 282, 290, 291

boron-doped diamonds 100, 226, 279
Bose 218, 228, 230, 263
boson 218, 239, 263
bound states 196, 198, 201, 202, 216, 232, 261, 264, 280

carbon dioxide 2–4, 6, 10, 13
carbon dots 21, 23, 24
carbon nanotubes 10, 11, 17, 19, 38, 43, 53, 55, 64, 66–70, 76, 104, 196, 198, 222, 223, 228, 265, 272–275
carbon spheres 23, 30
CDW 159
charcoal 3, 11, 21
charges 21, 49, 52, 74, 216, 217, 277
chemical vapor deposition 12, 31, 32, 36, 45, 57
chirality 10, 16–18, 21, 41, 46, 47–53, 83, 84, 91, 161, 273, 277
CNOT 284
computation 256, 261, 262, 264, 265, 273
condensate 192, 217, 283
condensation 218, 220
conductivity 66, 77, 80, 83, 85, 87, 105, 111–113, 122, 137, 139, 143, 157, 158, 162, 163–181, 190, 191, 203, 217
cooper pair 188, 190, 201, 208, 224, 225, 230, 270, 272
correlated 8, 15, 18, 60, 76, 98, 121, 135, 143–145, 148, 151, 152, 166, 170, 187, 188, 191, 224, 292
coulomb blockade 65, 70, 82, 90, 105, 107, 114, 187, 217
current-voltage 70, 110, 112, 142, 144, 149
curvature 14, 18, 43, 51, 67, 90, 132, 198, 199, 240
curved graphene 17, 130, 199
cycle 3, 4, 26, 27, 46, 256

dark states 252, 260, 261, 262
de Broglie 136
decoherence 98, 107, 170, 245, 249–255, 257, 261, 268, 273
defects 9, 16, 34, 35, 38, 39, 41, 60, 64, 75, 86, 102, 115, 120, 127, 128, 129, 133, 150, 198, 199, 211, 227, 228, 245, 246, 254, 256, 275, 279
delocalized 137, 157, 158, 173, 174–180, 217, 221, 263
delocalized transport 173, 180
device fabrication 85, 88, 105

DFS 169–172, 174
diamond(s) 3, 10, 13, 14, 54, 56, 57, 90, 116, 117,
    126, 181, 191, 192, 208, 211, 227, 240,
    244, 246, 248, 257, 262, 264, 266,
    272, 278, 292, 294
diamond-like carbon 26, 54
diffusive transport 75, 157, 168, 176
dimensions 6, 12, 24, 25, 26, 65, 75, 76, 77, 104,
    128, 136, 217
dipole 5, 49–51, 52, 100, 289, 300
Dirac cones 67, 83, 85, 129, 161, 194, 230, 240
disordered carbon 10, 52, 101, 109, 137, 139, 145,
    150, 152, 153, 168, 173, 176, 180, 181
DLC 26, 54, 60, 79, 137, 138, 142, 144, 150, 157, 158
DNA 8, 18–21, 266, 267, 277, 278
doping 9, 10, 13, 14, 24, 33, 35, 39–41, 54, 57, 59,
    60, 74, 80, 86, 100, 101, 116, 120, 121,
    123, 144, 145, 150, 157, 168, 169, 171,
    181, 187, 190, 191, 194, 203, 217, 220,
    264, 282, 293
DoS 68–70, 97, 132, 148, 157, 163, 164, 173, 176,
    177, 178, 180, 222
doublet 191, 200, 202, 237, 238
duality 6, 7, 10, 21, 216–218, 232, 236, 245, 272,
    280
DWNT 37, 53, 55, 105

electric field 18, 24, 27, 39, 49, 79, 80, 129, 159,
    180, 196, 199, 201, 238, 246, 248, 273,
    274, 290
electrodynamics 84, 270, 275
electronic properties 7, 11, 19, 24, 38, 64, 66,
    68, 70, 79, 81, 90, 116, 128, 129, 143,
    157, 180
electronic transport 38, 64, 70, 97, 141, 157, 158,
    172, 181, 196, 292
encapsulation 39, 43, 45, 47, 52, 108
energy diagram 153, 258, 260, 287, 295
energy gap 9, 10, 71, 79, 85, 151, 180, 188, 189,
    216, 263, 283
energy-level 245, 248, 294
entangled states 254
entanglement 69, 118, 248, 251, 252, 254–255,
    266, 273, 284
equation 85, 89, 109, 111, 163–167, 210, 238,
    259, 287
equilibrium 57, 128, 218
evolution 45, 58, 76, 110, 172, 205, 229, 231,
    253–256, 258, 260
exclusion principle 188, 189, 237, 265

Fano 113
Faraday 4, 216, 217
feedback 24, 287
Fermi level 9, 39, 60, 71–75, 92, 97, 117, 118, 129,
    130, 132, 139, 143, 149, 158, 163, 164,
    178, 188, 222, 226, 235

fermions 8, 76, 81, 83, 84, 85, 87, 88, 99, 131, 161,
    194, 218, 240, 283, 290
ferromagnetism 100, 115–117, 118, 121
FET 70, 76, 79, 119
FFLO 27, 211, 237, 238, 239, 290–292
field effect transistor 70
filled 39, 81, 109, 110, 113, 150, 153, 295
fire 1, 3, 56
flux 159, 187, 189, 210, 218, 235, 246, 253–254,
    256, 262–265, 272, 281, 286
force 7, 11, 49, 51, 126, 202, 291
four levels 287–289
fullerenes 1, 10, 17, 35, 36, 42, 43, 101, 104, 115,
    126, 187, 221, 222
functionalization 38–39, 120–124
FWHM 143, 149, 232

Gd-DTPA 38, 107, 109, 110, 114
geometric phase 158, 159, 180, 256, 257, 258,
    260, 261
geometry 6, 7, 50, 64, 74, 81, 142, 160, 171, 190,
    195, 196, 198, 199, 222, 257, 279
GMR 95–97, 105
GNR 117, 121
graphene 10–12, 15–17, 31–35, 40–42, 46–48,
    51, 57, 79–90, 102–104, 111, 115,
    120, 121, 126–132, 161, 162, 198,
    199, 224
graphene-based 34, 80, 90, 120, 274, 278
graphene oxide 12, 35
graphite 4–6, 11–15, 21, 22, 24, 27, 31, 32, 35,
    36, 38, 40, 43, 46, 49, 54, 57, 60,
    78, 80, 81, 86, 92, 100, 103, 116,
    121, 139, 153, 174, 178, 194, 240,
    299, 300
growth mechanism 5, 6, 41, 42, 44, 46, 53

Hall effect 76, 81–88, 117, 118, 121, 187, 279
Heisenberg 218, 219, 221
HFCVD 57
HF devices 76
high frequency 45, 137
holonomic 180, 256–261, 286–290, 295
hopping conduction 8, 9, 18, 75, 108, 130, 131,
    134, 139, 140–145, 147, 148, 150, 157,
    163, 164, 166, 173–180, 199, 240, 241,
    266, 292, 294
HRTEM 46, 55, 107–109, 133
hybrid circuits 90, 272
hybridization 7, 10, 11, 24, 25, 39, 40, 64, 101,
    103, 176, 196, 199
hydrocarbon 3, 8, 37, 46, 47, 53

IBM QE/IBM 21, 253–259, 260, 277, 284–286
impurity 87, 98, 99, 123, 130, 133, 144, 163, 193,
    220, 288
ion implantation 133

Josephson Junction 201, 207, 218, 270, 271, 275, 279

kinetic energy 69, 284
Klein 84, 118, 161, 187
Kondo 98, 99, 106, 109, 111–114, 190, 202, 221, 229, 288
Kramer doublet 191, 200, 202

Landau 85, 87–90, 129, 167, 218, 239
laser ablation 23, 31, 35, 36, 37, 44, 54, 137, 226
LDOS 41, 129, 141, 145, 147–152
lifetime 24, 90, 103, 126, 135, 136, 138, 152, 251
linear magnetoresistance 85, 87
localization 41, 64, 65, 75, 89, 115, 142, 143, 149, 152, 160, 162, 164, 173, 192, 203, 206, 228, 261, 284
Luttinger liquid theory 77

magnet 104, 105, 282
magnetic fields 17, 51, 70, 84, 85, 87, 88, 117, 129, 17, 179, 202, 216, 263, 279
magnetic moment 51, 90, 95, 101, 103, 120, 200
magnetic tunnel junction 95
magnetism 101, 102, 103, 115, 116, 118–121, 124, 204, 221, 229, 240, 294
magnetoresistance 85, 87, 95–97, 102, 108, 110, 114, 161, 168–171, 209, 216, 228–232, 238, 288, 294
Majorana 99, 236
massless 81–85, 87
Maxwell 217
mesoscopic 75, 114, 157–159, 178
metals 6, 24, 35, 45, 76, 79, 98, 144, 181, 222, 224, 270, 284
microstructure 22, 31, 133, 140, 145, 160, 170, 175, 197, 198, 201, 211, 216, 230, 237, 239, 279, 280, 291, 293–295
mobility 169, 173–176, 178
Moiré 48, 121, 122
MOSFET 73, 79, 127
motion 66, 105, 159
MRAM 97
multilayer 32, 48, 81, 85, 87–90, 97, 126, 132, 137, 138, 153, 283
MWCVD 46, 57
MWNT 44, 70, 107–114

nanodiamonds 13, 14, 165, 266, 268
nanomanipulation 85, 86, 88
nanostructure 10, 14, 16, 20, 21, 31, 38, 49, 64, 95, 99, 104, 107, 121, 132, 175, 181, 277, 278
nanotubes 11, 12, 16, 17, 19, 20, 36, 39, 41, 44, 46, 51, 52, 57, 64, 66–77, 104–108, 116, 119, 174, 187, 196, 198, 223, 264–266, 272–275, 278

NDR 132–135, 137, 138, 142, 152
nitrogen-doping 41, 60, 145, 158, 168, 171
nitrogen-vacancy center 246, 247, 269
noise 180, 29, 254, 257, 273, 274, 287
nonmagnetic 95, 96, 98, 250
NV centers 90, 91, 163, 244–257, 261–268, 273, 282, 283, 292

odd frequency 202, 204, 236, 237, 239, 292
ODMR 247
orbital 6, 7, 17, 20, 25, 62, 92, 103, 116, 121–125, 131, 132, 196, 198–199, 202, 236, 238, 273, 281, 287, 291
orbital magnetism 116, 121
order parameter 189, 194, 198, 201, 202, 209, 211, 216, 218, 235–238, 280, 281, 291–293
organic conductors 9, 222
oscillations 14, 75–76, 83, 87–90, 96, 159, 190, 210, 214, 224, 248, 272, 278, 286, 287, 296

particles 4, 5, 16, 18, 23, 43, 46, 49, 51, 56, 64, 111, 112, 130, 136, 161, 211, 218, 219, 244, 282, 286
Pauli spin 282, 289
pentagon 41, 42, 43, 47, 50, 59, 198, 222, 300
phase(s) 5, 43, 44, 46, 75, 126, 136, 145, 151, 153, 155, 159, 160, 190, 194, 205, 217, 224, 228, 229, 238, 245, 256, 260, 263, 281, 291, 300
phase-slip 236, 280–281
phase transition 8, 9, 120, 126, 150, 160, 189, 205, 228
photoluminescence 14, 19, 24, 246–248
photon 64, 80, 180, 220, 245, 249, 250, 252, 253, 255, 261, 274
photosynthesis 3, 4, 27
pi-electrons 10, 100
planet 2, 26
plasma 3, 23, 40, 44, 46, 49, 56–57, 61, 86, 158, 168, 170
polarization 90, 115, 117, 118, 236, 240, 248, 253
polyacetylene 9, 187, 194, 240, 292, 295
polymers 8, 27, 108, 126
potential 70, 126, 127, 135, 236

q-carbon 21, 23, 25, 26, 226
Qdots 24, 25
quadrupole 50, 189
quanta 217, 271
quantum dot 50, 64, 70, 82, 90, 98, 106, 127, 135, 157, 221, 264, 266, 277, 279, 285
quantum mechanics 159, 244, 283
quantum-phase 224
quantum well 22, 69, 91, 122, 136, 144, 149, 153, 181, 268, 283, 289, 295

quantum wires 66
qubit 27, 179, 180, 244, 245, 248, 249, 252–254,
    256–266, 268

radiation 2, 3, 4, 181, 246, 248, 281
Raman spectroscopy 41, 42, 86, 90, 144, 193, 294
Rashba 180, 190, 200, 201, 204, 208, 234, 241,
    286, 293
repulsion 7, 8, 49, 114, 217, 220
resistance 24, 31, 43, 65, 74, 75, 76, 78, 79, 82,
    85, 87, 88, 90, 96–99, 105, 108, 109,
    111, 112, 114, 115, 132–134, 141, 143,
    150, 153, 160–162, 164, 178, 179,
    181, 187, 188, 190, 191, 197, 205–211,
    216–218, 222, 224, 227–229, 231–233,
    244, 261, 289
resonance 8, 24, 69, 79, 98, 102, 110, 112–114,
    117, 142, 148, 153, 179, 190, 211, 218,
    234, 237, 248, 250, 255, 256, 262, 274
rotation 8, 160, 180, 198, 209, 256–260, 283, 287,
    288, 290, 291, 295
RVB model 218, 219, 221

scattering 20, 64, 67, 70, 74, 75–78, 81, 87–90,
    96–99, 102, 106, 111, 122, 138, 158–162
Schrödinger 259
SEM 31–33, 40, 92, 102, 104, 106, 152
shell 10, 23, 36, 47, 53, 200
Shockley model 195, 201, 202, 216, 235, 240, 291,
    293, 296
Shubnikov de Hass Oscillation 86
simulation 26, 153, 177, 179, 211, 245, 253, 254,
    255, 258, 259, 262, 268, 270, 271, 273,
    275, 278, 283, 284–286, 291, 295,
    297, 298
single electron 9, 65, 78, 82, 131, 276
SMM 105, 107
space 1, 2, 4, 5, 10, 11, 14, 18, 27, 49, 51, 70, 81,
    85, 91, 196, 217–219, 240, 244, 257,
    282, 292
space-time 91, 129, 158, 160, 161, 153, 181, 189,
    196, 198, 300
spin-charge 76, 160, 189, 219, 239
spin flip 114, 238, 287
spin liquid 8, 189, 211, 218, 219, 221, 224
spin-orbit coupling 50, 67, 160, 161, 181, 190,
    196, 199, 203, 261, 287, 288, 293
spin qubit 98, 126, 244, 246, 273, 273, 277, 288
spin triplet 239, 258, 260, 294
spin valve 95–97, 102, 105, 110, 235
spin zero 27, 189, 299
spinors 283, 283, 286, 287
spintronics 18, 95–96, 100, 104, 278, 281
SSH model 27, 194, 195, 211, 239–240, 300
STM 48, 49, 86, 114, 116, 131, 152
Stone-Wales 34, 128, 199
strong localization 64, 75, 91, 160

Sun 1, 3
superconductivity 187, 188, 189–194, 196,
    199–205, 211, 216–225, 228–240
supercurrent 188, 201, 217, 224, 244, 281
superlattice 122, 126, 129, 136, 137, 144, 153,
    165, 167, 168, 171, 224, 227, 283
superposition 5, 6, 27, 160, 178, 190, 201,
    202, 216, 219, 220, 236, 237, 244,
    245, 251, 264, 272, 275, 281, 283,
    284, 293
SWAP gates 276
SWNT 45, 66, 71, 77, 78, 108, 223
SYK model 298
synthesis 8, 11, 12, 24, 31, 33, 35, 36, 37, 39, 41,
    43, 44–46, 49, 51–53, 54, 57, 59, 60,
    70, 78, 117, 119, 162, 226, 246, 254

telescope 2
TEM 23, 34, 37, 44, 46–48, 55, 131, 137, 165, 210
theory 43, 53, 98, 99, 103, 112, 116, 128, 136, 152,
    153, 160, 162, 179, 188–190, 203, 216,
    218, 220, 223, 231, 238, 280
tight-binding model 9, 81, 129
time-reversal 51, 88, 162, 189, 191, 196, 198, 200,
    202, 208, 240, 261, 287
TMR 96, 97, 98, 114
topological 117, 128–130, 194, 195, 198, 199, 201,
    202, 208, 216, 218–220, 226–228, 230,
    233, 236–241, 255, 257, 264, 279, 280,
    281, 283, 290–295
torque 19, 50
triangle 135, 149, 233, 235, 288, 290–292, 295
triplet 50, 102, 118–120, 189, 190, 196, 201,
    202, 206, 207, 209, 221, 232, 234,
    236, 237, 239, 260, 261, 264, 282,
    291, 292
tunnel 26, 65, 70, 74, 84, 91, 95–97, 99, 108, 108,
    124, 134, 136, 137, 144, 150, 152, 153,
    179–180, 209, 216, 235–237, 270, 272,
    283–287, 291, 295, 297
twisted graphene 15, 187, 224

UCF 162
UHRTEM 291, 295
UNCD 158–160, 163–176
uncertainty 244
universe 1, 4, 21

vacuum 40, 51, 71, 74, 101, 189, 221
vapor 44, 46, 57
vapor-solid-solid 43
vector 17, 48, 54, 66, 67, 104, 132, 141, 160,
    200–202, 206, 241, 250, 257, 281,
    287, 293
velocity 13, 76, 78, 84, 121, 126
viral diseases 265

vortex 5, 8, 17, 51–52, 188–190, 198, 201–203,
         206, 207, 211, 216–218, 221–222,
         227–230, 232, 235, 237–240, 291, 292,
         295, 299
VRH conduction 149, 157, 158, 163, 164, 175,
         179–180, 190

wave 13, 20, 64, 237, 240, 270, 288
    d-wave 236
    p-wave 239
    s-wave 288
wavefunctions 151, 152, 160, 161, 181, 293
weak antilocalization 203, 206, 300
weak localization 64, 75, 88, 89, 132, 158, 161,
         206, 261, 283, 284

Wigner 179, 187
wormhole 153, 237, 245, 295

YSR states 117

Zak phase 240, 294
ZBCP 65, 75, 190, 195, 209, 216, 232, 233, 234,
         235, 236, 237, 238, 240, 279, 291
Zeeman 191, 200, 202, 225, 236, 239, 281, 292
zero-bias 65, 72, 113, 190, 196, 234, 241, 256, 281
zero-field 107, 113, 162, 200, 231, 248, 250
zigzag 16, 17, 35, 45, 47, 54, 66–68, 77, 102, 117,
         118, 119–120, 195, 240, 294
ZPL 248

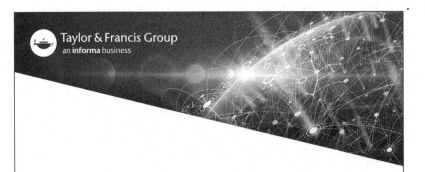